Static Analysis of Shells

Static Analysis of Shells

A Unified Development of Surface Structures

Phillip L. Gould
Washington University

Lexington Books
D. C. Heath and Company
Lexington, Massachusetts
Toronto

Library of Congress Cataloging in Publication Data
Gould, Phillip L
    Static analysis of shells.
    Includes bibliographical references and index.
    1. Shells (Engineering) I. Title.
TA660.S5G65   624'.1776   76–47142
ISBN 0–669–00966–0

To my parents, Belle and
David Gould.

Contents

List of Figures

xii

List of Tables

Preface

The study of three-dimensional continua has been a traditional part of graduate education in solid mechanics for some time. With rational simplifications to the three-dimensional theory of elasticity, the engineering theories of medium-thin plates and of thin shells may be derived and applied to a large class of engineering structures distinguished by a characteristically small dimension in one direction. Often, these theories are developed somewhat independently due to their distinctive geometrical and load-resistance characteristics. On the other hand, the two theories are frequently components of the same academic course, suggesting that the similarities and common bases be exploited as fully as possible.

A substantial portion of many traditional approaches to this subject has been devoted to constructing classical and approximate solutions to the governing equations of the system in order to proceed with applications. Within the context of analytical, as opposed to numerical, approaches, the limited generality of many such solutions has been a formidable obstacle to applications involving complex goemetry, material properties, and/or loading. It is now relatively routine to obtain computer-based solutions to quite complicated situations. However, the choice of the proper problem to solve through the selection of the mathematical model remains a human rather than a machine task and requires a basis in the theory of the subject.

With the requirement of a strong grounding in the engineering theories of shells and plates remaining firm, this book offers a unified development with emphasis on the fundamental engineering apsects. The basic material is designed to be covered in a single semester graduate course or through equivalent self-study; also, ample enrichment is provided for further independent study. Initially, the geometrical relationships are developed on a somewhat general level, with specific applications to frequently encountered forms. Following the geometric description of the surface and the consideration of equilibrium, we introduce a first logical simplification, the membrane theory of shells. After a further theoretical exposition of linear deformations, constitutive relationships, and energy principles, we present the flexural theory of plates and the bending theory of shells, including elastic instability aspects. Although this sequence postpones the introduction of the complete theory of plates until much of the theory of shells is covered, we believe it is logical and consistent with the unifying objective of the text.

The fundamental geometric and static relationships are considered from an integrated mathematical and physical point of view. Orthogonal curvilinear coordinates and vector calculus are used to provide a concise general derivation of the field equations. Early on, however, we present the physical resolution and interpretation of forces and deformations for specialized geometric forms such as rotational shells and flat plates. We believe that the physical notions are more meaningful in the specialized geometric context, whereas the mathematical formulation yields a conciseness not attainable with a strictly physical viewpoint. Also, we stress the energy aspects of the formulations because of the importance of energy methods in modern computational techniques.

With regard to applications, our focus is on classical examples that illustrate the basic resistance mechanisms of selected configurations without undue mathematical complications. The availability of numerically based, computer-implemented solution algorithms has diminished the need to rely on cumbersome, sometimes oversimplified, analytical solutions for complex problems; hence, they are paid scant attention in this text. Rather, we emphasize the fundamental aspects of equilibrium and compatibility, as they apply to the resistance of surface loading and the satisfaction of boundary constraints. We believe that a firm grasp of these fundamental aspects is necessary to perform the vital critical interpretation required of the analyst when computer-based solutions are employed. This book is dedicated to developing in the engineer the physical and mathematical understanding required to perform analysis and design in an interactive computer-assisted environment. The book is also designed to provide a foundation in the basic aspects of the subject which will enable the interested reader to progress to more advanced texts and technical papers.

Of special interest in the book is a thorough treatment of hyperboloidal shells of revolution. This topic is of current interest because of the wide use of this form in cooling tower applications.

Regarding the background of the reader, this book is written primarily for advanced students and practitioners in Civil, Mechanical, and Aeronautical Engineering and Engineering Mechanics. We presume that the reader has an elementary knowledge of vector calculus and matrix algebra, and some familiarity with the linear theory of elasticity.

Finally, the reader should note that the phrase "surface structures" may be used to describe both shells and plates. This term is a literal translation of the German word *Flächentragwerke*, and we hope that this book will further the unified consideration of plates and shells under the single heading "surface structures" in the English-language community as well.

Acknowledgments

I was introduced to and guided in the study of thin shells by my doctoral advisor, Professor S. L. Lee, then of Northwestern University. Also, the lectures of Professors J. Dundurs and G. Herrmann strongly influenced the approach taken in this book. I hope that some of the vivid insights of these gentlemen have penetrated into the text. The basic vectorial approach to the theoretical development of the theory of surfaces was inspired by the lucid writings of the Soviet applied mathematician V. V. Novozhilov. The English translations of his works have strongly influenced my approach and remain a continued source of inspiration.

I also wish to recognize two groups of professional colleagues. My many associates at the Institut für Konstruktiven Ingenieurbau of the Ruhr University–Bochum in West Germany have demonstrated to me the efficiency of a general mathematical approach in both the research and teaching aspects of this field. On the other hand, the practice-oriented members of the ACI–ASCE Committee on Concrete Thin Shell Design and Construction have helped to remind us that our efforts as engineers are ultimately manifested as physical structures rather than mathematical models. The consideration of several well-taken points of view has helped shape the somewhat balanced direction taken in the book between the highly mathematical and strongly physical possibilities.

I have also been aided by a number of dedicated students. Special acknowledgments are due to P. Basu, L. Brombolich, A. Cataloglu, H. Cecen, M-T. Chen, P. Chod, O. El-Shafee, R. Lowrey, R. Minkoff, A. Peano, S. Sen, H. Suryoutomo, R. Wang, T. Yang, and A. Zahoor, as well as to those colleagues who contributed photographs that are used in the text.

The manuscript was typed by A. Bletch, D. Kelleher, K. Jones, Z. Henkel, P. Ross, and L. Simons, and the index was compiled by E. Gould. The drawings were prepared by B. Sweeney.

Finally, thanks are due to the Engineering Mechanics Section of the National Science Foundation, whose support has enabled me to sustain and broaden my interest in shell structures, and the the Department of Civil Engineering at Washington University, which has provided a pleasant working environment.

Static Analysis of Shells

1 Introduction

1.1 Role of the Theory of Elasticity

The theory of elasticity is the basis for several engineering theories which, in turn, are applied to mechanical and structural design. The basic components of the elasticity problem are often designated as *equilibrium, compatibility*, and a *constitutive law*. The equilibrium equations represent a statement of Newton's Laws, which are restricted here to the static case. The compatibility conditions express the kinematic relationships between strains and displacements, and the constitutive law embodies the stress-strain behavior of the material which is presumed here to be linear elastic. In general, this set of basic components may be collected as a set of differential equations or as an energy principle. The simultaneous satisfaction of each component of the elasticity problem often is foreboding from the mathematical standpoint; therefore, engineers have naturally looked to simplifications and approximations.

One common simplification often imposed on elasticity problems has been to formulate less restrictive theories based on distinctive geometric characteristics. Among the relaxed theories, we find (1) the *theory of beams*, which is concerned with flexural members having one dimension characteristically far greater than the other two; (2) the *theory of plates*, which treats initially flat components having two dimensions far greater than the third; and (3) the *theory of shells*, which deals with curved bodies having one small dimension. Within these three theories, there are a variety of subtheories. For the purposes of this introduction it is sufficient to refer to the most common subtheory for each case; e.g., (1) *shallow* beam theory; (2) *medium-thin* plate theory, and (3) *thin* shell theory.

Although a beam may be considered as a one-dimensional member, whereas plates and shells are two-dimensional, the similarities in the theories are numerous, and the beam serves as a useful analogue for the exposition of the higher theories.

1.2 Engineering Theories

Sections of a beam, a plate and a shell are shown in figure 1–1. On each figure, the characteristic dimensions in the transverse and lateral directions are denoted as h and l, respectively. Also, a reference position is

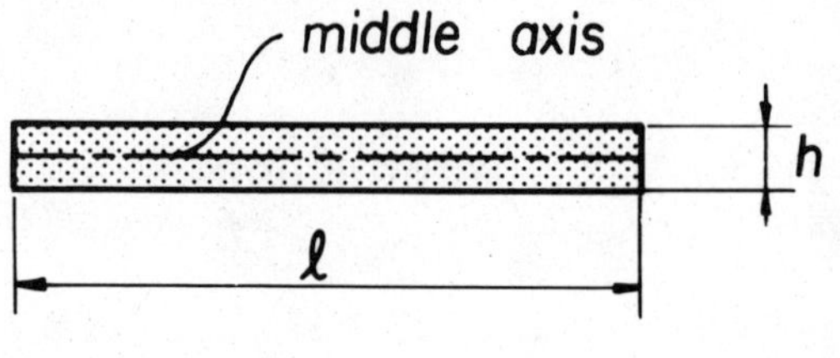

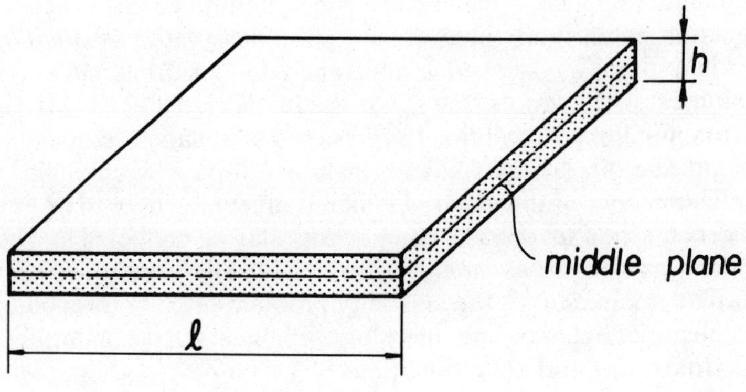

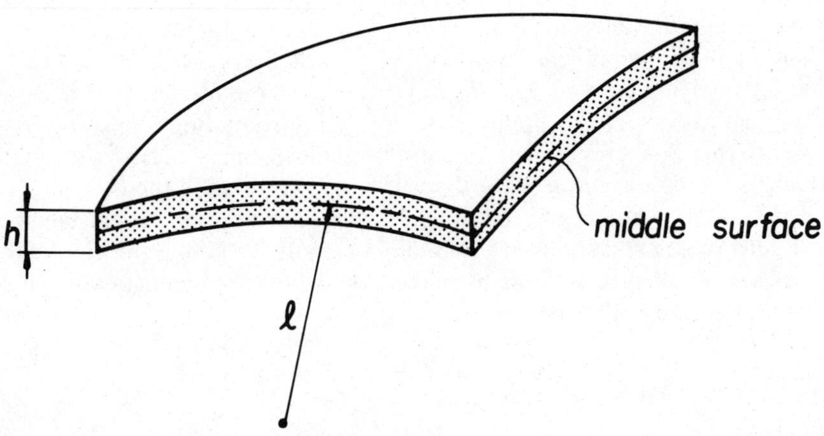

Figure 1–1. Characteristic Dimensions of Structural Forms

identified in the transverse direction, midway between the boundaries for symmetric cross sections. We call this reference position the *neutral* or *middle axis* for a beam, the *middle plane* for a plate and the *middle surface* for a shell. In the ensuing treatment, the terms *plane* and *plate*, and *surface* and *shell* are used synonymously and interchangeably to avoid repetitive distinction between the mathematical and physical objects.

The initial step in the derivation of each of the simplified theories usually consists of a set of assumptions with respect to the ratio of the characteristic dimensions, the relative magnitude of the deflection under the applied loading, the rotation of a normal to the undeformed reference position, and the stresses in the transverse directions. The statement and justification of these assumptions are largely attributable to several distinguished mathematicians and scientists of the eighteenth and nineteenth centuries. Thus, we have the Navier hypothesis and the Bernoulli–Euler theory for beams, the Kirchhoff theory of plates, and Love's first approximation to the theory of shells.[a,1]

These assumptions are collected in table 1–1. A number of comments with respect to the assumptions and justifications are appropriate:

[1] This assumption may be regarded as the most fundamental, since it clearly delineates the class of problems with which we are concerned from a physical standpoint. Also, [1] is the justification for [3] and [4]. The bounds on h/l are only approximate and are subject to considerable latitude depending on loading, geometry, etc.

[2] This assumption is independent, although the cases for which it would be violated might well coincide with the lower range of h/l stated in [1]. If [2] is not justified, geometrically nonlinear theories can be formulated retaining the remaining assumptions. (An illustration of such a formulation is presented in section 8.3). We may also accommodate material nonlinearities within the range of [2]. However, as the magnitude of the admissible displacements increases, the possibility of exceeding the limits of linear elastic material behavior increases proportionally.

[3] This assumption enables the mathematical formulation to be simplified. When only classical solution techniques were available, the suppression of transverse shear permitted many otherwise intractable problems to be approached. With powerful numerical procedures now well developed, the necessity of this assumption has diminished, although it still remains popular. It has been suggested in an indepen-

---

a Readers who are interested in the historical development of solid mechanics and in the distinguished personalities who contributed to this development are referred to Todhunter and Pearson; A. E. H. Love; S. A. Timoshenko; and, H. M. Westergaard.

4

Table 1–1
Basic Assumptions

Assumption	Theory	Consequence	Justification
[1] Transverse characteristic dimension is small in comparison to lateral characteristic dimension	Shallow beam	Beam is shallow in comparison to length, $h < l$	$0.01 \leq h/l \leq 0.5$ cable deep beam
	Medium-thin plate	Plate is thin in comparison to lateral dimension, $h < l$	$0.001 \leq h/l \leq 0.4$ membrane thick plate
	Thin shell	Shell is thin in comparison to minimum radius of curvature, $h \ll l$	$0.001 \leq h/l \leq 0.05$ curved thick membrane shell
[2] Displacements are small in comparison to transverse characteristic dimension	All	Equilibrium may be formulated with respect to the initial undeformed geometry. Products of deformation parameters may be neglected. The system may be described by a system of geometrically linear equations	Validity may be established by calculation in the course of the solution
[3] Shearing strains acting on planes parallel to middle (section, plane, surface) are neglected	Shallow beam	Plane sections before deformation remain plane after deformation	$h < l$
	Medium-thin plate	Straight fibers which are perpendicular to the middle plane before deformation remain perpendicular to the middle plane after deformation	$h < l$
	Thin shell	Straight fibers which are perpendicular to the middle surface before deformation remain perpendicular to the middle surface after deformation	$h \ll l$
[4] Normal stresses acting on planes parallel to middle (section, plane, surface) are neglected	Shallow beam	Beam depth does not change during deformation	$h < l$
	Medium-thin plate	Plate thickness does not change during deformation	$h < l$
	Thin shell	Shell thickness does not change during deformation	$h \ll l$

dent observation that the retention of this assumption, originally conceived to facilitate analytical solutions, indeed often complicates numerically based solutions.[2] Relaxation of [3] enables the upper limit on h/l in [1] to be extended in many cases.

[4] Situations for which this assumption would not be justified, apart from the immediate vicinity of concentrated loads, would coincide with the upper limits of [1].

The assumptions and consequences collected in table 1–1 are of utmost importance in what follows and are referred to frequently.

1.3 Load Resistance Mechanisms

The common basis of shallow beam, medium-thin plate and thin shell theories is illustrated by the unified set of underlying assumptions. However, the means for resisting applied loading among these structural forms may be quite different. Idealized free-body diagrams of the three cases, along with an additional form, the arch, are shown in figure 1–2. For simplicity, only a vertical loading is shown in each case but the following observations are generally valid for any distributed loading.

The beam, being straight, depends on the shear V to resist transverse loading. In turn, V acting together with the loading and reaction R, requires a moment M for equilibrium. The beam may be classified as a *one-dimensional flexural* member.

The arch, because it is curved, can develop a thrust N to resist the applied loading in addition to the shear V. Although V and M are still present in the general case, the efficiency of the arch form lies primarily in resisting the loading with N and minimizing V and M. The arch may be called a *one-dimensional extensional* member.

The plate, being flat, relies on the transverse shear V to resist transverse loading in the same manner as the beam. The two-dimensional configuration results in bending moments M and, additionally, twisting moments T on each internal face. Because the loading is generally carried in both directions and because the twisting rigidity in isotropic plates is quite significant, a plate is considerably stiffer than a team of comparable span and thickness. The plate may be considered a *two-dimensional flexural* member.

The shell, being curved, can develop thrusts N to form the primary resistance mechanism in addition to those forces and moments present in

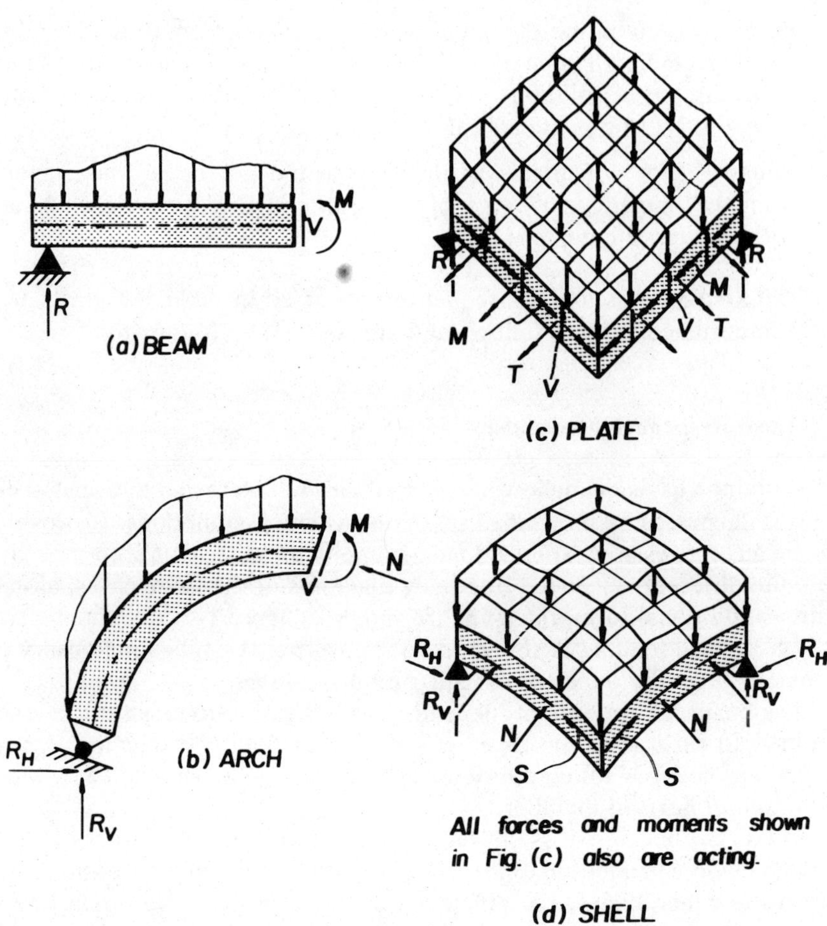

Figure 1–2. Means of Load Resistance

the plate. Also, the two-dimensional curved configuration mobilizes an in-plane shear S in each direction. Although V, M, and T are still present in the general case, the efficiency of the shell form rests with the reliance on N and S as the primary means of resistance with V, M, and T minimized. The shell may be termed a *two-dimensional extensional* member. The classification of these forms is summarized in table 1–2.

Of course, there is some overlap within the classification in table 1–2. One common situation is the presence of axial loading in a beam and, correspondingly, in-plane loading in a plate. Within the limitations of assumption [2], these effects, called *rod* and *diaphragm* action, respectively, are uncoupled from the primary action and may be combined with the

Table 1–2
Classification of Structural Forms

Primary Resistance Mechanism	Configuration	
	1-dimensional	*2-dimensional*
Flexural	Beam	Plate
Extensional	Arch	Shell

flexural behavior by superposition. The study of instability, however, involves a coupling between extensional and flexural effects.

Another possibility with respect to the beam or plate is the presence of axial or in-plane forces which develop a vertical component through a curvature induced by the flexural action. This case may be treated only with a relaxation of assumption [2].

A third possibility which is of considerable practical importance is a form curved in only one direction, such as a cylinder or cone. Although these so-called *developable* surfaces are generally regarded as shells, the resistance mechanism in the uncurved direction can be basically flexural, whereas that in the curved direction may be primarily extensional.

As a closing introductory remark, it should be pointed out that structural materials are generally far more efficient in an extensional rather than in a flexural mode, making the arch and shell preferable over the beam and plate. The extensional mode, within the scope of small deformations, can be developed only through curvature; this limits the applications of arches and shells both from a fabrication and from a utilization standpoint. The structural form is subject to many constraints apart from the most efficient use of the material and will be considered as specified a priori in most of this text.

It is of present as well as historical interest to recognize that significant structures were erected utilizing the efficient doubly curved form of the dome long before the development of modern engineering analysis.[3] Several of these structures survive today. The Pantheon of ancient Rome (figure 1–3), attributed to Marcus Agrippa and Emperor Hadrian, has stood for about two thousand years, while the Hagia Sophia in Istanbul (figure 1–4), originally completed by Isidorus, Jr., has epitomized Byzantine architecture for fifteen centuries. Increased geometrical refinement is exhibited in the Renaissance cathedrals of Santa Maria del Fiore in Florence (figure 1–5), constructed without shoring by Brunelleschi, and St. Peter's in Rome (figure 1–6), designed by Michelangelo. Also, Sir Christopher Wren's St. Paul's Cathedral (figure 1–7) remains to grace the London skyline.

Although the domes atop these structures are not thin or engineered in

Figure 1–3. Pantheon, Rome, Italy. Dome Span = 43.4 m; Dome Rise = 21.6 m. (Author's Photo)

Figure 1–4. Hagia Sophia, Istanbul, Turkey. Dome Span = 31.9 m; Dome Rise = 13.8 m (Courtesy Dr. I. Mungan)

Figure 1–5. S. Maria del Fiore, Florence, Italy. Dome Span = 42.4 m; Dome Rise = 36.6 m. (Author's Photo)

the modern sense, they exhibit the unique capability of the curved surface to bridge considerable space without intermediate supports utilizing construction materials capable of resisting only compressive forces.

The connection between the modern developments in thin shell technology and the corresponding significant scientific events in the post-industrial revolution are documented in chapter 1 of Fung and Sechler.[4] The remaining chapters of this modern collection of papers (which is devoted to the unification of theory, experiment, and design of thin-shell structures) is probably best appreciated after the fundamental material presented in this book has been mastered.

Figure 1–6. S. Peter's, Rome, Italy. Dome Span = 41.6 m; Dome Rise = 35.1 m

Notes

1. See I. Todhunter and K. Pearson, *A History of the Theory of Elasticity and of the Strength of Materials*. 2 vols. (Cambridge: Cambridge University Press, 1886 and 1893). Also see A. E. H. Love, *A Treatise on the Mathematical Theory of Elasticity*, 4th ed. (New York: Dover Publications, 1944), Introduction. And see S. A. Timoshenko, *History of Strength of Materials* (New York: McGraw-Hill, 1953). Finally, see H. M. Westergaard, *Theory of Elasticity and Plasticity* (New York: Dover Publications, 1964), chap. II.

Figure 1–7. St. Paul's, London, England. Dome Span = 30.8 m; Dome Rise = 33.5 m. (Author's Photo)

2. J. T. Oden, *Finite Elements of Nonlinear Continua* (New York: McGraw-Hill, 1952), pp. 144–145.

3. See C. T. Grimm, "Brick Masonry Shells," *Journal of the Structural Division, ASCE* 101, no. ST1 (January 1975): 79–95; and the discussion by K. Anadol and E. Arioylu, no. ST11 (November 1975):2451–2455.

4. Y. C. Fung and E. E. Sechler, eds., *Thin Shell Structures* (Englewood Cliffs, N.J.: Prentice-Hall, 1971).

2

Geometry of Middle Surface

2.1 Curvilinear Coordinates

Consider a portion of the middle surface of a shell as shown in figure 2–1.

The surface is defined with respect to the X–Y–Z global Cartesian coordinates by

$$Z = f(X,Y) \tag{2.1}$$

Then, a set of coordinates α and β that are related to the Cartesian system by

$$\left. \begin{array}{ll} X = f_1(\alpha,\beta) & \text{(a)} \\[2mm] Y = f_2(\alpha,\beta) & \text{(b)} \\[2mm] Z = f_3(\alpha,\beta) & \text{(c)} \end{array} \right\} \tag{2.2}$$

in which f_1, f_2, and f_3 are continuous, single-valued function, is chosen. Since each ordered pair (α,β) corresponds to only one point on the surface, the surface is uniquely described in terms of α and β, which are called *curvilinear* coordinates. If one of the coordinates, e.g., β, is incremented $\beta = \beta_1, \beta = \beta_2, \ldots, \beta = \beta_n$, then we define a series of parametric curves on the surface, along which only α varies. These curves are termed the s_α *coordinate lines*. Similarly, if α takes on the values $\alpha = \alpha_1, \alpha = \alpha_2, \ldots, \alpha = \alpha_m$, we get the s_β coordinate lines. The coordinate lines are shown in figure 2–1.

If the coordinate lines s_α and s_β are mutually perpendicular at all points on the surface, the curvilinear coordinates are said to be *orthogonal*. Orthogonal curvilinear coordinates are used exclusively in this book.

2.2 Description of the Middle Surface in Orthogonal Curvilinear Coordinates and First Quadratic Form of the Theory of Surfaces

Equation (2.1), which describes the middle surface, may be written in terms of a radius vector emanating from the origin as shown in figure 2–2:

$$\mathbf{r} = X\mathbf{u} + Y\mathbf{v} + Z\mathbf{w} \tag{2.3}$$

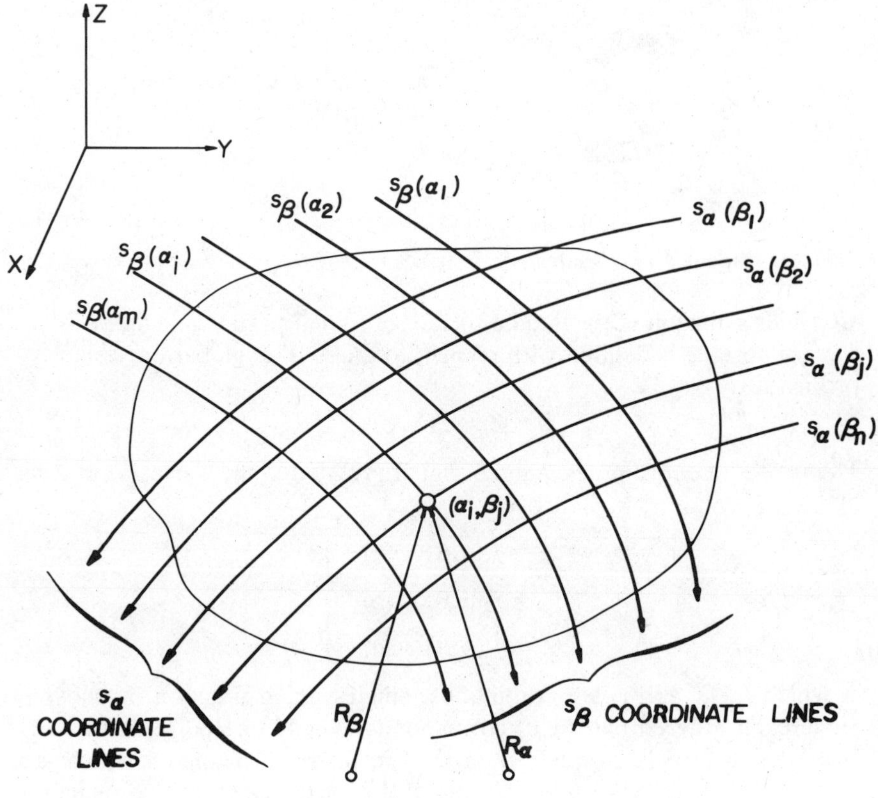

Figure 2–1. Middle Surface of a Shell

in which **u**, **v**, and **w** are unit vectors along the X, Y, and Z axes, respectively.

Substituting equation (2.2) into (2.3), the radius vector is defined in terms of the curvilinear coordinates by

$$\mathbf{r}(\alpha,\beta) = f_1(\alpha,\beta)\mathbf{u} + f_2(\alpha,\beta)\mathbf{v} + f_3(\alpha,\beta)\mathbf{w} \tag{2.4}$$

The derivatives of **r** with respect to the curvilinear coordinates are considered next:

$$\frac{\partial \mathbf{r}}{\partial \alpha} = \mathbf{r}_{,\alpha} \tag{2.5a}$$

and

$$\frac{\partial \mathbf{r}}{\partial \beta} = \mathbf{r}_{,\beta} \tag{2.5b}$$

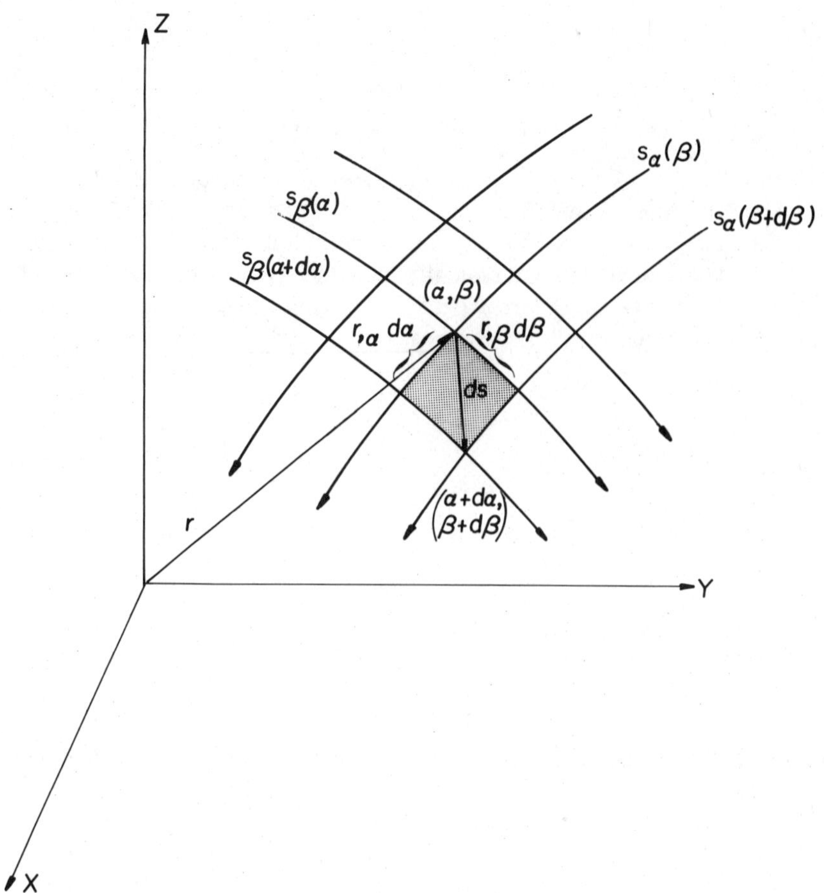

Figure 2–2. Radius Vector to a Point on the Middle Surface

are vectors that are tangent to the s_α and s_β coordinate lines, respectively. Since the coordinate lines are orthogonal, these tangent vectors are orthogonal as well, and their scalar product $\mathbf{r},_\alpha \cdot \mathbf{r},_\beta = 0$.

The vector joining the two points on the middle surface (α,β) and $(\alpha + d\alpha, \beta + d\beta)$ shown in figure 2–2 is

$$\mathbf{ds} = \mathbf{r},_\alpha \, d\alpha + \mathbf{r},_\beta \, d\beta \tag{2.6}$$

If we form the scalar product of $\mathbf{ds}$ with itself, we have

$$\mathbf{ds} \cdot \mathbf{ds} = ds^2 = (\mathbf{r},_\alpha \cdot \mathbf{r},_\alpha) \, d\alpha^2 + (\mathbf{r},_\beta \cdot \mathbf{r},_\beta) \, d\beta^2 \tag{2.7}$$

Defining

$$A^2 = \mathbf{r},_\alpha \cdot \mathbf{r},_\alpha \tag{2.8a}$$

16

and

$$B^2 = \mathbf{r},_\beta \cdot \mathbf{r},_\beta \tag{2.8b}$$

$$ds^2 = A^2\,d\alpha^2 + B^2\,d\beta^2 \tag{2.9}$$

which is known as the *first quadratic form of the theory of surfaces*.

The quantities A and B are called *Lamé parameters* or *measure numbers* and are fundamental to the understanding of curvilinear coordinates. To interpret their meaning physically, consider two cases in which each of the coordinates α and β are varied individually and independently. For these cases, equation (2.9) becomes

$$ds_\alpha = A\,d\alpha \tag{2.10a}$$

and

$$ds_\beta = B\,d\beta \tag{2.10b}$$

Thus, ds_α is the change in arc length along coordinate line s_α when α is incremented by $d\alpha$, and ds_β is the change in arc length along coordinate line s_β when β is incremented by $d\beta$. We see that the Lamé parameters are quantities which relate the *change in arc length* on the surface to the corresponding *change in the curvilinear coordinate*; hence the alternate name, measure number.

As a simple example, consider a circular arc of radius R in the Y–Z plane shown in figure 2–3. If the curvilinear coordinate is chosen as the polar angle β,

$$ds = R\,d\beta \tag{2.11}$$

and the Lamé parameter is R. Alternately, the curvilinear coordinate could be chosen as the Z coordinate, in which case we have (from figure 2–3)

$$ds^2 = dY^2 + dZ^2 \tag{2.12}$$

From the equation of the circle $Y^2 + Z^2 = R^2$,

$$2Y\,dY = -2Z\,dZ$$

or

$$dY^2 = \frac{Z^2}{Y^2}\,dZ^2$$

so that equation (2.12) becomes

$$ds = \left(1 + \frac{Z^2}{Y^2}\right)^{1/2} dZ \tag{2.13}$$

and the Lamé parameter is

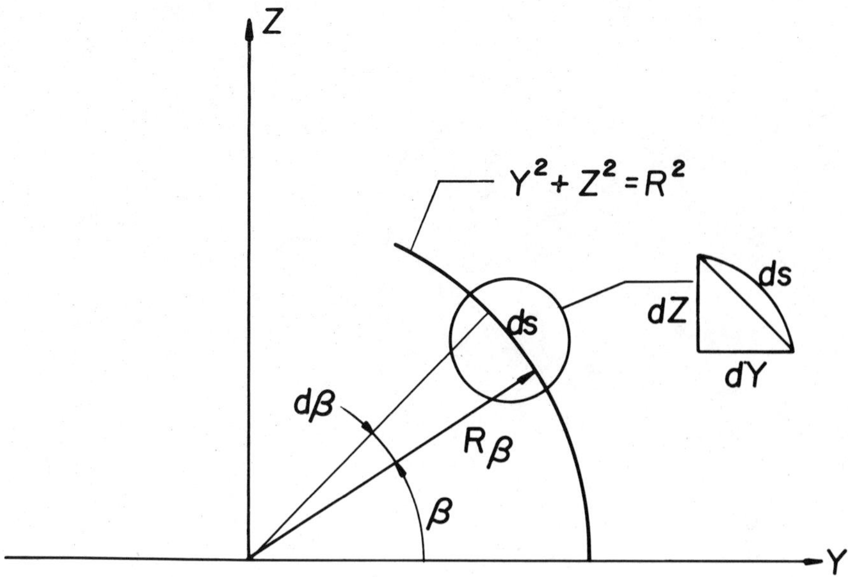

Figure 2–3. Lamé Parameter for a Circular Arc

$$\left(1 + \frac{Z^2}{Y^2}\right)^{1/2}$$

A third possibility is to choose the arc length itself as the curvilinear co-ordinate. Then

$$ds = 1 \cdot ds \tag{2.14}$$

and the Lamé parameter is the constant 1. It is apparent from this elementary example that the Lamé parameter may be a constant or a rather involved function. For a particular problem, the choice of curvilinear coordinates which correspond to the simplest possible expressions for the Lamé parameter can serve to greatly expedite the mathematics of the solution.

The preceding example illustrates that the Lamé parameters may sometimes be found by the geometric representation of equations (2.9) and (2.10). In more complicated cases, we may compute the Lamé parameters from equations (2.8) and (2.3) as

$$A^2 = (X_{,\alpha})^2 + (Y_{,\alpha})^2 + (Z_{,\alpha})^2 \tag{2.15a}$$

$$B^2 = (X_{,\beta})^2 + (Y_{,\beta})^2 + (Z_{,\beta})^2 \tag{2.15b}$$

Finally, with respect to the first quadratic form, note that it generally pertains to the *measurement of distances* on the surface, but does not involve the specific *shape* of the surface.

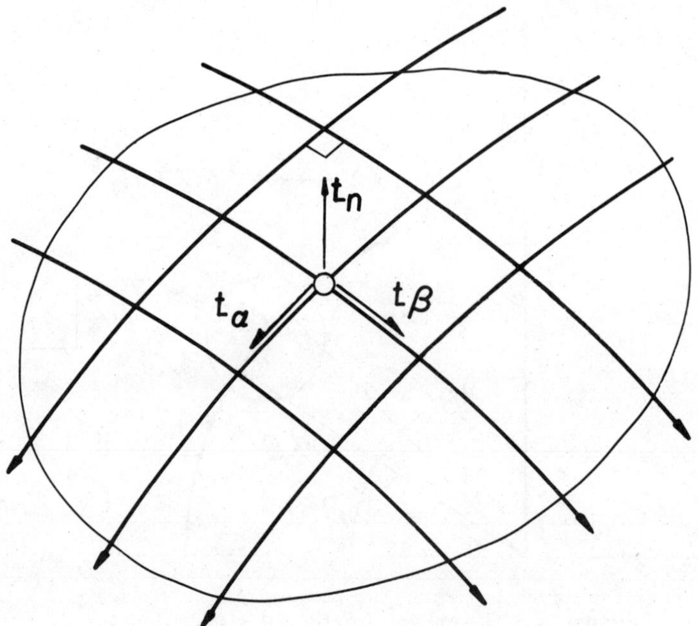

Figure 2–4. Unit Tangent Vectors

2.3 Unit Tangent Vectors and Principal Directions

2.3.1 Definition of Unit Tangent Vectors and Normal Section

It is convenient to refer all vector point functions to a triad of *unit* vectors composed of tangent vectors to the coordinate lines and a normal vector which define a right hand system as shown in figure 2–4.

We have already defined tangent vectors to the coordinate lines in equation (2.5). Hence, in view of equation (2.8),

$$t_\alpha = \frac{\mathbf{r}_{,\alpha}}{\pm|\mathbf{r}_{,\alpha}|} = \frac{\mathbf{r}_{,\alpha}}{A} \qquad\qquad (2.16a)$$

$$t_\beta = \frac{\mathbf{r}_{,\beta}}{\pm|\mathbf{r}_{,\beta}|} = \frac{\mathbf{r}_{,\beta}}{B} \qquad\qquad (2.16b)$$

and the normal vector is found by forming the vector product of $\mathbf{t}_\alpha$ and $\mathbf{t}_\beta$:

$$t_n = t_\alpha \times t_\beta = \frac{1}{AB}(\mathbf{r}_{,\alpha} \times \mathbf{r}_{,\beta}) \qquad\qquad (2.16c)$$

A *normal section* of the surface is defined as a plane curve obtained by cutting the surface with a plane containing the normal to the surface $\mathbf{t}_n$.

2.3.2 Principal Directions

Now, consider the general point (α_i, β_j) on the middle surface as shown in figure 2–1. At this point, each coordinate line, $s_\beta(\alpha_i)$ and $s_\alpha(\beta_j)$, may be regarded as a normal section with a corresponding radius of curvature $R_\beta(\alpha_i, \beta_j)$ and $R_\alpha(\alpha_i, \beta_j)$, respectively, which is directed from the center of curvature to the point (α_i, β_j) along $\mathbf{t}_n$. Obviously, there are an infinite number of possible R_α and R_β at any point, since an infinite number of orientations for the curvilinear coordinates exist.

From the theory of surfaces, it may be shown that there is a system of orthogonal curvilinear coordinates (α^*, β^*) oriented such that one radius of curvature, $R_\alpha^* = |\mathbf{R}_i^*|$ $(i = \alpha$ or $\beta)$, is the maximum of all possible $|\mathbf{R}_i|$, whereas the other radius of curvature, $R_\beta^* = |\mathbf{R}_j^*|$ $(j = \beta$ or $\alpha)$, is the minimum of all possible $|\mathbf{R}_j|$.[1] We call this system the *principal orthogonal curvilinear coordinates* corresponding to the *principal directions*, α^* and β^*. The associated coordinate lines are known as the *lines of principal curvature*, and R_α^* and R_β^* as the *principal radii of curvature*. In the subsequent derivation of unit tangent vector derivatives, principal directions will be used exclusively, so that α and β imply α^* and β^*.

2.3.3 Derivatives of Unit Tangent Vectors

To establish the relationships between the Lamé parameters and the principal radii of curvature for a surface, it is necessary to derive a set of relationships for the derivatives of the tangent vectors, $\mathbf{t}_\alpha$, $\mathbf{t}_\beta$, $\mathbf{t}_n$, with respect to α and β.[2] The derivatives of the unit tangent vectors are expressed in terms of the unit tangent vectors themselves:

$$
\begin{Bmatrix} \mathbf{t}_{\alpha,\alpha} \\[6pt] \mathbf{t}_{\alpha,\beta} \\[6pt] \mathbf{t}_{\beta,\alpha} \\[6pt] \mathbf{t}_{\beta,\beta} \\[6pt] \mathbf{t}_{n,\alpha} \\[6pt] \mathbf{t}_{n,\beta} \end{Bmatrix}
=
\begin{bmatrix}
0 & \dfrac{-A,_\beta}{B} & \dfrac{-A}{R_\alpha} \\[10pt]
0 & \dfrac{B,_\alpha}{A} & 0 \\[10pt]
\dfrac{A,_\beta}{B} & 0 & 0 \\[10pt]
\dfrac{-B,_\alpha}{A} & 0 & \dfrac{-B}{R_\beta} \\[10pt]
\dfrac{A}{R_\alpha} & 0 & 0 \\[10pt]
0 & \dfrac{B}{R_\beta} & 0
\end{bmatrix}
\begin{Bmatrix} \mathbf{t}_\alpha \\[6pt] \mathbf{t}_\beta \\[6pt] \mathbf{t}_n \end{Bmatrix}
\tag{2.17}
$$

We now consider the verification of equation (2.17). It is convenient to write the equation once again with the elements grouped as shown:

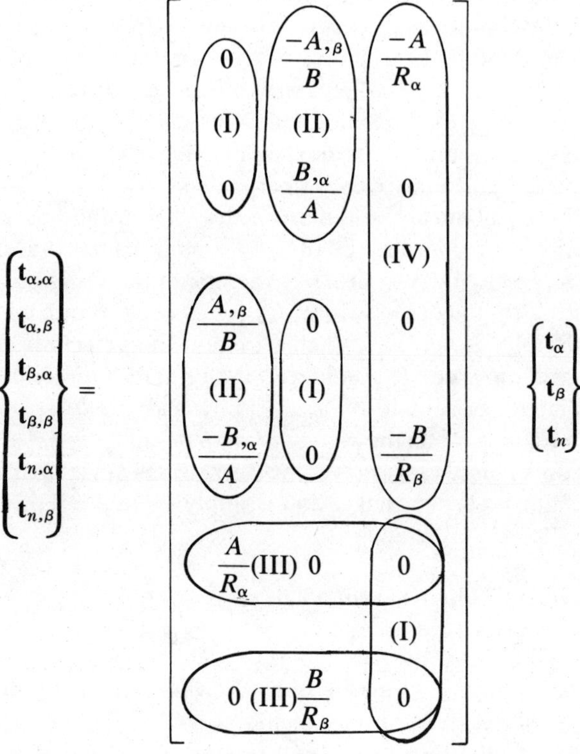

The grouping refers to the arguments in the following sections.

2.3.3.1 (I) Component in Direction of Differentiated Vector. The derivative of any unit tangent vector is normal to the vector itself so that there are no components in the direction of the vector being differentiated, e.g., there is no t_α component for $t_{\alpha,\alpha}$ or $t_{\alpha,\beta}$.

This assertion is well known from elementary vector calculus. As a proof, consider the dot product of two unit tangent vectors in the curvilinear coordinate system:

$$\mathbf{t}_i \cdot \mathbf{t}_j = \delta_{ij} \begin{cases} \delta_{ij} = 1, \ i = j; \ \delta_{ij} = 0, \ i \neq j \\ \\ i,j = \alpha, \beta, n \end{cases}$$

Differentiating the dot product with respect to $k(k = \alpha$ or $\beta)$

$$(\mathbf{t}_i \cdot \mathbf{t}_j)_{,k} = 0$$

from which

$$\mathbf{t}_j \cdot \mathbf{t}_{i,k} + \mathbf{t}_i \cdot \mathbf{t}_{j,k} = 0$$

Then set $j = i$ to get

$$2(\mathbf{t}_i \cdot \mathbf{t}_{i,k}) = 0$$

so that $\mathbf{t}_{i,k}$ has no component along $\mathbf{t}_i$. Taking $i = \alpha, \beta$ and n and $k = \alpha$ and β, in turn, verifies the indicated zero terms in this group.

2.3.3.2 (II) Components of Derivatives of $\mathbf{t}_\alpha$ and $\mathbf{t}_\beta$ in α and β Directions. We multiply equations (2.16a) and (2.16b) by A and B, respectively, and form the mixed second partial derivatives,

$$\mathbf{r}_{,\alpha\beta} = \mathbf{r}_{,\beta\alpha} \tag{2.18a}$$

With equation (2.16) substituted into equation (2.18a), we obtain

$$(A\mathbf{t}_\alpha)_{,\beta} = \mathbf{t}_\alpha A_{,\beta} + A\mathbf{t}_{\alpha,\beta} = (B\mathbf{t}_\beta)_{,\alpha} = \mathbf{t}_\beta B_{,\alpha} + B\mathbf{t}_{\beta,\alpha} \tag{2.18b}$$

from which

$$\mathbf{t}_{\beta,\alpha} = \frac{1}{B}[-\mathbf{t}_\beta B_{,\alpha} + \mathbf{t}_\alpha A_{,\beta} + A\mathbf{t}_{\alpha,\beta}] \tag{2.18c}$$

Now consider the component of $\mathbf{t}_{\alpha,\alpha}$ in the $\mathbf{t}_\beta$ direction given by $\mathbf{t}_\beta \cdot \mathbf{t}_{\alpha,\alpha}$. We take the derivative of the product $\mathbf{t}_\alpha \cdot \mathbf{t}_\beta$ with respect to α

$$(\mathbf{t}_\alpha \cdot \mathbf{t}_\beta)_{,\alpha} = \mathbf{t}_\beta \cdot \mathbf{t}_{\alpha,\alpha} + \mathbf{t}_\alpha \cdot \mathbf{t}_{\beta,\alpha} \tag{2.18d}$$

Since $\mathbf{t}_\alpha \cdot \mathbf{t}_\beta = 0$,

$$\mathbf{t}_\beta \cdot \mathbf{t}_{\alpha,\alpha} = -\mathbf{t}_\alpha \cdot \mathbf{t}_{\beta,\alpha} \tag{2.18e}$$

We replace $\mathbf{t}_{\beta,\alpha}$ by equation (2.18c) to get

$$\mathbf{t}_\beta \cdot \mathbf{t}_{\alpha,\alpha} = -\frac{1}{B}\mathbf{t}_\alpha \cdot [-\mathbf{t}_\beta B_{,\alpha} + \mathbf{t}_\alpha A_{,\beta} + A\mathbf{t}_{\alpha,\beta}]$$

Since $\mathbf{t}_\alpha \cdot \mathbf{t}_\beta = \mathbf{t}_\alpha \cdot \mathbf{t}_{\alpha,\beta} = 0$, and $\mathbf{t}_\alpha \cdot \mathbf{t}_\alpha = 1$,

$$\mathbf{t}_\beta \cdot \mathbf{t}_{\alpha,\alpha} = \frac{-A_{,\beta}}{B} \tag{2.18f}$$

as given in the first row of equation (2.17). The other components of the derivatives of $\mathbf{t}_\alpha$ and $\mathbf{t}_\beta$ in the α and β directions may be verified in a similar manner.

2.3.3.3 (III) Derivatives of $\mathbf{t}_n$. Consider the normal section at the point p_1 on the s_α coordinate line as shown in figure 2–5. The vector $\mathbf{t}_n$ is shown at the point p_1 and also at point p_2, a small distance Δs_α away. The vector construction at p_1 shows that the change in $\mathbf{t}_n$ is approximately parallel to the tangent to the curve at p_1 and the chord $\overline{p_1p_2}$. Therefore

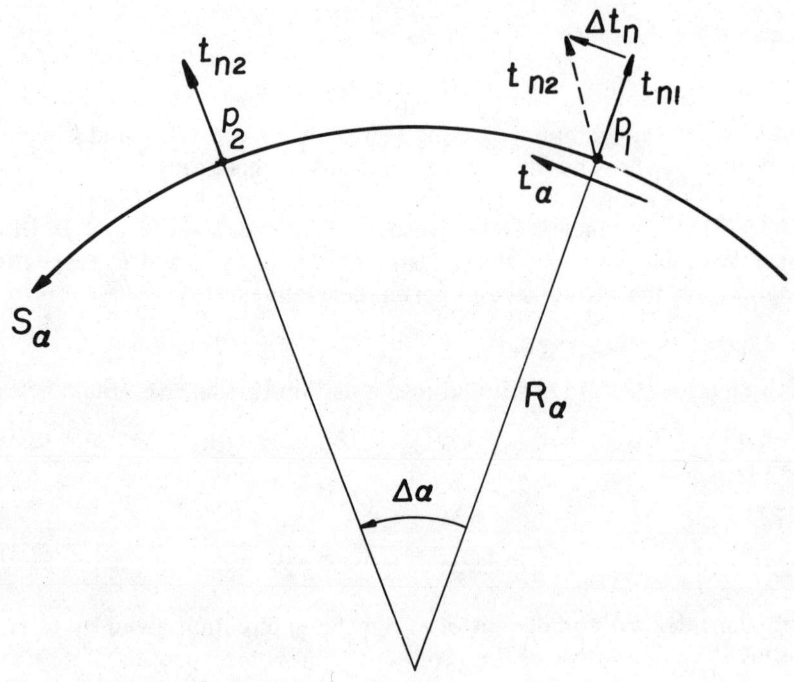

Figure 2–5. Normal Section

$$\Delta\mathbf{t}_n = |\Delta\mathbf{t}_n|\mathbf{t}_\alpha \tag{2.19a}$$

By similar triangles

$$\frac{|\Delta\mathbf{t}_n|}{|\mathbf{t}_{n1}|} = \frac{\overline{p_1 p_2}}{R_\alpha} \tag{2.19b}$$

Considering figure 2–5 and equation (2.10a),

$$\overline{p_1 p_2} \simeq \Delta s_\alpha = A\,\Delta\alpha \tag{2.19c}$$

Recognizing $|\mathbf{t}_{n1}| = 1$, and substituting equations (2.19a) and (2.19c) into (2.19b), we have

$$\frac{\Delta\mathbf{t}_n}{\Delta\alpha} = \frac{A}{R_\alpha}\mathbf{t}_\alpha \tag{2.19d}$$

Taking the limit of both sides as $\Delta\alpha \to 0$,

$$_{-n,\alpha} = \frac{A}{R_\alpha}\mathbf{t}_\alpha \tag{2.19e}$$

as given in the fifth row of equation (2.17). $t_{n,\beta}$ is evaluated in a similar manner. Note that the argument employed in (III) is more general than that used in (II), since the entire derivative is computed instead of just one component.

2.3.3.4 (IV) Components of Derivatives of t_α and t_β in Normal Direction.

Consider the normal component of $t_{\alpha,\alpha}$ given by $t_n \cdot t_{\alpha,\alpha}$. Proceeding as before,

$$(t_n \cdot t_\alpha)_{,\alpha} = t_\alpha \cdot t_{n,\alpha} + t_n \cdot t_{\alpha,\alpha} \qquad (2.20a)$$

Since $t_n \cdot t_\alpha = 0$,

$$t_n \cdot t_{\alpha,\alpha} = -t_\alpha \cdot t_{n,\alpha} \qquad (2.20b)$$

But we have already evaluated $t_{n,\alpha}$ in equation (2.19e); hence

$$t_n \cdot t_{\alpha,\alpha} = \frac{-A}{R_\alpha} \qquad (2.20c)$$

as given in the first row, third column of equation (2.17). The remaining normal components of the derivatives of t_α and t_β may be verified in a similar manner.

2.4 Second Quadratic Form of the Theory of Surfaces

Recall that in section 2.2, we derived the first quadratic form of the theory of surfaces, which pertains to distance on the surface but not specifically to the shape of the surface. In this section we seek information with respect to the latter property, the shape.

Consider a normal section which traces a plane curve with arc length coordinate s_i. An example is the normal section along s_α shown in figure 2–5; however, s_i is not necessarily restricted to only the principal direction coordinate lines in the ensuing development. The curvature of such a section is known as the *normal curvature* κ_i and is defined as a function of the radius vector $\mathbf{r}$, figure 2–2, by the Frenet–Serret formula[3] as

$$\kappa_i t_n = -\frac{1}{R_i} t_n = \mathbf{r}_{,s_i s_i} \qquad (2.21)$$

We want to express the normal curvature terms of the curvilinear coordinates α and β. Starting with

$$\mathbf{r}_{,s_i s_i} = (\mathbf{r}_{,s_i})_{,s_i} = (\mathbf{r}_{,\alpha}\alpha_{,s_i} + \mathbf{r}_{,\beta}\beta_{,s_i})_{,s_i}$$

we first evaluate

$$(\mathbf{r}_{,\alpha}\alpha_{,s_i})_{,s_i} = \alpha_{,s_i}\mathbf{r}_{,\alpha s_i} + \mathbf{r}_{,\alpha}\alpha_{,s_i s_i}$$

$$= \alpha_{,s_i}(\mathbf{r}_{,\alpha\alpha}\alpha_{,s_i} + \mathbf{r}_{,\alpha\beta}\beta_{,s_i}) + \mathbf{r}_{,\alpha}\alpha_{,s_i s_i}$$

and

$$(\mathbf{r}_{,\beta}\beta_{,s_i})_{,s_i} = \beta_{,s_i}\mathbf{r}_{,\beta s_i} + \mathbf{r}_{,\beta}\beta_{,s_i s_i}$$

$$= \beta_{,s_i}(\mathbf{r}_{,\beta\alpha}\alpha_{,s_i} + \mathbf{r}_{,\beta\beta}\beta_{,s_i}) + \mathbf{r}_{,\beta}\beta_{,s_i s_i}$$

from which

$$\mathbf{r}_{,s_i s_i} = \mathbf{r}_{,\alpha\alpha}(\alpha_{,s_i})^2 + 2\mathbf{r}_{,\alpha\beta}\alpha_{,s_i}\beta_{,s_i} + \mathbf{r}_{,\beta\beta}(\beta_{,s_i})^2$$

$$+ \mathbf{r}_{,\alpha}\alpha_{,s_i s_i} + \mathbf{r}_{,\beta}\beta_{,s_i s_i} \tag{2.22}$$

Next, we scalar multiply each side of equation (2.22) by $\mathbf{t}_n$. The last two terms on the r.h.s. (right hand side) vanish, since $\mathbf{t}_n$ is normal to $\mathbf{r}_{,\alpha}$ and $\mathbf{r}_{,\beta}$ by equation (2.16). In order to simplify the first three terms of equation (2.22), we first consider

$$\mathbf{t}_n \cdot \mathbf{r}_{,\alpha} = 0 \tag{2.23a}$$

$$\mathbf{t}_n \cdot \mathbf{r}_{,\beta} = 0 \tag{2.23b}$$

and differentiate both with respect to α and then β, which gives

$$\mathbf{t}_n \cdot \mathbf{r}_{,\alpha\alpha} = -\mathbf{r}_{,\alpha} \cdot \mathbf{t}_{n,\alpha} \tag{2.23c}$$

$$\mathbf{t}_n \cdot \mathbf{r}_{,\alpha\beta} = -\mathbf{r}_{,\alpha} \cdot \mathbf{t}_{n,\beta} \tag{2.23d}$$

$$\mathbf{t}_n \cdot \mathbf{r}_{,\beta\alpha} = -\mathbf{r}_{,\beta} \cdot \mathbf{t}_{n,\alpha} \tag{2.23e}$$

$$\mathbf{t}_n \cdot \mathbf{r}_{,\beta\beta} = -\mathbf{r}_{,\beta} \cdot \mathbf{t}_{n,\beta} \tag{2.23f}$$

We then compute the scalar product of $\mathbf{t}_n$ with the first three terms of equation (2.22). In view of equations (2.21) and (2.23c–f), we have

$$\kappa_i = -\mathbf{r}_{,\alpha} \cdot \mathbf{t}_{n,\alpha}(\alpha_{,s_i})^2 - 2\mathbf{r}_{,\alpha} \cdot \mathbf{t}_{n,\beta}\alpha_{,s_i}\beta_{,s_i}$$

$$-\mathbf{r}_{,\beta} \cdot \mathbf{t}_{n,\beta}(\beta_{,s_i})^2$$

$$= -A\mathbf{t}_\alpha \cdot \mathbf{t}_{n,\alpha}(\alpha_{,s_i})^2 - 2A\mathbf{t}_\alpha \cdot \mathbf{t}_{n,\beta}\alpha_{,s_i}\beta_{,s_i}$$

$$-B\mathbf{t}_\beta \cdot \mathbf{t}_{n,\beta}(\beta_{,s_i})^2 \tag{2.24}$$

We may evaluate the scalar products in equation (2.24) from equation (2.17). Therefore

$$\kappa_i = \frac{-A^2}{R_\alpha}(\alpha_{,s_i})^2 + 0(\alpha_{,s_i}\beta_{,s_i}) - \frac{B^2}{R_\beta}(\beta_{,s_i})^2 \tag{2.25}$$

It is convenient to multiply equation (2.25) by

$$\frac{ds_i^2}{ds_i^2}$$

whereupon

$$\kappa_i = \frac{L \cdot (\alpha_{,s_i})^2 \, ds_i^2 + N \cdot (\beta_{,s_i})^2 \, ds_i^2}{ds_i^2} \qquad (2.26a)$$

where

$$L = \frac{-A^2}{R_\alpha} \quad \text{and} \quad N = -\frac{B^2}{R_\beta} \qquad (2.26b)$$

Using equation (2.9), we may consolidate equation (2.26a) to get

$$\kappa_i = \frac{L \, d\alpha^2 + N \, d\beta^2}{A^2 \, d\alpha^2 + B^2 \, d\beta^2} \qquad (2.27)$$

In equation (2.27), $d\alpha$ and $d\beta$ represent $(\partial\alpha/\partial s_i) \cdot ds_i$ and $(\partial\beta/\partial s_i) \cdot ds_i$, respectively, for a particular direction i on the surface, as shown on figure 2–2.

Finally, we write equation (2.26) as

$$\kappa_i = \frac{\mathrm{II}}{\mathrm{I}} \qquad (2.28a)$$

where

$$\mathrm{II} = L \, d\alpha^2 + N \, d\beta^2 \qquad (2.28b)$$

and

$$\mathrm{I} = A^2 \, d\alpha^2 + B^2 \, d\beta^2 \qquad (2.28c)$$

$$= \text{First quadratic form (equation 2.9).}$$

The second quadratic form II, thus relates to the shape of the curve through the presence of radii of curvature R_α and R_β.

2.5 Evaluation of Principal Radii of Curvature

We thus far have assumed that the principal radii of curvature of the shell R_α and R_β are known or easily found. We now examine the details of this calculation.

First, consider the case where the section is a plane curve $Z = f(Y)$ in the Y–Z plane, which also contains the coordinate β, as shown in figure 2–3.

26

$$R_\beta = \frac{-[1 + (Z_{,Y})^2]^{3/2}}{Z_{,YY}} \tag{2.29a}$$

For a plane curve in the X–Z plane,

$$R_\alpha = \frac{-[1 + (Z_{,X})^2]^{3/2}}{Z_{,XX}} \tag{2.29b}$$

When the surface is specified parametrically in terms of the curvilinear coordinates as in equation (2.4), we may compute the radius of curvature using equation (2.21). If we take the s_i direction as one of the coordinate lines, e.g., s_α, then

$$-\frac{1}{R_\alpha} = \kappa_\alpha = \mathbf{t}_n \cdot \mathbf{r}_{,s_\alpha s_\alpha} = \mathbf{t}_n \cdot [\mathbf{r}_{,\alpha\alpha}(\alpha_{,s_\alpha})^2 + \mathbf{r}_{,\alpha}\alpha_{,s_\alpha s_\alpha}] \tag{2.30}$$

Equation (2.30) is evaluated in an identical fashion to equation (2.22), recognizing that $\beta_{,s_\alpha} = 0$. Since $\mathbf{t}_n \cdot \mathbf{r}_{,\alpha} = 0$, equation (2.30) reduces to

$$R_\alpha = \frac{-1}{\mathbf{t}_n \cdot \mathbf{r}_{,\alpha\alpha}(\alpha_{,s_\alpha})^2}$$

Noting that $\alpha_{,s_\alpha} = 1/A$ from equation (2.10a), and $\mathbf{t}_n = (1/AB)(\mathbf{r}_{,\alpha} \times \mathbf{r}_{,\beta})$ from equation (2.16c),

$$R_\alpha = \frac{-A^3 B}{(\mathbf{r}_{,\alpha} \times \mathbf{r}_{,\beta}) \cdot \mathbf{r}_{,\alpha\alpha}} \tag{2.31}$$

Similarly,

$$R_\beta = \frac{-B^3 A}{(\mathbf{r}_{,\alpha} \times \mathbf{r}_{,\beta}) \cdot \mathbf{r}_{,\beta\beta}} \tag{2.32}$$

Equations (2.31) and (2.32) are evaluated from equations (2.4), (2.15), and (2.2).

Another technique for computing the radii of curvature that is useful for shells of revolution is illustrated in section 4.3.2.3.

2.6 Gauss-Codazzi Relations

Thus far, no connective relationships between the Lamé parameters, A and B, and the principal radii of curvature, R_α and R_β, have been set forth. To explore this further, consider the equality of the second mixed partials

$$\mathbf{t}_{n,\alpha\beta} = \mathbf{t}_{n,\beta\alpha} \tag{2.33}$$

or, from equation (2.17),

$$\left(\frac{A}{R_\alpha}\mathbf{t}_\alpha\right)_{,\beta} = \left(\frac{B}{R_\beta}\mathbf{t}_\beta\right)_{,\alpha} \tag{2.34a}$$

$$\left(\frac{A}{R_\alpha}\right)_{,\beta}\mathbf{t}_\alpha + \frac{A}{R_\alpha}\mathbf{t}_{\alpha,\beta} = \left(\frac{B}{R_\beta}\right)_{,\alpha}\mathbf{t}_\beta + \frac{B}{R_\beta}\mathbf{t}_{\beta,\alpha} \tag{2.34b}$$

Again, using the differential relationships in equation (2.17), we find

$$\left[\left(\frac{A}{R_\alpha}\right)_{,\beta} - \frac{A_{,\beta}}{R_\beta}\right]\mathbf{t}_\alpha = \left[\left(\frac{B}{R_\beta}\right)_{,\alpha} - \frac{B_{,\alpha}}{R_\alpha}\right]\mathbf{t}_\beta \tag{2.34c}$$

With $\mathbf{t}_\alpha$ and $\mathbf{t}_\beta$ being mutually orthogonal, equation (2.34c) may only be satisfied if

$$\left(\frac{A}{R_\alpha}\right)_{,\beta} = \frac{A_{,\beta}}{R_\beta} \tag{2.35}$$

and

$$\left(\frac{B}{R_\beta}\right)_{,\alpha} = \frac{B_{,\alpha}}{R_\alpha} \tag{2.36}$$

If we now consider one of the other second mixed partial derivative identities

$$\mathbf{t}_{\alpha,\alpha\beta} = \mathbf{t}_{\alpha,\beta\alpha}$$

or

$$\mathbf{t}_{\beta,\alpha\beta} = \mathbf{t}_{\beta,\beta\alpha}$$

and manipulate those equations using equation (2.17), a third differential relationship

$$\left(\frac{B_{,\alpha}}{A}\right)_{,\alpha} + \left(\frac{A_{,\beta}}{B}\right)_{,\beta} = -\frac{AB}{R_\alpha R_\beta} \tag{2.37}$$

results. Equations (2.35), (2.36), and (2.37) are known as the Gauss–Codazzi relations and define the connectivity between A, B, R_α, and R_β, such that these parameters define a surface. Equation (2.37) is particularly useful in the derivation of the equations of equilibrium, and is discussed later.

Although we will not pursue the details, note that a parallel set of equations may be derived for the *deformed* middle surface by starting with the radius vector to the deformed surface in place of $\mathbf{t}_n$ in equation (2.33).[2] The resulting equations, properly termed Gauss–Codazzi relations for the deformed middle surface, are the conditions for the *continuity* of the middle surface displacements and the *compatibility* between the strains and displacements. They serve the same role as the St. Venant equations in the Theory of Elasticity. We refer to such equations in connection with the bending of shells in chapter 9.

2.7 Gaussian Curvature

On the right side of equation (2.37), note the fraction $1/R_\alpha R_\beta$, which is the product of the principal curvatures. This is known as the Gaussian curvature and plays an important role in the characterization of shells. Although the Gaussian curvature may be readily computed using the equations of section 2.5, such a calculation is seldom required for purposes of classification; it is often sufficient to know the algebraic sign of the Gaussian curvature.

If we consider the normal sections corresponding to the principal directions, the Gaussian curvature is positive if both centers of curvature lie on the same side of the surface, and is negative if the centers lie on opposite sides. If one of the radii of curvatures is equal to infinity, the Gaussian curvature is zero. Representative cases are shown in figure 2–6. A plate is the degenerate case of a shell with zero Gaussian curvature, since both radii are infinite.

A classification of shells by Gaussian curvature is given in figure 2–7. With respect to this classification, note that only the zero curvature shells are developable, and hence may be formed from flat material. This property of zero curvature shells contributes to their wide usage. Also note that the negative curvature shells have two sets of straight or ruled lines on the surface which correspond to the two sets of real characteristics associated with hyperbolic surfaces,[4] so that the formwork for such a shell can be fabricated from straight materials. This feature of negative curvature shells is largely responsible for the economical usage of reinforced concrete hyperbolic paraboloid (HP) shells in a variety of applications.[5]

The mathematical classification refers to the type of partial differential equation associated with quadratic surfaces having the indicated Gaussian curvature. If a shell is or can be approximated logically as a quadratic surface $Z = f(X,Y)$, then the algebraic sign of the discriminant

$$\delta = Z_{,XX}Z_{,YY} - (Z_{,XY})^2 \tag{2.38}$$

determines the type of partial differential equation and, further, coincides with the sign of the Gaussian curvature,[4] as indicated in figure 2–7.

A variety of thin shell and plate structures are shown in figure 2–8 and will be referred to frequently throughout the book.

2.8 Specialization of Shell Geometry

Because of the wide variety of plate and shell structures encountered in engineering practice, several geometrical classes are of particular interest.

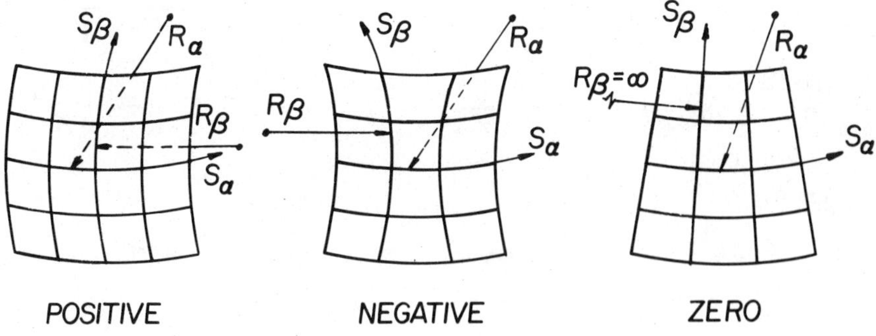

Figure 2–6. Gaussian Curvature

2.8.1 Shallow Shells

The theory of shallow shells has wide application for roof shells that have a relatively small rise as compared to their spans. Considering the surface specified in Cartesian coordinates as in equation (2.1), the shell is said to be shallow if, in the subsequent mathematical analysis, $(Z_{,X})^2$ and $(Z_{,Y})^2$ may be neglected by virtue of smallness in comparison to unity. The effect of this simplification is seen by considering figure 2–9.

Consider a differential element of the middle surface bound by the intersections with two planes parallel to the Y–Z plane and separated by a distance dX apart, and two planes parallel to the X–Z plane separated by a distance dY. As illustrated by the inset in figure 2–9,

$$ds_X \simeq dX[1 + (Z_{,X})^2]^{1/2} \tag{2.39a}$$

$$ds_Y \simeq dY[1 + (Z_{,Y})^2]^{1/2} \tag{2.39b}$$

which, when the geometric simplification is applicable, become

$$ds_X \simeq dX \tag{2.40a}$$

$$ds_Y \simeq dY \tag{2.40b}$$

The practical interpretation of equations (2.40) is that the curvilinear coordinates α and β may be selected as the Cartesian coordinates X and Y with the Lamé parameters $A = B = 1$. Additional approximations are introduced into shallow shell theory with respect to the equilibrium equations in chapter 9.

The practical range of the theory of shallow shells is restricted to shells with a central rise of one-fifth or less than the span.[6] However, this theory has been applied to shells that may not fit the criteria in a global

Gaussian Curvature Classification	Positive	Negative	Zero
Surface	Doubly Curved Synclastic	Doubly Curved Anticlastic	Singly Curved
Developability	Nondevelopable	Nondevelopable	Developable
Type of Equation (Discriminant)	Elliptic (Positive)	Hyperbolic (Negative)	Parabolic (Zero)
Straight or Ruled Lines on Surface	None	Two Sets (A and B Below)	One Set (C Below)
Examples	 Sphere	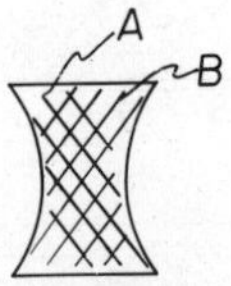 Hyperboloid of Revolution	 Cylinder
	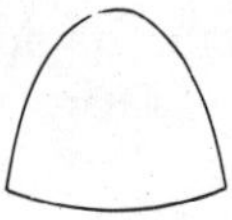 Paraboloid of Revolution	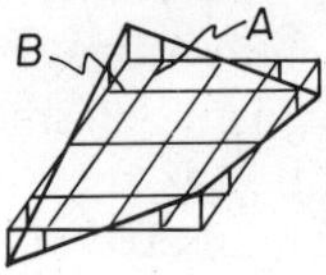 Hyperbolic Paraboloid	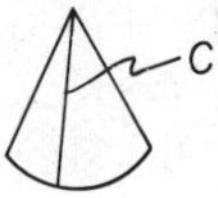 Cone
			Flat Plate (Degenerate Case)

Figure 2–7. Classification of Shells by Gaussian Curvature

sense, by treating pieces or finite elements of the shell which can be considered shallow, and then assembling these elements to satisfy the global geometry.[7]

2.8.2 Shells of Revolution

A surface of revolution is generated by rotating a plane curve *generator* around an *axis of rotation* to form a closed surface, as illustrated in figure 2–10. The lines of principal curvature are called the *meridians* (normal

Figure 2–8(a). Open Hyperbolic Paraboloid Roof, Colosseum San Juan, Puerto Rico (Courtesy Professor A. Scordelis)

sections formed by planes containing the axis of rotation) and *parallel circles* (normal sections traced by planes perpendicular to the axis of rotation).

In figure 2–11, we show the meridian of a shell of revolution illustrating both positive and negative Gaussian curvature. The equation of the meridian is given by

$$Z = Z(R) \tag{2.41a}$$

where

$$R = \sqrt{(X^2 + Y^2)} \tag{2.41b}$$

Consider a reference point on the surface. The angle formed by the extended normal to the surface at this point and the axis of rotation is defined as the *meridional angle* ϕ; and, the angle between the radius of the parallel circle at the point and the X axis is designated as the *circumferential angle* θ. Correspondingly, the meridians are taken as the α coordinate lines, and $R_\alpha = R_\phi$, the meridional radius of curvature. The parallel circles are the β coordinate line, with $R_\beta = R_\theta$, the circumferential radius of curvature. The radius of the parallel circle, which is equal to R as defined in equation (2.41b), is termed the *horizontal radius* and is denoted by R_0. Note that R_0 is not a principal radius of curvature, since it is not normal to the surface. Rather, it is the projection of R_θ on the horizontal plane, i.e.

Figure 2–8(b). Hyperbolic Paraboloid Roof, Theater, Northbrook, Ill. (Author's Photo)

$$R_0 = R_\theta \sin \phi \qquad (2.42)$$

A closed shell of revolution is frequently called a *dome*, and the peak of such a shell is termed the *pole*. A pole introduces certain mathematical complications because, at this point, $R \rightarrow 0$.

In most applications, the curvilinear coordinate in the β or θ direction is chosen as the circumferential angle θ. Therefore, from equation (2.10b),

$$ds_\beta = Bd\beta = ds_\theta = R_0 \, d\theta \qquad (2.43a)$$

and thus

$$B = R_0 = R_\theta \sin \phi \qquad (2.43b)$$

Figure 2–8(c). Vertical Hyperbolic Paraboloids, Church, San Francisco, Calif. (Author's Photo)

In the α or ϕ direction, there are at least three useful choices for the curvilinear coordinate: a) the meridional angle ϕ; b) the axial coordinate Z; and c) the arc length s_ϕ. The respective Lamé parameter A for each case is:

a) *Meridional angle, ϕ:*

$$ds_\phi = R_\phi \, d\phi \tag{2.44a}$$

and from equation (2.10)

$$A = R_\phi \tag{2.44b}$$

b) *Axial coordinate, Z:*

Figure 2–8(d). Hyperbolic Paraboloids, Church, Mexico City, Mexico (Author's Photo)

Figure 2–8(e). Cantilevered Hyperbolic Paraboloid, Canopy, Miami Beach, Fla. (Author's Photo)

Figure 2–8(f). Hyperbolic Paraboloid Vaults, Church, St. Louis County, Mo. (Author's Photo)

Considering an element of arc length similar to that shown on the inset of figure 2–9,

$$ds_\phi^2 = dZ^2 + dR_0^2$$

$$= dZ^2 + (R_{0,z})^2\, dZ^2 \tag{2.45a}$$

$$ds_\phi = dZ[1 + (R_{0,z})^2]^{1/2} \tag{2.45b}$$

and therefore

$$A = [1 + (R_{0,z})^2]^{1/2} \tag{2.45c}$$

c) *Arc length* s_ϕ:

$$A = 1 \tag{2.45d}$$

Figure 2–8(g). Hyperboloid of Revolution, Planetarium, St. Louis, Mo. (Author's Photo)

Figure 2–8(h). Open Cylindrical Roof, Airport, Barcelona, Spain (Author's Photo)

Figure 2–8(i). Spherical Roof, Auditorium, Cambridge, Mass. (Author's Photo)

Figure 2–8(j). Folded Dome, Assembly Hall, Urbana, Ill. (Author's Photo)

Figure 2–8(k). Intersecting Barrel Shells, Airport, St. Louis, Mo. (Author's Photo)

Figure 2–8(l). Shallow Spherical Roof with Cutouts, Restaurant, Miami Beach, Fla. (Author's Photo)

Figure 2–8(m). Shell-Like Spherical Roof, Astrodome, Houston, Tex. (Author's Photo)

Figure 2–8(n). Inside of Astrodome (Authors Photo)

Figure 2–8(o). Cooling Towers, Schmeehausen, West Germany (Author's Photo)

Figure 2–8(p). Supporting Columns for Hyperbolic Cooling Tower (Courtesy Prof. W. Schnobrich)

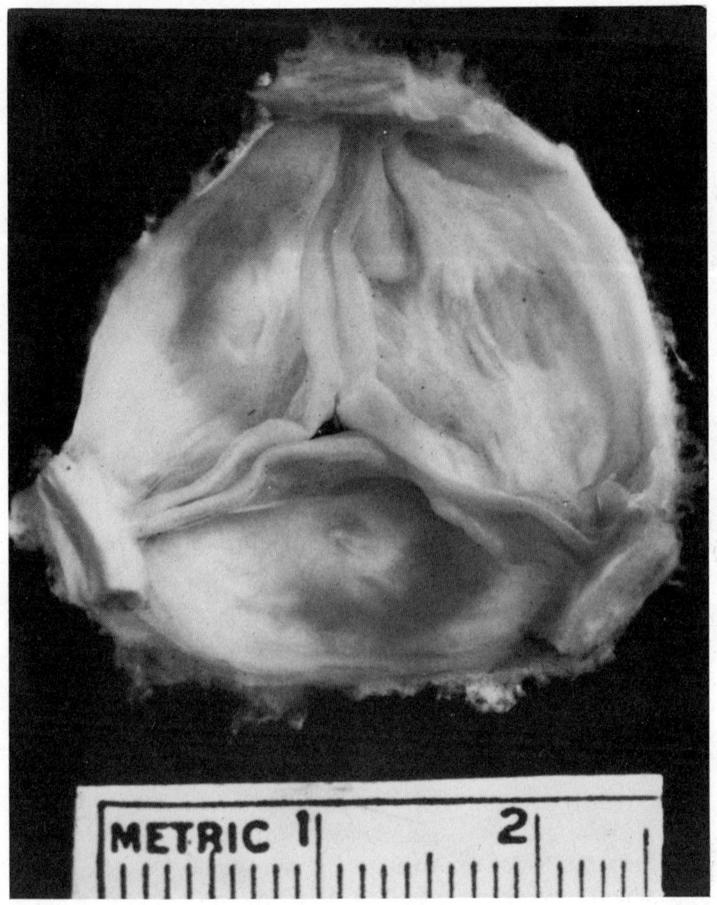

Source: P. L. Gould et al., "Stress Analysis of the Human Aortic Valve," *Journal of Computers and Structures* 3 (1973): 379.

Figure 2–8(q). Human Aortic Heart Valve (Courtesy Dr. R. Clark. Reprinted with permission of Pergamon Press.)

but since the meridian is given by $Z = Z(R_0)$, the arc length coordinate must be computed by integrating equation (2.45b). Therefore,

$$s_\phi = \int_{Z(0)}^{Z(\phi)} [1 + (R_{0,z})^2]^{1/2}\, dZ \qquad (2.46)$$

The limits of the integral indicate the coordinate $Z(\phi)$ at the origin for s_ϕ and at the section where the coordinate is being evaluated, respectively.

The choice of an appropriate curvilinear coordinate for the meridional direction is problem-dependent and involves several considerations. In tower-type shells, which are essentially vertical structures, the axial co-

Figure 2–8(r). Stiffened Shell (Courtesy Chicago Bridge & Iron Co.)

ordinate Z has the greatest significance with respect to the physical construction of the shell. For relatively flat domes, however, the axial coordinate approaches zero even for points relatively far from the pole. The meridional angle ϕ may behave in a similar manner, so the arc length will be the most stable coordinate for such cases. Also, if the meridian has an inflection point, the coordinate ϕ might cause difficulties because it may not provide a one-to-one correspondence with all points on the shell surface.

The most popular, but by no means universal, preference of theoreticians in the field of shells of revolution has been to choose the merdional angle ϕ as the basis for the development of the governing equations. This may be due somewhat to historical precedent, since the early work on shells of revolution focused on spherical shells[8] for which $A = R_\phi = $ constant, thereby greatly simplifying the ensuing treatment.

We noted in section 2.6 that the parameters A, B, R_α, and R_β must satisfy the three Gauss–Codazzi conditions to define a surface. These condi-

Figure 2–8(s). Water Tower (Courtesy Chicago Bridge & Iron Co.)

tions, as given by equations (2.35), (2.36), and (2.37), can be checked for the ϕ–θ curvilinear coordinate system:

Equation (2.35):
Satisfied identically, since none of the parameters are functions of $\beta(\theta)$.

Equation (2.36):

$$\left(\frac{R_\theta \sin \phi}{R_\theta}\right)_{,\phi} = \frac{(R_\theta \sin \phi)_{,\phi}}{R_\phi}$$

$$\cos \phi = \frac{(R_\theta \sin \phi)_{,\phi}}{R_\phi}$$

Figure 2–8(t). Column-Supported Spherical Tanks (Courtesy Chicago Bridge & Iron Co.)

or

$$(R_\theta \sin \phi)_{,\phi} = R_{0,\phi} = R_\phi \cos \phi \tag{2.47}$$

Equation (2.37):

$$\left[\frac{(R_\theta \sin \phi)_{,\phi}}{R_\phi}\right]_{,\phi} = -\frac{R_\phi R_\theta \sin \phi}{R_\phi R_\theta} \tag{2.48}$$

Substituting equation (2.47) for the numerator of equation (2.48) gives an identity. Thus, the Gauss–Codazzi conditions for a shell of revolution described by the coordinates ϕ and θ are satisfied provided equation (2.47) is valid. This equations is very useful in what follows, and it is instructive to derive it from a purely geometric argument. Consider the meridian shown in figure 2–12:

Referring to points B and D

At B, $R_0(\phi) = AB$

At D, $R_0(\phi + \Delta\phi) = CD$

$\Delta R_0 = CD - AB$

$\qquad = BD \sin\left(\dfrac{\pi}{2} - \phi\right)$

$\qquad = BD \cos \phi$

Figure 2–8(u). Column-Supported, Stiffened Water Tower (Courtesy Chicago Bridge & Iron Co.)

Since $BD = R_\phi \Delta\phi$

$$\Delta R_0 = R_\phi \cos \phi \, \Delta\phi$$

$$\Delta R_0 = \Delta(R_\theta \sin \phi) = R_\phi \cos \phi \, \Delta\phi$$

Or,

$$\frac{\Delta(R_\theta \sin \phi)}{\Delta\phi} = R_\phi \cos \phi$$

$\lim \Delta\phi \to 0$ gives equation (2.47).

One often desires to calculate the value of Z corresponding to a par-

Figure 2–8(v). Steel Hyperbolic Paraboloid Roof, Aircraft Hangar (Courtesy Lev Zetlin Associates, Inc.)

Figure 2–8(w). Waffle Slab, Library, St. Louis, Mo. (Author's Photo)

48

Figure 2–8(x). Folded Plate Roof, Law School, St. Louis, Mo. (Author's Photo)

ticular value of ϕ. R_θ and R_ϕ may be computed from the equation of the curve using equations (2.29a and b). Once R_θ is computed, R_0 can be calculated from equation (2.42) and substituted into equation (2.41a) to find the corresponding value of $Z(\phi)$. If the arc length also is required, equation (2.46) should be evaluated. A direct transformation between s_ϕ and ϕ can be derived by integrating equation (2.44a).

Also, it is expedient to be able to transform derivatives with respect to ϕ to derivatives with respect to Z, i.e. $(\ ),_z = (\ ),_\phi \cdot (d\phi/dZ)$. If we consider equation (2.45c) with $R_{0,z}$ given by $R_{0,\phi} \cdot (d\phi/dZ)$ and $R_{0,\phi}$ given by equation (2.47), we have

$$A = \left[1 + R_\phi^2 \cos^2\phi \left(\frac{d\phi}{dZ}\right)^2 \right]^{\frac{1}{2}} \tag{2.49}$$

for the axial coordinate.

Next, we write ds_ϕ for both the ϕ and Z coordinates

$$ds_\phi = R_\phi d\phi = \left[1 + R_\phi^2 \cos^2\phi \left(\frac{d\phi}{dZ}\right)^2 \right]^{\frac{1}{2}} dZ$$

or

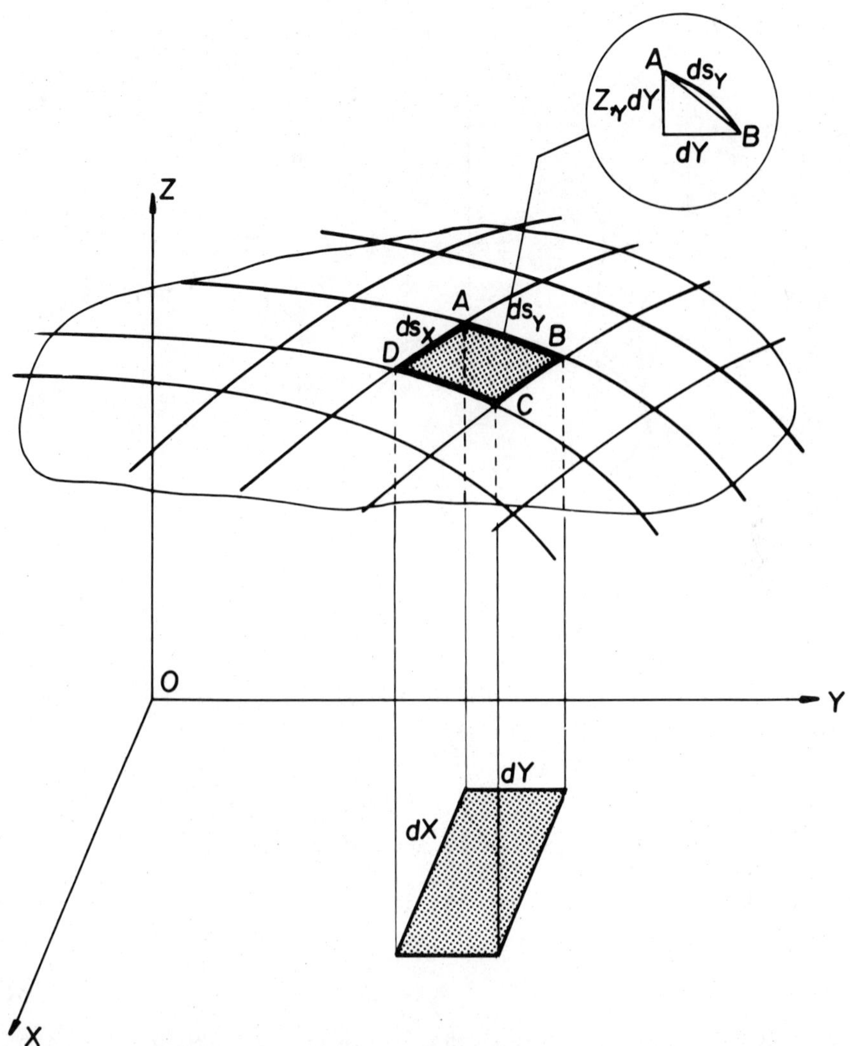

Figure 2–9. Shallow Shell Geometry

$$R_\phi\left(\frac{d\phi}{dZ}\right) = \left[1 + R_\phi^2 \cos^2\phi\left(\frac{d\phi}{dZ}\right)^2\right]^{\frac{1}{2}}$$

Squaring both sides, we find

$$\frac{d\phi}{dZ} = \frac{1}{R_\phi(1 - \cos^2\phi)^{\frac{1}{2}}}$$

$$= \frac{1}{R_\phi \sin\phi} \tag{2.50}$$

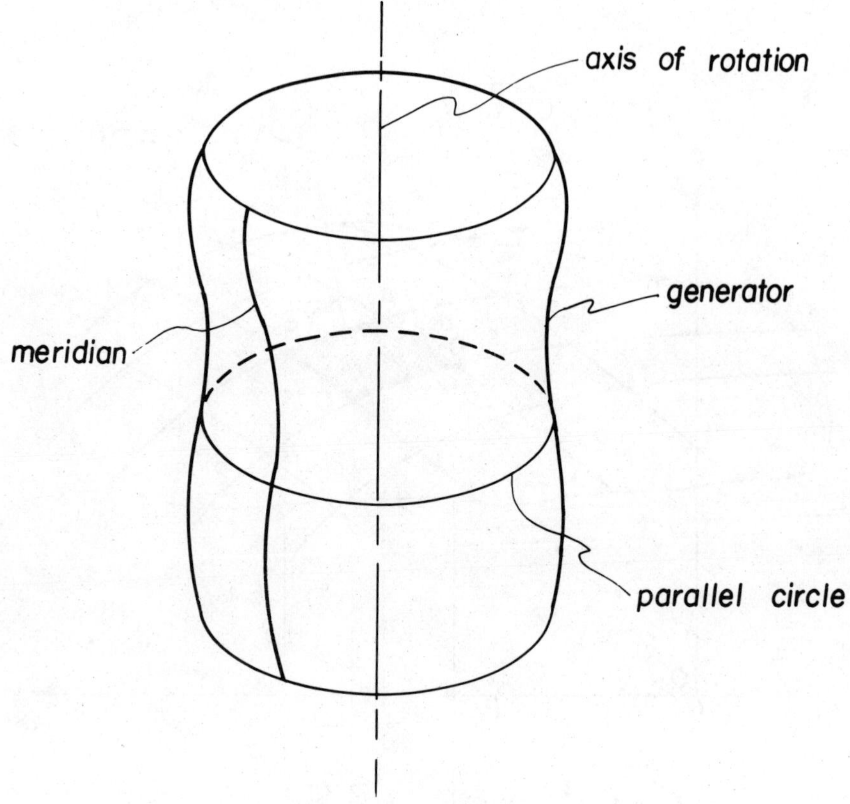

Figure 2–10. Shell of Revolution

Substituting equation (2.50) into (2.49) gives

$$A = \csc \phi \qquad (2.51)$$

Equation (2.50) is quite useful, since it is often preferable to differentiate with respect to ϕ first in order to utilize equation (2.47).

Notes

1. H. Kraus, *Thin Elastic Shells* (New York: Wiley, 1967), pp. 12–14.
2. The complete derivation requires a rather tedius set of calculations which are presented in specialized shell texts such as Kraus's *Thin Elastic*

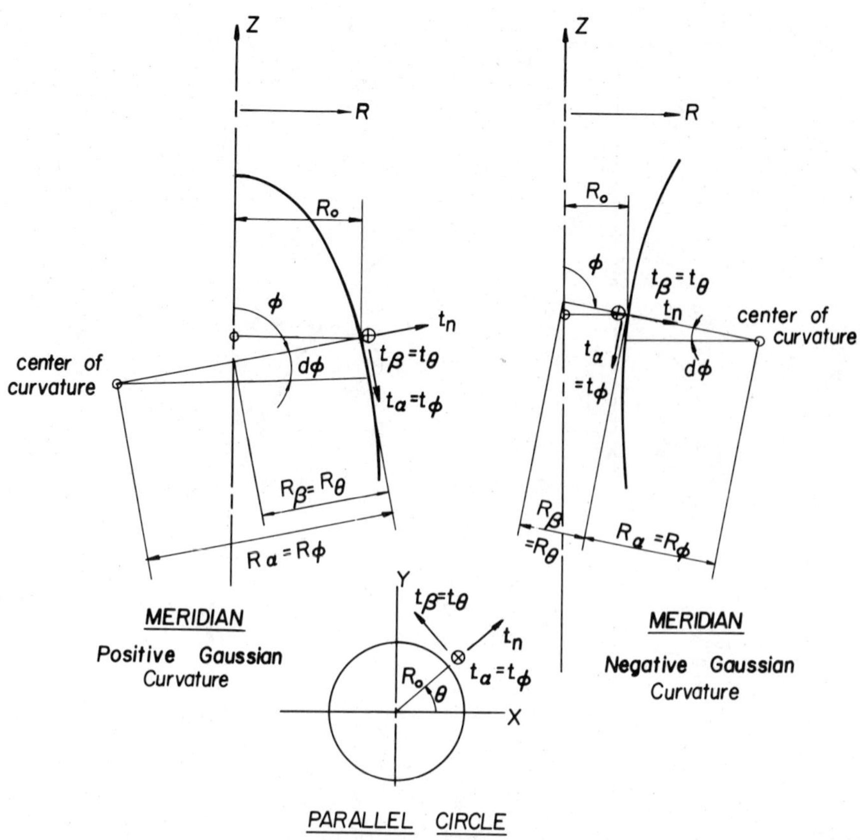

Figure 2–11. Meridians of a Shell of Revolution

Shells, pp. 14–16, and V. V. Novozhilov's *Thin Shell Theory* [translated from 2nd Russian ed. by P. G. Lowe (Groningen, The Netherlands: Noordhoff, 1964), pp. 9–30]. Here, we find it adequate to list the relationships, initially without proof, and then to selectively derive some of the terms to illustrate the mathematical procedure. Our derivations basically follow those presented in Novozhilov's *Thin Shell Theory*.

3. A. L. Gol'denveizer, *Theory of Elastic Thin Shells* [translated from Russian by G. Herrmann (New York: Pergamon Press, 1961), p. 13].

4. A. M. Haas, *Design of Thin Concrete Shells*, vol. 2 (New York: Wiley, 1967), pp. 5–11.

5. C. Faber, *Candela: The Shell Builder* (New York: Reinhold, 1963).

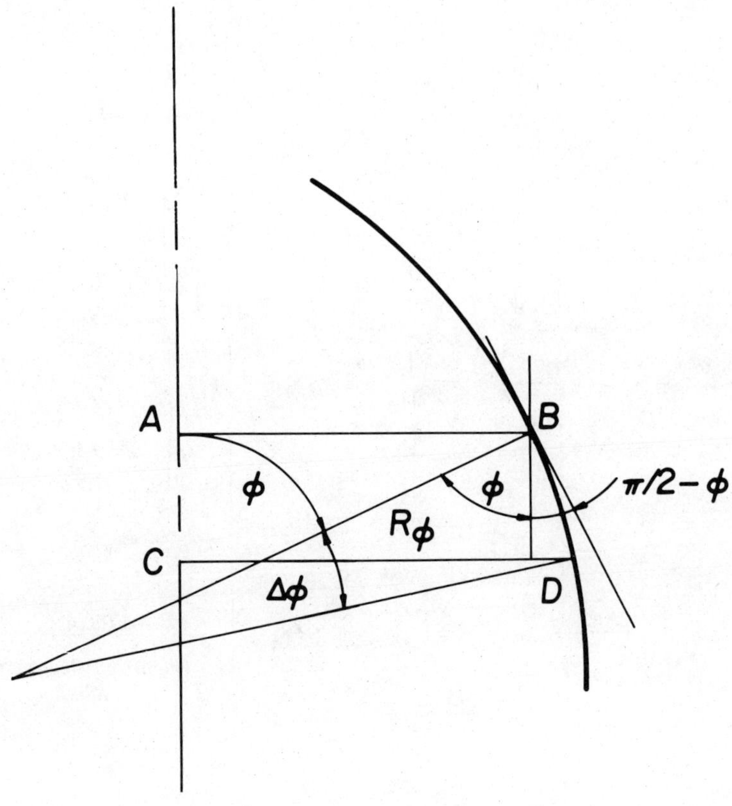

Figure 2–12. Geometrical Interpretation of Gauss–Codazzi Condition

6. Novozhilov, *Thin Shell Theory*, pp. 94–99.

7. G. R. Cowper, G. M. Lindberg, M. D. Olson, "A Shallow Shell Finite Element of Triangular Shape," *International Journal of Solids and Structures* 6, no. 8 (1970): 1133–1156.

8. H. Reissner, *Spannungen in Kugelschalen* (Leipzig: Muller-Breslau-Festschrift, 1912), pp. 181–193.

3   Equilibrium

3.1 Stress Resultants and Couples

Consider the element of the shell shown in figure 3–1(a) that is bounded by the normal sections α, $\alpha + d\alpha$, β, and $\beta + d\beta$. The geometry of the middle surface of such an element was considered in chapter 2 (see figures 2–2 and 2–4). Here we show the entire thickness h with the coordinate ζ defined in the direction of t_n and depict a differential volume element $d\alpha\ d\beta\ d\zeta$ parallel to and displaced from the middle surface a distance ζ with thickness $d\zeta$.

In figure 3–1(b), the stresses acting on the volume element are indicated. The sign convention is that of the theory of elasticity, with the first subscript indicating the *surface* on which the stress acts, as identified by the normal to that surface, and the second subscript indicating the *direction* in which the stress acts. The solution for the various *stresses*, shown in figure 3–1(b) at a particular *point* of the continuum (α, β, ζ), is the fundamental problem of the thoery of elasticity. Plate and shell theories deal with a simplification of the *stress at a point* problem. Instead of the stresses, it is considered sufficient to solve for the *total forces and moments* per length of middle surface, which are known as *stress resultants* and *stress couples*, respectively.[1] The stress at any point presumably can then be computed by back-substitution, although we will find that this can sometimes be done only approximately.

We have thus alluded to the major simplification present in the plate and shell problem as compared to the theory of elasticity problem: the plate or shell problem is formulated and solved for the forces and moments per unit length of the middle surface, rather than for the stresses at each point of the continuum. Toward this end, we now consider the definition of the stress resultants and couples in terms of the stresses.

In figure 3–2, the stress resultants and couples are shown on an element of the middle surface. The extensional forces N_α and N_β, transverse shearing forces Q_α and Q_β, and bending moments M_α and M_β are represented by simple subscripted variables indicating the direction of the vector, whereas the in-plane shearing forces $N_{\alpha\beta}$ and $N_{\beta\alpha}$, and the twisting moments $M_{\alpha\beta}$ and $M_{\beta\alpha}$ carry a double subscript. The first subscript corre-

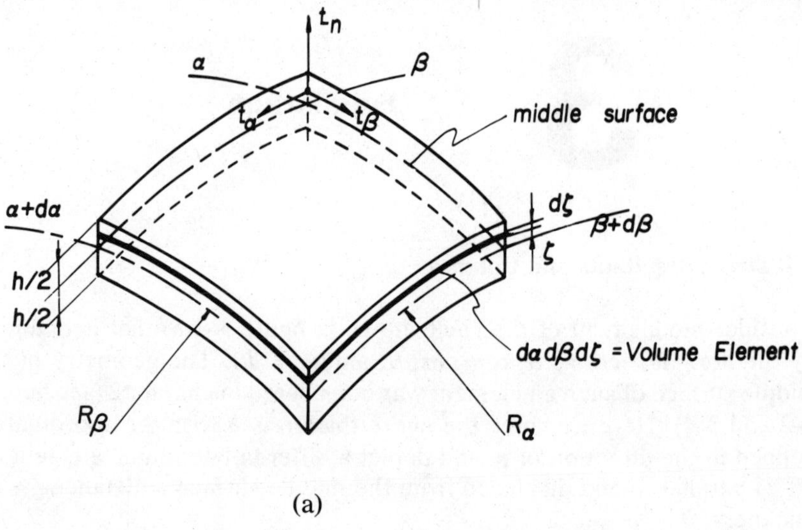

(a)

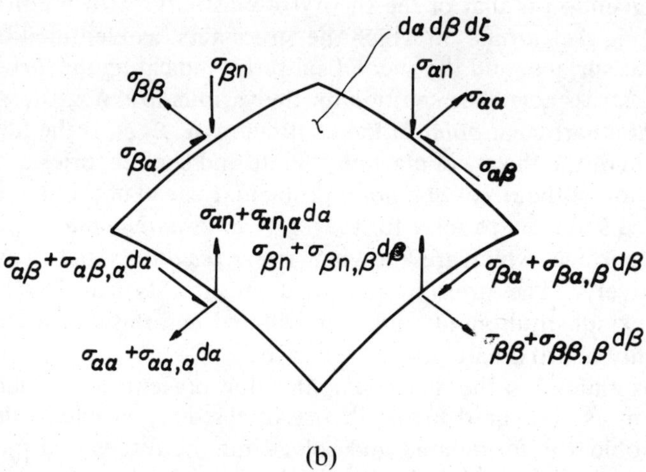

(b)

Figure 3–1. Shell Volume Element and Stresses

sponds to the surface on which the force or moment acts, whereas the second indicates the direction in which the force or moment acts. Though this notation is not completely consistent with the theory of elasticity definitions cited in connection with figure 3–1, it has been used almost exclusively in classical texts and references, and a departure here is felt to be inappropriate.

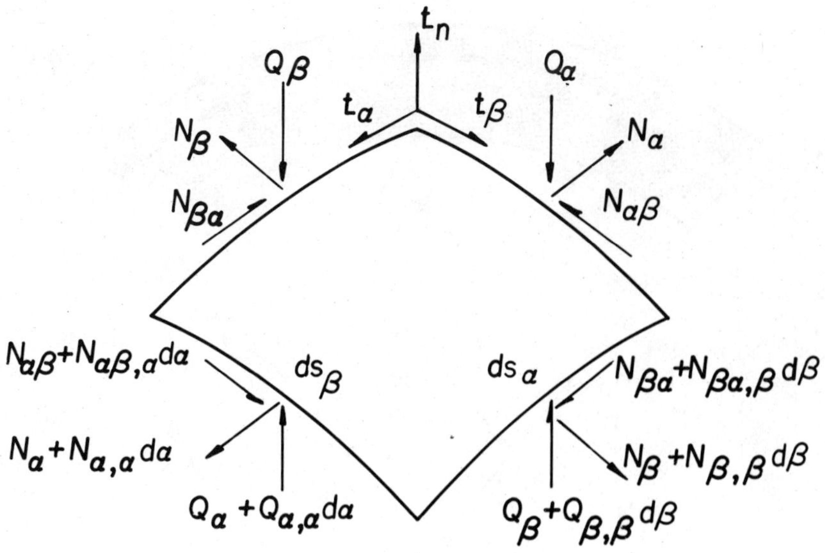

(a) Stress Resultants

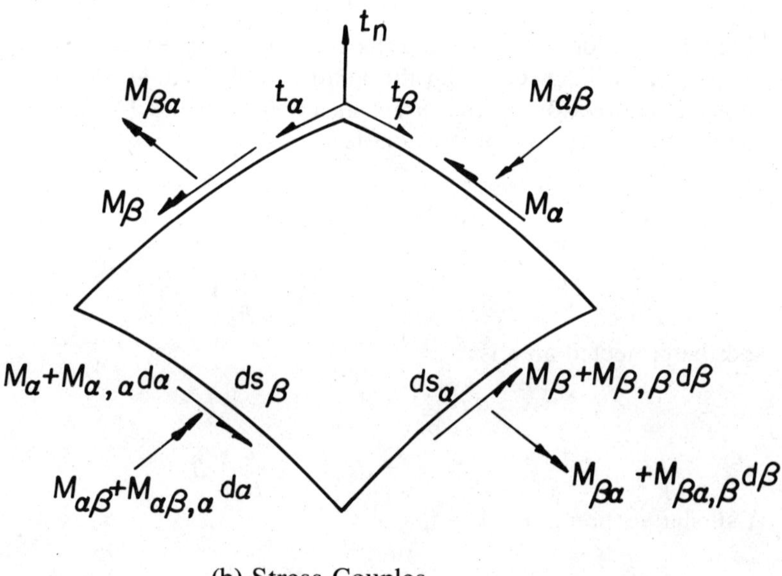

(b) Stress Couples

Figure 3–2. Stress Resultants and Stress Couples

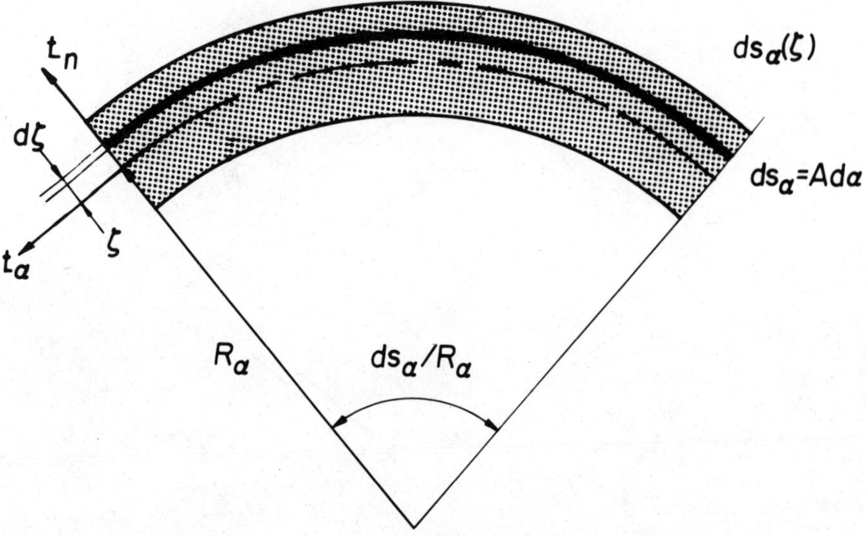

Figure 3–3. Normal Section on a Shell Element

We now focus on an arc of the middle surface

$$ds_\alpha = A\,d\alpha \tag{3.1}$$

along the s_α coordinate line and the corresponding arc of the volume element lying a distance ζ from the middle surface along the normal $\mathbf{t}_n$ as shown in figure 3–3, which represents a view of figure 3–1(a) normal to $\mathbf{t}_\beta$. The length of the volume element is

$$ds_\alpha(\zeta) = \frac{R_\alpha + \zeta}{R_\alpha}\,ds_\alpha$$

$$= ds_\alpha\left(1 + \frac{\zeta}{R_\alpha}\right) \tag{3.2}$$

and the projected area is

$$da_\beta(\zeta) = ds_\alpha(\zeta)\,d\zeta$$

$$= ds_\alpha\left(1 + \frac{\zeta}{R_\alpha}\right)d\zeta \tag{3.3}$$

A similar section normal to $\mathbf{t}_\alpha$ gives

$$ds_\beta(\zeta) = ds_\beta\left(1 + \frac{\zeta}{R_\beta}\right) \tag{3.4}$$

$$da_\alpha(\zeta) = ds_\beta\left(1 + \frac{\zeta}{R_\beta}\right)d\zeta \tag{3.5}$$

The magnitudes of the stress resultants acting in the section normal to $\mathbf{t}_\alpha$ are

$$\begin{Bmatrix} N_\alpha \\ N_{\alpha\beta} \\ Q_\alpha \end{Bmatrix} = \frac{1}{ds_\beta} \int_{-h/2}^{h/2} \begin{Bmatrix} \sigma_{\alpha\alpha} \\ \sigma_{\alpha\beta} \\ \sigma_{\alpha n} \end{Bmatrix} da_\alpha(\zeta)$$

$$= \int_{-h/2}^{h/2} \begin{Bmatrix} \sigma_{\alpha\alpha} \\ \sigma_{\alpha\beta} \\ \sigma_{\alpha n} \end{Bmatrix} \left(1 + \frac{\zeta}{R_\beta} \right) d\zeta \qquad (3.6)$$

On the section normal to $\mathbf{t}_\beta$ we have

$$\begin{Bmatrix} N_\beta \\ N_{\beta\alpha} \\ Q_\beta \end{Bmatrix} = \frac{1}{ds_\alpha} \int_{-h/2}^{h/2} \begin{Bmatrix} \sigma_{\beta\beta} \\ \sigma_{\beta\alpha} \\ \sigma_{\beta n} \end{Bmatrix} da_\beta(\zeta)$$

$$= \int_{-h/2}^{h/2} \begin{Bmatrix} \sigma_{\beta\beta} \\ \sigma_{\beta\alpha} \\ \sigma_{\beta n} \end{Bmatrix} \left(1 + \frac{\zeta}{R_\alpha} \right) d\zeta \qquad (3.7)$$

The magnitudes of the stress couples are

$$\begin{Bmatrix} M_\alpha \\ M_{\alpha\beta} \end{Bmatrix} = \frac{1}{ds_\beta} \int_{-h/2}^{h/2} \begin{Bmatrix} \sigma_{\alpha\alpha} \\ \sigma_{\alpha\beta} \end{Bmatrix} \zeta \, da_\alpha(\zeta)$$

$$= \int_{-h/2}^{h/2} \begin{Bmatrix} \sigma_{\alpha\alpha} \\ \sigma_{\alpha\beta} \end{Bmatrix} \zeta \left(1 + \frac{\zeta}{R_\beta} \right) d\zeta \qquad (3.8)$$

and

$$\begin{Bmatrix} M_\beta \\ M_{\beta\alpha} \end{Bmatrix} = \frac{1}{ds_\alpha} \int_{-h/2}^{h/2} \begin{Bmatrix} \sigma_{\beta\beta} \\ \sigma_{\beta\alpha} \end{Bmatrix} \zeta \, da_\beta(\zeta)$$

$$= \int_{-h/2}^{h/2} \begin{Bmatrix} \sigma_{\beta\beta} \\ \sigma_{\beta\alpha} \end{Bmatrix} \zeta \left(1 + \frac{\zeta}{R_\alpha} \right) d\zeta \qquad (3.9)$$

58

We reiterate at this point that once the plate or shell problem has been solved for the stress resultants and couples, the recovery of the elasticity problem would necessitate the inversion of equations (3.6)–(3.9), which might prove difficult in a mathematically exact sense. In a way, this is the price paid to realize a simplified theory—the sacrifice of mathematical precision to achieve a solution that is satisfactory from an engineering standpoint.

In practice, the stresses at a particular point in the continuum are usually recovered from the stress resultants and couples by (a) neglecting the terms $(1 + \zeta/R_\alpha)$ and $(1 + \zeta/R_\beta)$ in equations (3.6)–(3.9); and (b) assuming that all stresses, except for the transverse shearing stress $\sigma_{\alpha n}$ and $\sigma_{\beta n}$, vary linearly across the cross section. The preceeding arguments imply that

$$\sigma_{ii}(\zeta) = \frac{N_i}{h} + \sigma_{ii}\left(\frac{h}{2}\right)\frac{\zeta}{h/2}$$

$$\sigma_{ij}(\zeta) = \frac{N_{ij}}{h} + \sigma_{ij}\left(\frac{h}{2}\right)\frac{\zeta}{h/2} \qquad (i,j = \alpha,\beta)$$

(3.10a)

It is consistent with elementary theory to assume that the transverse shearing stresses are distributed through the thickness as a second-degree parabola with a maximum at the middle surface. Thus,

$$\sigma_{in}(\zeta) \simeq \sigma_{in}(0)\left[1 - \left(\frac{\zeta}{h/2}\right)^2\right] (i = \alpha,\beta) \qquad (3.10b)$$

When these relationships are introduced into equations (3.6)–(3.9), the maximum stresses in the cross section are

$$\sigma_{ii}\left(\pm\frac{h}{2}\right) = \frac{N_i}{h} \pm \frac{6M_i}{h^2} \qquad (i = \alpha,\beta) \qquad (3.10c)$$

$$\sigma_{ij}\left(\pm\frac{h}{2}\right) = \frac{N_{ij}}{h} \pm \frac{6M_{ij}}{h^2} \qquad \left(\begin{matrix} i = \alpha,\beta \\ j = \beta,\alpha \end{matrix}\right) \qquad (3.10d)$$

$$\sigma_{in}(0) = \frac{3}{2}\frac{Q_i}{h} \qquad (i = \alpha,\beta) \qquad (3.10e)$$

Then, the stresses at any other point within the cross section are found from equations (3.10a) and (3.10b).

3.2 Equilibrium of the Shell Element

Now that all the stresses acting within the shell are referred to quantities defined on the middle surface, we have reduced the problem to two dimensions and seek a set of relationships between the stress resultants and

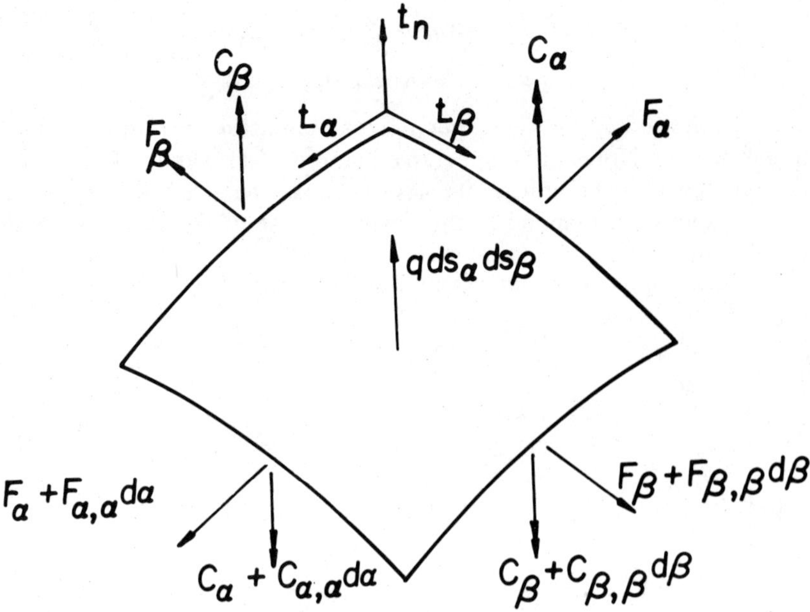

Figure 3–4. Force and Couple Vectors

couples that reflects the equilibrium of the middle surface. In this derivation, we basically follow the approach of Novozhilov,[2] which is based on vector algebra. More concise mathematical approaches based principally on tensor calculus,[3] as well as more physical approaches relying mainly on the free body diagram and trigonometry,[4] are available; in selecting the vector derivation, we hope to achieve an accommodation between mathematical elegance and sometimes cumbersome physical reasoning. The physical interpretation of the resulting equations will be explored later for selected geometrical forms.

The stress resultants and couples on each face of the middle surface element, shown in figure 3–2, are combined into force and couple vectors in figure 3–4. Also shown is the load vector **q**. In terms of the stress resultants and couples, the resultant vectors are

$$\mathbf{F}_\alpha = (N_\alpha \mathbf{t}_\alpha + N_{\alpha\beta}\mathbf{t}_\beta + Q_\alpha \mathbf{t}_n)\,ds_\beta \qquad (3.11a)$$

$$\mathbf{F}_\beta = (N_{\beta\alpha}\mathbf{t}_\alpha + N_\beta \mathbf{t}_\beta + Q_\beta \mathbf{t}_n)\,ds_\alpha \qquad (3.11b)$$

$$\mathbf{C}_\alpha = (-M_{\alpha\beta}\mathbf{t}_\alpha + M_\alpha \mathbf{t}_\beta)\,ds_\beta \qquad (3.11c)$$

$$\mathbf{C}_\beta = (-M_\beta \mathbf{t}_\alpha + M_{\beta\alpha}\mathbf{t}_\beta)\,ds_\alpha \qquad (3.11d)$$

and the load vector is

$$\mathbf{q}\, ds_\alpha\, ds_\beta = [q_\alpha(\alpha, \beta)\mathbf{t}_\alpha + q_\beta(\alpha, \beta)\mathbf{t}_\beta$$

$$+ q_n(\alpha, \beta)\mathbf{t}_n]\, ds_a\, ds_\beta \tag{3.11e}$$

where q_α, q_β, and q_n are the respective load intensities per unit area of the middle surface. Note with respect to equations (3.11e) and (3.11d) that the negative signs on the first terms arise because the stress couple vectors were chosen to correspond to the positive sense of the stresses (see figures 3–1 and 3–2).

We now apply the equations of static equilibrium

$$\sum \mathbf{F} = 0 \tag{3.12a}$$

$$\sum \mathbf{C} = 0 \tag{3.12b}$$

Referring to figure 3–4, the force equilibrium equation, equation (3.12a), becomes

$$(\mathbf{F}_\alpha + \mathbf{F}_{\alpha,\alpha}\, d\alpha - \mathbf{F}_\alpha) + (\mathbf{F}_\beta + \mathbf{F}_{\beta,\beta}\, d\beta - \mathbf{F}_\beta) + \mathbf{q}\, ds_\alpha\, ds_\beta = 0$$

or

$$\mathbf{F}_{\alpha,\alpha}\, d\alpha + \mathbf{F}_{\beta,\beta}\, d\beta + \mathbf{q}AB\, d\alpha\, d\beta = 0 \tag{3.13}$$

Substituting equations (3.11a), (3.11b), and (3.11e) into equation (3.13), and dividing through by $d\alpha\, d\beta$, we have

$$[(N_\alpha \mathbf{t}_\alpha + N_{\alpha\beta}\mathbf{t}_\beta + Q_\alpha \mathbf{t}_n)B]_{,\alpha} + [(N_{\beta\alpha}\mathbf{t}_\alpha + N_\beta \mathbf{t}_\beta + Q_\beta \mathbf{t}_n)A]_{,\beta}$$

$$+ (q_\alpha \mathbf{t}_\alpha + q_\beta \mathbf{t}_\beta + q_n \mathbf{t}_n)AB = 0 \tag{3.14}$$

The differentiations indicated in equation (3.14) are carried out in a straightforward manner using equation (2.17) to evaluate the derivatives of the unit tangent vectors. The resulting vector equations may be factored into the form

$$F_\alpha \mathbf{t}_\alpha + F_\beta \mathbf{t}_\beta + F_n \mathbf{t}_n = 0 \tag{3.15}$$

Since the unit tangent vectors are independent, equation (3.15) can only be satisfied if

$$F_\alpha = 0; \quad F_\beta = 0; \quad F_n = 0 \tag{3.16}$$

which gives the three scalar equations of *force* equilibrium

$$F_\alpha = [(BN_\alpha)_{,\alpha} + (AN_{\beta\alpha})_{,\beta} + A_{,\beta}N_{\alpha\beta} - B_{,\alpha}N_\beta]$$

$$+ Q_\alpha \frac{AB}{R_\alpha} + q_\alpha AB = 0 \tag{3.17a}$$

$$F_\beta = [(BN_{\alpha\beta})_{,\alpha} + (AN_\beta)_{,\beta} + B_{,\alpha}N_{\beta\alpha} - A_{,\beta}N_\alpha]$$

$$+ Q_\beta \frac{AB}{R_\beta} + q_\beta AB = 0 \tag{3.17b}$$

$$F_n = [(BQ_\alpha)_{,\alpha} + (AQ_\beta)_{,\beta}] - N_\alpha \frac{AB}{R_\alpha} - N_\beta \frac{AB}{R_\beta} + q_n AB = 0 \tag{3.17c}$$

The individual terms in equations (3.17a–c) will be interpreted in section 3.3 for the shell of revolution geometry. However, note that there is considerable coupling of the equations; i.e., N_α appears in F_β and F_n as well as in F_α as expected. This coupling is due to two general sources: (a) the curvature of the surface; and (b) the possibility that the Lamé parameters A and B vary with both α and β.

If, for example, $A = A(\alpha)$ and $B = B(\beta)$ only, many of the coupling terms drop out. This emphasizes the importance of choosing the curvilinear coordinates so as to achieve the simplest mathematical formulation consistent with the problem.

The moment equilibrium equation, (3.12b), is evaluated about axes through point O in figure 3–4.

1. First, we have the contributions of the stress couple vectors

$$\mathbf{C}_\alpha + \mathbf{C}_{\alpha,\alpha}\, d\alpha - \mathbf{C}_\alpha + \mathbf{C}_\beta + \mathbf{C}_{\beta,\beta}\, d\beta - \mathbf{C}_\beta \tag{3.18a}$$

or from equations (3.11c) and (3.11d),

$$[(-M_{\alpha\beta}\mathbf{t}_\alpha + M_\alpha \mathbf{t}_\beta)B]_{,\alpha}\, d\alpha\, d\beta + [(-M_\beta \mathbf{t}_\alpha + M_{\beta\alpha}\mathbf{t}_\beta)A]_{,\beta}\, d\alpha\, d\beta \tag{3.18b}$$

2. Next, we investigate the contributions of the stress resultants. Observe in equation (3.18b) that the stress couple terms are of second differential order, $d\alpha\, d\beta$, so that terms of third or higher differential order will drop out in the limit. From figure 3–2, we see that the extensional stress resultants $\mathbf{N}_\alpha$ and $\mathbf{N}_\beta$ contribute moments about $\mathbf{t}_n$ of

$$-N_{\alpha,\alpha}\, d\alpha\, ds_\beta \cdot \frac{ds_\beta}{2}\, \mathbf{t}_n \rightarrow \text{third order}$$

and

$$N_{\beta,\beta}\, d\beta \cdot ds_\alpha \cdot \frac{ds_\alpha}{2}\, \mathbf{t}_n \rightarrow \text{third order}$$

while the in-plane shearing stress resultants $\mathbf{N}_{\alpha\beta}$ and $\mathbf{N}_{\beta\alpha}$ give

$$N_{\alpha\beta}\, ds_\beta\, ds_\alpha \mathbf{t}_n \rightarrow \text{second order}$$

and

$$-N_{\beta\alpha}\, ds_\alpha\, ds_\beta \mathbf{t}_n \rightarrow \text{second order}$$

The transverse shear resultant $\mathbf{Q}_\alpha$ contributes

$$-Q_\alpha\, ds_\beta\, ds_\alpha\, \mathbf{t}_\beta \to \text{second order}$$

about $\mathbf{t}_\beta$ while $\mathbf{Q}_\beta$ gives

$$Q_\beta\, ds_\alpha\, ds_\beta\, \mathbf{t}_\alpha \to \text{second order}$$

about $\mathbf{t}_\alpha$. Collecting the second order contributions of the force vector to moment equilibrium, we have

$$[Q_\beta \mathbf{t}_\alpha - Q_\alpha \mathbf{t}_\beta + (N_{\alpha\beta} - N_{\beta\alpha})\mathbf{t}_n]\, ds_\alpha\, ds_\beta \qquad (3.19)$$

3. Then we note from figure 3–4 that the contribution of $\mathbf{q}$ is of third differential order and therefore negligible.

4. Finally, we expand equation (3.18b) in accordance with equation (2.17), combine the resulting terms with equation (3.19), factor in terms of the unit vectors, and divide by $d\alpha\, d\beta$, giving

$$G_\alpha \mathbf{t}_\alpha + G_\beta \mathbf{t}_\beta + G_n \mathbf{t}_n = 0 \qquad (3.20)$$

which is satisifed if and only if

$$G_\alpha = 0; \quad G_\beta = 0; \quad G_n = 0 \qquad (3.21)$$

Thus, we have the three scalar equations of *moment* equilibrium

$$G_\alpha = -(BM_{\alpha\beta}),_\alpha - (AM_\beta),_\beta - B,_\alpha M_{\beta\alpha} + A,_\beta M_\alpha + Q_\beta AB = 0 \qquad (3.22a)$$

$$G_\beta = (BM_\alpha),_\alpha + (AM_{\beta\alpha}),_\beta + A,_\beta M_{\alpha\beta} - B,_\alpha M_\beta - Q_\alpha AB = 0 \qquad (3.22b)$$

$$G_n = N_{\alpha\beta} - N_{\beta\alpha} + \frac{M_{\alpha\beta}}{R_\alpha} - \frac{M_{\beta\alpha}}{R_\beta} = 0 \qquad (3.22c)$$

The set of six equations of static equilibrium, equations (3.17a–c) and (3.22a–c), may immediately be reduced to five, since it is easily shown by the substitution of equations (3.6)–(3.9) into equation (3.22c) that the later equation, commonly known as the *sixth* equation of equilibrium, is satisfied *identically* if the symmetry of the stress tensor $\sigma_{\beta\alpha} = \sigma_{\alpha\beta}$ is invoked.[2] Thus equation (3.22c) may be discarded. There are several modified shell theories which attempt to redefine the stress resultants and couples so that the sixth equation can be satisifed in the form of equation (3.22c).[5] None of these alters the fact that the sixth equation is identically satisfied using a proposition from a higher theory, the theory of elasticity, and consequently cannot be germane in any further development. Therefore, five equilibrium equations in ten unknowns remain. Such a problem is classified as statically indeterminate, indicating that the solution awaits the introduction of additional relationships between the stress resultants and couples, and the deformations of the shell.

3.3 Equilibrium Equations for Shells of Revolution

3.3.1 Specialization of General Equations

The shell of revolution is a widely used shell geometry and affords a sufficiently general, yet easily visualized model for the physical interpretation of the equilibrium equations.

For the shell of revolution, shown in figure 2–10, we set $\alpha = \phi$ and $\beta = \theta$. From equations (2.43b) and (2.44b), $A = R_\phi(\phi)$ and $B = R_0(\phi) = R_\theta(\phi) \sin \phi$. Also, recall the Gauss–Codazzi condition, equation (2.47), which gives

$$B_{,\alpha} = R_{0,\phi} = (R_\theta \sin \phi)_{,\phi} = R_\phi \cos \phi \tag{3.23}$$

Writing the equilibrium equations including these relations, we have from equations (3.17a–c), (3.22a), and (3.22b)

$$(R_0 N_\phi)_{,\phi} + R_\phi N_{\theta\phi,\theta} - R_\phi \cos \phi \, N_\theta + R_0 Q_\phi + q_\phi R_\phi R_0 = 0 \tag{3.24a}$$

$$(R_0 N_{\phi\theta})_{,\phi} + R_\phi N_{\theta,\theta} + R_\phi \cos \phi \, N_{\theta\phi}$$

$$+ R_\phi Q_\theta \sin \phi + q_\theta R_\theta R_\phi = 0 \tag{3.24b}$$

$$(R_0 Q_\phi)_{,\phi} + R_\phi Q_{\theta,\theta} - R_0 N_\phi - R_\phi \sin \phi \, N_\theta + q_n R_\phi R_0 = 0 \tag{3.24c}$$

$$(R_0 M_{\phi\theta})_{,\phi} + R_\phi M_{\theta,\theta} + R_\phi \cos \phi \, M_{\theta\phi} - R_\phi R_0 Q_\theta = 0 \tag{3.24d}$$

$$(R_0 M_\phi)_{,\phi} + R_\phi M_{\theta\phi,\theta} - R_\phi \cos \phi \, M_\theta - R_\phi R_0 Q_\phi = 0 \tag{3.24e}$$

The middle surface element is shown for a shell of revolution geometry with positive Gaussian curvature in figure 3–5. Stress resultants and couples are shown on an enlarged view of this element normal to $\mathbf{t}_n$ in figure 3–6. Also shown in figure 3–6 are horizontal and meridional sections, $HV1$, $MV1$, and $MV2$, with selected resultants and couples indicated. In comparing the terms of equation (3.24) with the forces and moments on figure 3–6, recall that we divided through by $d\alpha \, d\beta$, or, in this case, $d\phi \, d\theta$ in the course of the derivation, and that we have retained terms of second differential order only. Also, for the differential element, $\sin d\phi/2 = d\phi/2$, $\sin d\theta/2 = d\theta/2$, and $\cos d\phi/2 = \cos d\theta/2 = 1$.

3.3.2 Physical Interpretation of Equilibrium Equations

3.3.2.1 Force Equilibrium in the ϕ Direction: Equation (3.24a). The first term represents the increment in the force $N_\phi R_0 \, d\theta$ as ϕ changes, whereas the second gives the increment in the force $N_{\theta\phi} R_\phi \, d\phi$ with respect to θ, as seen on the force element in figure 3–6. Since R_ϕ is not a function of θ, it is

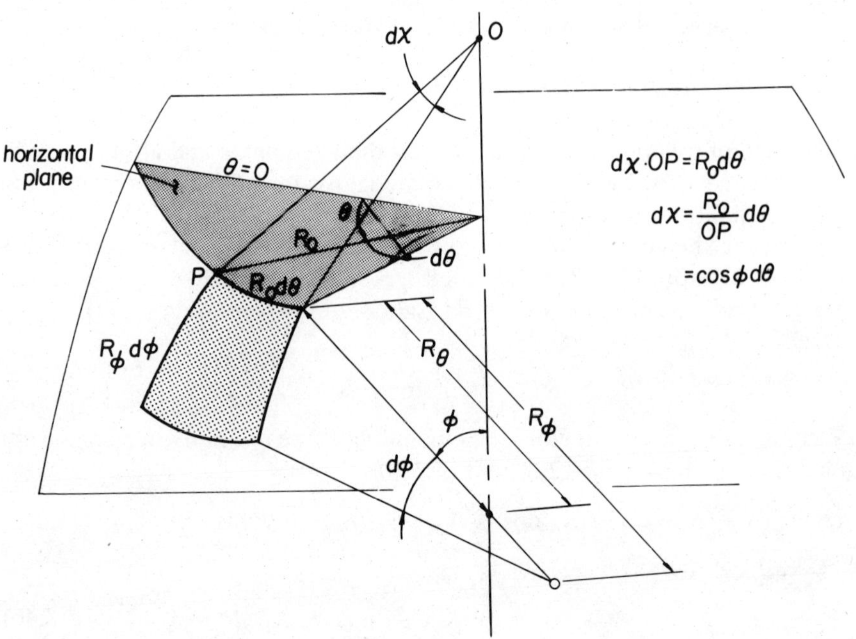

Figure 3–5. Middle Surface Element and Geometrical Relationships for a Shell of Revolution

treated as a constant in the second term. The third term indicates a contribution of the force $N_\theta R_\phi d\phi$ to the ϕ direction equilibrium and is explained by first referring to view $HV1$, where the radial component $2N_\theta R_\phi d\phi \cdot d\theta/2$ which is in the horizontal plane, is developed. This radial force is transferred to view $MV1$, and is further resolved in the ϕ and n directions, with the ϕ component being the third term in equation (3.24a). The negative sign indicates that the component acts in the negative ϕ direction as defined on figure 2–11. The fourth term is the contribution of $Q_\phi R_0 \, d\theta$ and is substantiated by considering view $MV2$. The last term is, of course, the applied loading.

3.3.2.2 Force Equilibrium in the θ Direction: Equation (3.24b). The first and second terms are analogous to the corresponding terms of equation (3.24a) and are easily verified on the force element. The third term is the circumferential contribution of force $N_{\theta\phi} R_\phi \, d\phi$ and arises because forces $N_{\theta\phi} R_\phi \, d\phi$ and $N_{\theta\phi} R_\phi \, d\phi + (N_{\theta\phi} R_\phi \, d\phi)_{,\theta} \, d\theta$ are not parallel to one another, but rather are directed along tangents to the meridians at θ and $\theta + d\theta$, respectively. As shown in figure 3–5, the extension of these tangents de-

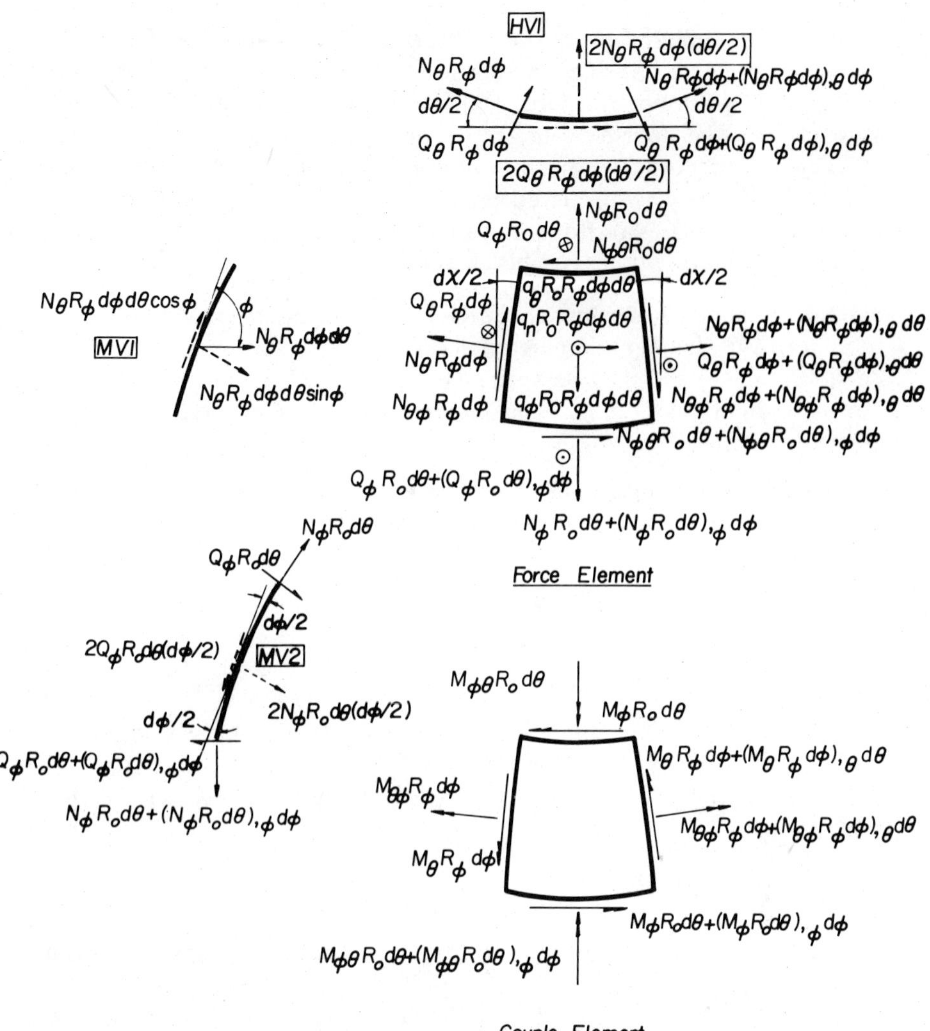

Figure 3–6. Forces and Moments Acting on a Shell of Revolution Element

fines an intersection on the axis of rotation at an angle $d\chi$, which is expressed as $d\theta \cos \phi$. Therefore, from the force element, we have $2N_{\theta\phi} R_\phi \, d\phi \cdot d\chi/2 = N_{\theta\phi} R_\phi \cos \phi \, d\phi \, d\theta$ in the $\mathbf{t}_\theta$ direction. The fourth term is the contribution of $Q_\theta R_\phi \, d\phi$ and is shown in view $HV1$, whereas the fifth term is the applied loading.

3.3.2.3 Force Equilibrium in the n Direction: Equation (3.24c). The first two terms are easily verified in views $MV2$ and $HV1$. The third term represents the contribution of force $\mathbf{N}_\phi R_0\, d\theta$ and is obtained by projection in view $MV2$. The fourth term, representing a similar contribution of $\mathbf{N}_\theta R_\phi\, d\phi$, is the normal projection of the radial force $\mathbf{N}_\theta R_\phi\, d\phi\, d\theta$ in view $MV1$. This force was discussed previously in connection with the third term in equation (3.24a). Both of these normal components are directed in the negative $\mathbf{t}_n$ direction. The fifth term again is the applied loading.

3.3.2.4 Moment Equilibrium about the $\mathbf{t}_\phi$ Axis: Equation (3.24d). The first three terms of this equation are analogous to the corresponding terms of equation (3.24e) and may be verified accordingly. The fourth term is the contribution of the force $\mathbf{Q}_\theta R_\phi\, d\phi$ multiplied by $R_0\, d\theta$ as seen on the force element.

3.3.2.5 Moment Equilibrium about the $\mathbf{t}_\theta$ Axis: Equation (3.24e). Again the first three terms are analogous to the force equilibrium equation, (3.24b), and the last term is the couple $\mathbf{Q}_\phi R_0\, d\theta \cdot R_\phi\, d\phi$, which is directed along the negative $\mathbf{t}_\theta$ direction.

3.4 Equilibrium Equations for Plates

Equations (3.17a–c) and (3.22a–c) may easily be reduced to sufficiently general equilibrium equations for medium-thin plates by allowing R_α and $R_\beta \to \infty$. After these simplifications, we have

$$[\,(BN_\alpha)_{,\alpha} + (AN_{\beta\alpha})_{,\beta} + A_{,\beta}N_{\alpha\beta} - B_{,\alpha}N_\beta\,] + q_\alpha AB = 0 \quad (3.25a)$$

$$[\,(BN_{\alpha\beta})_{,\alpha} + (AN_\beta)_{,\beta} + B_{,\alpha}N_{\beta\alpha} - A_{,\beta}N_\alpha\,] + q_\beta AB = 0 \quad (3.25b)$$

$$[\,(BQ_\alpha)_{,\alpha} + (AQ_\beta)_{,\beta}\,] + q_n AB = 0 \quad (3.25c)$$

$$(BM_{\alpha\beta})_{,\alpha} + (AM_\beta)_{,\beta} + B_{,\alpha}M_{\beta\alpha} - A_{,\beta}M_\alpha - Q_\beta AB = 0 \quad (3.25d)$$

$$(BM_\alpha)_{,\alpha} + (AM_{\beta\alpha})_{,\beta} + A_{,\beta}M_{\alpha\beta} - B_{,\alpha}M_\beta - Q_\alpha AB = 0 \quad (3.25e)$$

Further simplifications will be made for regular geometries and for bending in the absence of in-plane forces; for now, we note in equations (3.25a–c) that the in-plane stress resultants (N_α, N_β, $N_{\alpha\beta}$, $N_{\beta\alpha}$) are uncoupled from the stress couples (M_α, M_β, $M_{\alpha\beta}$, $M_{\beta\alpha}$) and the transverse shear stress resultants (Q_α, Q_β). This suggests that the plate resists in-plane loading (q_α, q_β) exclusively through extensional action and transverse loading (q_n) only by flexure. This is an important contrast to shell action as demonstrated by equation (3.17c), where the *in-plane* stress resultants provide resistance against transverse loading by virtue of the curvature of the shell.

3.5 Nature of the Applied Loading

Thus far we have assumed that the applied loading is expressible as a distributed force per unit area of middle surface, with components in the directions of the unit vectors. This leaves open the cases of concentrated forces, and distributed and concentrated moments.

In general, shells are most efficient when the loading is distributed over the surface, because—provided certain geometrical and support conditions are met—such loading can be resisted primarily by the extensional and the in-plane shear stress resultants, rather than by the transverse shear stress resultants and bending and twisting stress couples. In addition to the distributed surface loading included in the formulation by $\mathbf{q}$, distributed moments about the $\mathbf{t}_\phi$ and $\mathbf{t}_\theta$ axes can easily be accommodated by including appropriate terms in equations (3.22a) and (3.22b). Moments about the $\mathbf{t}_n$ axis, however, are not admissible, since the corresponding equilibrium equation, (3.22c), was suppressed. A plate or shell generally offers very great stiffness to twisting about the normal, provided rigid body motion is restrained.

Concentrated forces and moments require special attention in plate and shell theory. Often, these loadings produce singular points, in which case admissible solutions may be found only away from the point of application of the load. In any case, these forces are usually resisted primarily by the transverse shears that are directly related to the bending and twisting stress couples. We may view concentrated forces and moments as limits of the corresponding distributed effect. Consider the area Δa subjected to a uniformly distributed loading of intensity p in figure 3–7. We define a concentrated force $\mathbf{P}_c$ as

$$\mathbf{P}_c = \lim_{\Delta A \to 0} (p\Delta a) \qquad (3.26)$$

$$p\Delta a \text{ remains constant}$$

Now consider two concentrated coplanar forces as defined in equation (3.26) and shown in figure 3–7(b). The forces $\mathbf{P}_c$ are equal in magnitude and opposite in direction a distance $\Delta\xi$ apart. We define a concentrated moment $\mathbf{M}_c$ about an axis normal to the plane of the forces as

$$\mathbf{M}_c = \lim_{\Delta\xi \to 0} (\mathbf{P}_c\Delta\xi) \qquad (3.27)$$

$$\mathbf{P}_c\Delta\xi \text{ remains constant}$$

Equations (3.26) and (3.27) are somewhat abstract in this form, but are useful in obtaining many solutions for plates and shells.

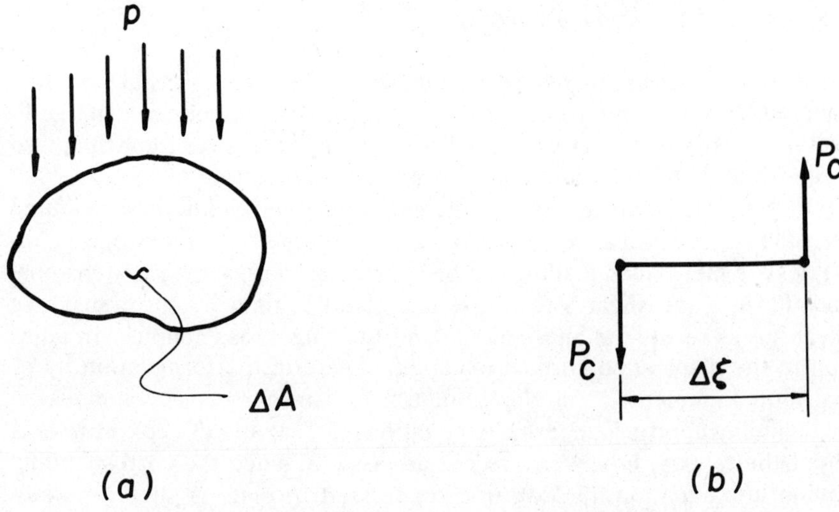

Figure 3–7. Concentrated Force and Moment

Notes

1. H. Kraus, *Thin Elastic Shells* (New York: Wiley, 1967), p. 33.

2. V. V. Novozhilov, *Thin Shell Theory* [translated from 2nd Russian ed. by P. G. Lowe (Groningen, The Netherlands: Noordhoff, 1964), pp. 34–39].

3. A. E. Green and W. Zerna, *Theoretical Elasticity* (Oxford: Oxford University Press, 1968).

4. See S. Timoshenko and S. Woinowsky-Krieger, *Theory of Plates and Shells*, 2nd ed. (New York: McGraw-Hill, 1959). Also see W. Flügge, *Stresses in Shells*, 2nd ed. (Berlin: Springer-Verlag, 1973).

5. Kraus, *Thin Elastic Shells*, pp. 58–65.

4   Membrane Theory

4.1 Simplification of the Equilibrium Equations

We now examine the individual terms of the force equilibrium equations, (3.17a–c), and the moment equilibrium equations, (3.22a–c). We see that the equations are coupled only through the transverse shear stress resultants, Q_α and Q_β. If we suppose for the moment that for a certain class of shells, the stress couples are an order of magnitude smaller than the in-plane stress resultants, we may deduce from equations (3.22a–c) that the transverse stress resultants are similarly small and thus may be neglected in the force equilibrium equations, (3.17). This implies that the shell may achieve *force equilibrium* through the action of *in-plane forces* alone. From a physical viewpoint, this possibility is evident for the first two equilibrium equations which reflect in-plane resistance to in-plane loading, a natural and obvious mechanism. On the other hand, the third equilibrium equation refers to the normal direction, and the possibility of resisting *transverse* loading with *in-plane* forces alone is not as apparent. It is evident from equation (3.17c) that this mode of resistance is possible only if at least one radius of curvature is finite; i.e., R_α and/or $R_\beta \neq \infty$. Thus, flat plates are excluded from resisting transverse loading in this manner within the limitations of small deformation theory (assumption [2], table 1–1.

The applications and imitations of this idealized behavior, termed the *membrane theory of thin shells*, is examined in this chapter for a variety of shell forms. At present, it is constructive to consider some of the consequences of membrane behavior. First, it is quite desirable from a material efficiency standpoint. Recall from chapter 1 that structural materials are generally far more efficient in extension rather than in flexure. Second, if we introduce an additional simplification with respect to the shearing stress resultants,

$$N_{\alpha\beta} \simeq N_{\beta\alpha} = S \tag{4.1}$$

the force equilibrium equations become

$$\frac{1}{AB}[(BN_\alpha)_{,\alpha} + (AS)_{,\beta} + A_{,\beta}S - B_{,\alpha}N_\beta] + q_\alpha = 0 \tag{4.2a}$$

$$\frac{1}{AB}[(BS)_{,\alpha} + (AN_\beta)_{,\beta} + B_{,\alpha}S - A_{,\beta}N_\alpha] + q_\beta = 0 \tag{4.2b}$$

69

$$\frac{N_\alpha}{R_\alpha} + \frac{N_\beta}{R_\beta} = q_n \qquad (4.2c)$$

In these three equations, there are but three unknowns, N_α, N_β, and S. Such a system is said to be *statically determinate* and thus independent of compatibility or constitutive considerations. This, of course, simplifies the subsequent mathematics considerably. The convenient assumption stated in equation (4.1) may be easily justified by considering the definition of $N_{\alpha\beta}$ and $N_{\beta\alpha}$ as given in equations (3.6) and (3.7). Since $\sigma_{\alpha\beta} = \sigma_{\beta\alpha}$ because the stress tensor is symmetric, $N_{\alpha\beta}$ and $N_{\beta\alpha}$ may differ only in the disparity of the terms $1 + \zeta/R_\beta$ and $1 + \zeta/R_\alpha$, respectively. Since ζ/R_β and ζ/R_α are themselves necessarily small in comparison to 1, the difference in $N_{\alpha\beta}$ and $N_{\beta\alpha}$ is obviously negligible. Thus the simplification introduced in equation (4.1) is justified apart from other considerations.

Equations (4.2a–c), together with the requisite boundary conditions, constitute the essential components of the membrane theory. Strictly speaking, the entire solution includes the evaluation of the displacements as well. This, of course, requires the constitutive and compatibility relationships that have not yet been developed here. However, just as in the analysis of statically determinate structures, the membrane displacements may be computed following the determination of the force field with only a few exceptions.[1] Consequently, consideration of the membrane displacements is deferred until chapter 6.

4.2 Applicability of Membrane Theory

The shells for which the membrane theory is applicable fall into two general classes: (a) absolutely flexible shells or true membranes, which by virtue of their thinness have a negligible bending stiffness; and, (b) shells with finite bending stiffnesses which still develop relatively small bending stresses.[2]

Concerning the absolutely flexible shells, there is little question of the dominance of membrane action, provided that the principal stresses remained tensile. The presence of compressive stresses in this type of shell would likely cause buckling of the membrane. For flexible shells, the requirement of small displacements as stated in assumption [2] of table 1–1 is often not attainable, and the equations as developed herein are not applicable without modification. An interesting example of a thin flexible curved object to which shell theory has been applied for the purpose of estimating the stresses present is the human aortic heart valve [figure 2–8(q)].[3]

A wide variety of shells with finite bending rigidities can be designed and constructed to resist external loading almost exclusively through membrane action. We should recognize that the establishment of the precise bounds of the membrane theory requires the consideration of the governing equations of the general theory of shells. Rigorously, a valid membrane theory solution must be a close approximation to the response that would be computed using the general shell equations. However, we may make some general observations regarding the *geometry, boundary conditions*, and *loading* generally consistent with membrane behavior.

With respect to the *geometry*, a continuous surface is conducive to membrane action. Also recall that the second fundamental form of the theory of surfaces, equation (2.28a), contains the principal radii of curvature, and a smooth variation of these parameters is essential if equilibrium is to be achieved without transverse shears and bending and twisting moments.

The *boundary conditions* must be specified with respect to the membrane stress resultants, N_α, N_β, and S, or the corresponding displacements, and must indicate either a *constraint* that will fully develop the force on the boundary, a *release* that will enable the shell to displace freely in the corresponding direction, or, possibly, a linear combination of both, as in the case of an elastic support. Moreover, the boundary should not provide constraints that would develop moments and transverse shears. This implies that the ideal boundary must, in some cases, permit rotations and transverse displacements while providing complete restraint in the plane of the shell.

A *loading* that can be resisted by membrane action must be distributed over the surface without severe variations or concentrations. In the membrane equilibrium equations, (4.2a–c), concentrated loadings are not admissible. From a physical standpoint, it is apparent that a transverse load concentrated over a small area of the shell surface must be resisted principally by transverse shears which, in turn, produce bending as demonstrated by equations (3.22a–c).

From this brief discussion of the conditions which correspond to ideal membrane behavior, it is obvious that to design and construct an actual shell that satisfies all of these conditions would be quite difficult. However, even shells which do not meet all of the requirements entirely may be designed so that membrane action predominates throughout most of the continuum, except for localized regions in which the required conditions are violated. In short, many shells exhibit basically membrane behavior augmented by locally prominent bending action. Of course, this is not to say that all shells may be made to act primarily as membranes or that all bending effects are localized, but such behavior is a desirable and often attainable result of good design and careful construction.

4.3 Shells of Revolution

4.3.1 Specialization of Equilibrium Equations

We now specialize the membrane theory equilibrium equations, (4.2), for the shell of revolution geometry, where the curvilinear coordinates are taken as the meridional angle, ϕ, and the circumferential coordinate, θ, and with $\alpha = \phi$ and $\beta = \theta$, $A = R_\phi(\phi)$ and $B = R_0(\phi) = R_\theta \sin\phi$ as previously established in equations (2.44b) and (2.43b). If we make these substitutions into equations (4.2a–c) and carry out the differentiations using the Gauss–Codazzi condition, equations (2.47), we find

$$\frac{1}{R_\phi}N_{\phi,\phi} + \frac{\cot\phi}{R_\theta}(N_\phi - N_\theta) + \frac{1}{R_0}S_{,\theta} + q_\phi = 0 \tag{4.3a}$$

$$\frac{1}{R_\phi}S_{,\phi} + \frac{2\cot\phi}{R_\theta}S + \frac{1}{R_0}N_{\theta,\theta} + q_\theta = 0 \tag{4.3b}$$

$$\frac{N_\phi}{R_\phi} + \frac{N_\theta}{R_\theta} = q_n \tag{4.3c}$$

We immediately note that equations (4.3a–c) are identical to equations (3.24a–c) divided through by $R_\phi R_0$, with the bending terms suppressed and the in-plane shearing stress resultants $N_{\phi\theta}$ and $N_{\theta\phi}$ set equal to S. As such, the physical interpretation of these equations is contained within section 3.3.2. We also observe that the third equation, (4.3c), is an algebraic equation. Hence, either N_θ or N_ϕ may be directly eliminated from equations (4.3a) and (4.3b), leaving a second order system of partial differential equations to be solved.

To simplify the solution of the equilibrium equations, it is convenient to introduce certain transformations at this stage. Timoshenko and Woinowsky-Krieger use integrating functions,[4] whereas Novozhilov suggests the auxiliary variables[5]

$$\psi = N_\phi R_\theta \sin^2\phi \tag{4.4a}$$

and

$$\xi = SR_\theta^2 \sin^2\phi \tag{4.4b}$$

We first solve equation (4.3c) for

$$N_\theta = R_\theta\left(q_n - \frac{N_\phi}{R_\phi}\right) \tag{4.5}$$

and then introduce equations (4.4) and (4.5) into equations (4.3a) and (4.3b) to get

$$\frac{R_\theta^2 \sin \phi}{R_\phi}\psi,_\phi + \xi,_\theta = (q_n \cos \phi - q_\phi \sin \phi)R_\theta^3 \sin^2 \phi \qquad (4.6a)$$

$$\xi,_\phi - \frac{R_\theta}{\sin \phi}\psi,_\theta = -(q_\theta \sin \phi + q_{n,\theta})R_\phi R_\theta^2 \sin \phi \qquad (4.6b)$$

Equations (4.6a) and (4.6b) are the transformed membrane equilibrium equations. Once ψ and ξ are determined, N_ϕ and S follow from equations (4.4a) and (4.4b), and N_θ may then be calculated from equation (4.5). The subsequent combination and the eventual solution of the transformed equilibrium equations are dependent on the circumferential distribution of the applied surface loading. If the loading is distributed uniformly around the circumference, it is termed *axisymmetric*, whereas if the loading is a variable function of θ, it is obviously *nonsymmetric*. Each possibility is considered further in the subsequent sections.

4.3.2 Axisymmetrical Loading

4.3.2.1 Integration of Equilibrium Equations. The case of axisymmetrically loaded shells of revolution is perhaps the most widely studied class of problems in the shell literature. From a standpoint of applications, this case is often the most important, since it corresponds to self dead weight and internal pressurization, as well as to many other practical loading situations.

We may drop the θ-dependent terms in equations (4.6a) and (4.6b) to derive the uncoupled set of ordinary differential equations

$$\psi,_\phi = (q_n \cos \phi - q_\phi \sin \phi)R_\phi R_\theta \sin \phi \qquad (4.7a)$$

$$\xi,_\phi = -q_\theta R_\phi R_\theta^2 \sin^2 \phi \qquad (4.7b)$$

Considering equation (4.7a) and integrating, we get

$$\psi(\phi) = \int (q_n \cos \phi - q_\phi \sin \phi)R_\phi R_\theta \sin \phi \, d\phi \qquad (4.8)$$

In many cases, an alternate form of equation (4.8) is convenient. We recognize that evaluating the integral for a particular shell geometry will produce a single integration constant, and that this constant must be determined from a boundary condition on ψ, or on N_ϕ. We designate the boundary where the condition is specified as $\phi = \phi'$ and the corresponding boundary condition as $\psi(\phi')$. Then, equation (4.8) may be written as

$$\psi(\phi) = \psi(\phi') + \int_{\phi'}^{\phi} (q_n \cos \phi - q_\phi \sin \phi)R_\phi R_\theta \sin \phi \, d\phi \qquad (4.9)$$

Of course, at $\phi = \phi'$, $\psi = \psi(\phi') = N_\phi(\phi')R_\theta(\phi') \sin^2 \phi'$. The solution for N_ϕ follows from equations (4.4a) and (4.9) as

$$N_\phi(\phi) = \frac{N_\phi(\phi')R_\theta(\phi') \sin^2 \phi'}{R_\theta \sin^2 \phi}$$

$$+ \frac{1}{R_\theta \sin^2 \phi} \int_{\phi'}^{\phi} (q_n \cos \phi - q_\phi \sin \phi)R_\phi R_\theta \sin \phi \, d\phi \qquad (4.10)$$

Then, N_θ is computed from equation (4.5).

The in-plane shearing stress resultant S is completely uncoupled from N_ϕ and N_θ, as is apparent from equations (4.7a) and (4.7b). For common axisymmetric loading cases, $q_\theta = 0$ and the shearing stress is also 0; however, this is not necessarily so, since equation (4.7b) may be integrated in an identical fashion to equation (4.7a), yielding

$$\xi(\phi) = \xi(\phi'') - \int_{\phi''}^{\phi} q_\theta R_\phi R_\theta^2 \sin^2 \phi \, d\phi \qquad (4.11)$$

or

$$S(\phi) = \frac{S(\phi'')R_\theta^2(\phi'') \sin^2 \phi''}{R_\theta^2 \sin^2 \phi}$$

$$- \frac{1}{R_\theta^2 \sin^2 \phi} \int_{\phi''}^{\phi} q_\theta R_\phi R_\theta^2 \sin^2 \phi \, d\phi \qquad (4.12)$$

In equations (4.11) and (4.12), $\xi(\phi'')$ and $S(\phi'')$ are the corresponding values of the functions specified at the boundary $\phi = \phi''$.

In summary, we observe that the membrane theory equilibrium equations for axisymmetrically loaded shells of revolution reduce to two uncoupled first order ordinary differential equations. This system admits the prescription of two boundary conditions, one on the meridional stress resultant N_ϕ, and the other on the in-plane shearing stress resultant S.

One special case is worthy of consideration before we turn to the integration of equations (4.10) and (4.12) for specific shell geometries. If we consider a closed shell as first discussed in section 2.8.2, we have the two possibilities illustrated in figure 4–1. In case (a), which we will call a *dome* [figures 2–(i) and (l)], the meridian remains continuous as ϕ and $R_0 \to 0$; whereas in case (b), which may be termed a *pointed* shell, $\phi = \phi_t$ at $R_0 = 0$. If we take the pole angle $\phi = \phi_p$ as the boundary ϕ' in equation (4.9), where $\phi_p = 0$ for case (a) and $\phi_p = \phi_t$ for case (b), and evaluate $\psi(\phi_p)$, we find

$$\psi(\phi_p) = N_\phi(\phi_p)R_\theta(\phi_p) \sin^2 \phi_p$$

$$= N_\phi(\phi_p)R_0(\phi_p) \sin \phi_p$$

$$= 0 \qquad (4.13)$$

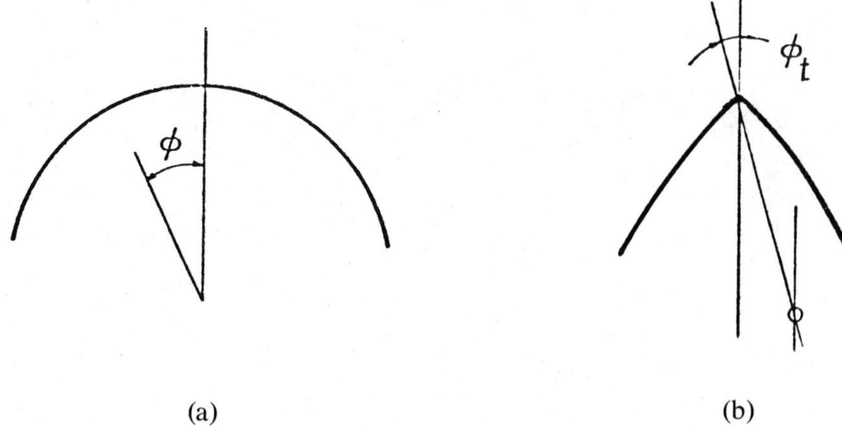

(a)   (b)

Figure 4–1. Closed Shells

since $R_0(\phi_p) = 0$. Similarly, considering equation (4.11) for ξ, $\xi(\phi_p) = 0$. Thus the solutions for closed shells simplify to

$$N_\phi = \frac{1}{R_\theta \sin^2 \phi} \int_{\phi_p}^{\phi} (q_n \cos \phi - q_\phi \sin \phi) R_\phi R_\theta \sin \phi \, d\phi \qquad (4.14a)$$

$$S = -\frac{1}{R_\theta^2 \sin^2 \phi} \int_{\phi_p}^{\phi} q_\theta R_\phi R_\theta^2 \sin^2 \phi \, d\phi \qquad (4.14b)$$

For the common case of a dome [figure 4–1(a)], $\phi_p = 0$.

We may observe several interesting points about the preceding solutions for closed shells:

1. The pole condition does not necessarily imply that $N_\phi(\phi_p)$ or $S(\phi_p) = 0$.

2. For the case of a dome, $\phi_p = 0$, the integrated expressions may produce an indeterminate form which must be evaluated by some limiting operation.

3. Since the available boundary conditions on N_ϕ and S are implied by the pole, there is no opportunity to specify any further boundary conditions on the stress resultants within the membrane theory. This means that the remaining boundary of the shell must develop whatever values of membrane stress resultants are computed from equations (4.14) or (4.15) to maintain membrane equilibrium. We recall that such an ideal boundary may not impose constraints which will develop bending forces or moments.

4. For a dome, all meridians meet at the pole, and any direction is parallel to one meridian and at right angles to another; that is, the curvilinear coordinates ϕ and θ are interchangeable and indistinguisable. There-

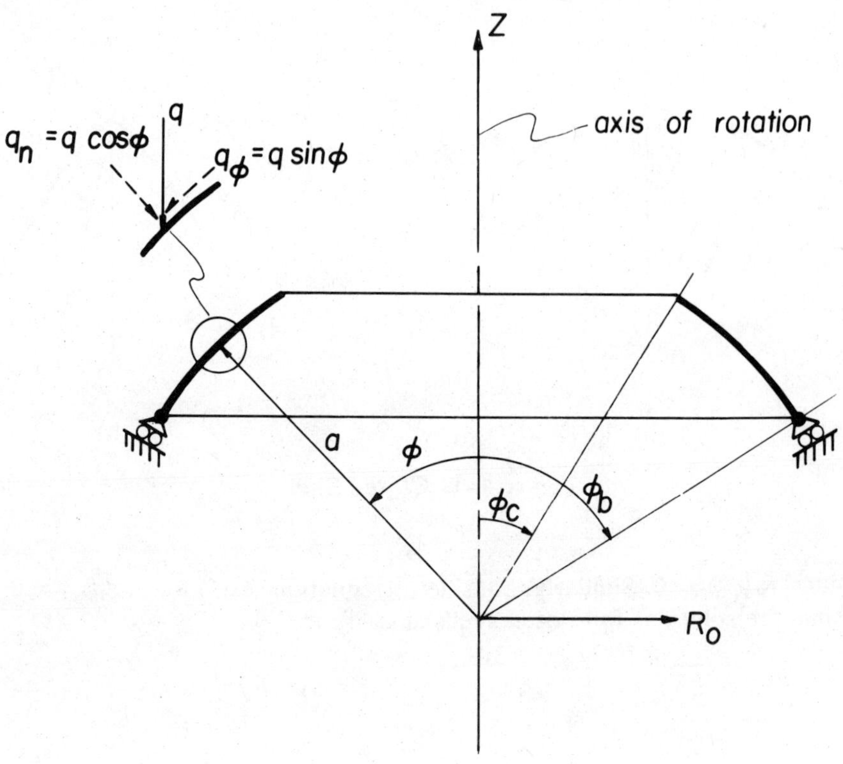

Figure 4–2. Spherical Shell under Self-Weight Load

fore, $R_\phi(0) = R'_\theta(0) = R_p$ and, correspondingly, $N_\phi(0) = N_\theta(0)$. Then, from equation (4.3c)

$$N_\phi(0) = N_\theta(0) = q_n(0)\frac{R_p}{2} \tag{4.15}$$

4.3.2.2 Spherical Shells. As a first example, consider the spherical shell illustrated in figure 4–2. The shell has radius a, constant thickness h, mass density ρ, and is bound by angles ϕ_t and ϕ_b. Initially the weight of the shell alone, known as the *self-weight*, is taken as the loading condition.

The principal radii of curvature are the definitive geometric parameters. The radius of curvature of the meridian is, of course, the radius of the generating circle a. This is easily verified by considering the equation of the meridian $Z^2 + R_0^2 = a^2$ and substituting into equation (2.29b) with α

$= \phi$ and $X = R_0$. With $Z_{,R_0} = -R_0/Z$ and $Z_{,R_0R_0} = -(1/Z)(1 + R_0^2/Z^2)$, we find

$$R_\phi = \frac{-(1 + [R_0^2/Z^2])^{3/2}}{-(1/Z)[1 + (R_0^2/Z^2)]}$$

$$= \sqrt{(Z^2 + R_0^2)}$$

$$= a \tag{4.16a}$$

The other principal radius may be computed from equation (2.42). We note from figure 4–2 that the horizontal radius $R_0 = a \sin \phi$, so that

$$R_\theta = \frac{a \sin \phi}{\sin \phi} = a \tag{4.16b}$$

Equations (4.16a and b) obviously satisfy the Gauss–Codazzi condition, equation 2.47.

Next, consider the applied loading. The self-weight of the shell per unit area of the middle surface is given by

$$q = \rho g h \tag{4.17}$$

in which $g =$ the acceleration of gravity. The load q acts vertically, of course. The governing equilibrium equations require the loading to be resolved in the ϕ, θ, and n directions. Since the vertical load has no component in the circumferential direction, $q_\theta = 0$. On the inset in figure 4–2, we have the resolution of q into q_ϕ and q_n. The signs are established by comparison with the unit vectors shown in figure 2–11.

At present, the boundaries are assumed to satisfy ideal membrane restrictions. The top boundary is presumed to be free of stresses, and the bottom boundary is taken to be unyielding in the ϕ or meridional direction, free to displace in the n or normal direction, and unrestrained against rotation about an axis along the θ or circumferential direction. The base conditions imply that the necessary N_ϕ to maintain equilibrium will be developed, but no Q_ϕ or M_ϕ can occur. The roller symbol used on the figure corresponds to a similar condition in plane structural analysis; however, here the boundary is a space curve, and for cases which have circumferential loading, a condition in the θ direction also must be specified. In the latter case, the ideal membrane boundary must be extended beyond the simple conceptual model of a roller.

We now substitute the geometry, loading, and boundary conditions into equation (4.10). These calculated quantities are $R_\phi = R_\theta = a$; $q_\phi = q \sin \phi$ and $q_n = -q \cos \phi$. We take $\phi' = \phi_t$ so that $N_\phi(\phi_t) = 0$. Then we have

78

$$N_\phi(\phi) = \frac{1}{a \sin^2\phi} \int_{\phi_t}^{\phi} (-q \cos^2 \phi - q \sin^2 \phi)a^2 \sin \phi \, d\phi$$

$$= \frac{qa}{\sin^2 \phi}(\cos \phi - \cos \phi_t) \tag{4.18}$$

From equation (4.5), the circumferential or hoop stress is given by

$$N_\theta = a\left(-q \cos \phi - \frac{N_\phi}{a}\right) \tag{4.19}$$

We now examine a spherical dome by letting $\phi_t = 0$. Using an elementary trigonometric identity, equation (4.18) becomes

$$N_\phi = \frac{qa}{(1 + \cos \phi)(1 - \cos \phi)}(\cos \phi - 1)$$

$$= \frac{-qa}{1 + \cos \phi} \tag{4.20}$$

and equation (4.19) follows as

$$N_\theta = a\left(-q \cos \phi + \frac{q}{1 + \cos \phi}\right)$$

$$= qa\left(\frac{1}{1 + \cos \phi} - \cos \phi\right) \tag{4.21}$$

At $\phi = 0$, we have

$$N_\phi = N_\theta = \frac{-qa}{2} \tag{4.22}$$

which is in accordance with equation (4.15). It is also notable that the indeterminate form mentioned as point (2.) in section 4.3.2.1 was resolved by evaluating the integral first and then going to the limit $\phi = 0$.

It is quite instructive to consider equations (4.20) and (4.21) for increasing values of ϕ as shown in figure 4–3, where nondimensional plots for N_ϕ and N_θ are provided. These graphs actually represent the self-weight stresses for *all* constant thickness spherical domes. For any particular dome for which the lower boundary is located by a given value of ϕ_b, the relevant parts are the regions $\phi \leq \phi_b$. A shallow dome ($\phi_b << 90°$), a hemispherical dome ($\phi_b = 90°$), and a deep dome ($\phi_b > 90°$) are illustrated in the insets of the graphs. An interesting feature of the graphs is that N_ϕ is always negative or compressive, whereas N_θ changes from negative to positive. This exact transition point may be computed from equation (4.21) as the value of ϕ satisfying

$$\frac{1}{1 - \cos \phi} - \cos \phi = 0$$

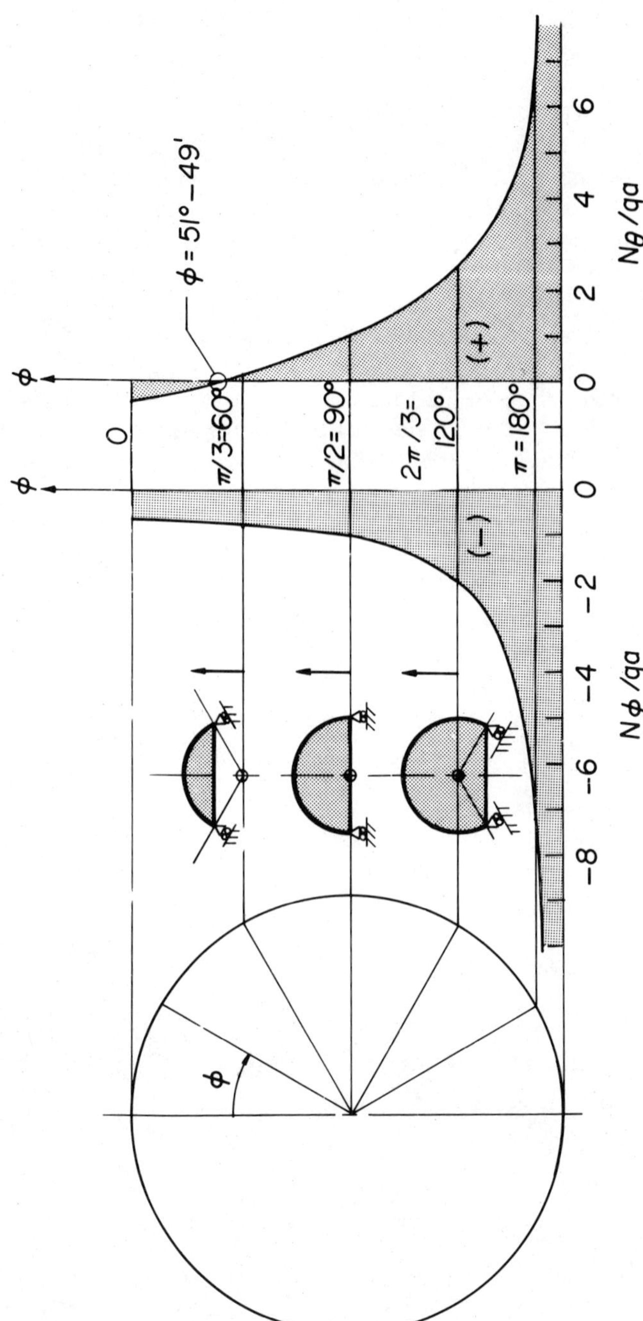

Figure 4–3. Self-Weight Stress Resultants for Spherical Domes of Constant Thickness

which has the solution $\phi = 51°49'$. The transition angle has some general practical ramifications, since a shell constructed such that $\phi_b < 51°$ will be entirely in compression under gravity loading, which is especially desirable for shells constructed of concrete.

With reference to the ancient masonry domes described in chapter 1, the necessity for maintaining the entire dome in compression is obvious and is apparently reflected in the change from the spherical to the more parabolic profile used in the Renaissance.

The apparent desirability of providing an entirely compressive state of stress is countered by another practical consideration. Recall that the support for the spherical shell must develop the calculated value of N_ϕ at $\phi = \phi_b$. An idealized typical support is shown in figure 4–4(a), where a circumferential ring beam is employed to resist the thrust. The shell is assumed to extend to the centroid of the ring beam to eliminate the introduction of eccentricity.[6] A section through the ring beam is shown in the inset. The vertical component of the thrust, $\mathbf{V} = N_\phi(\phi_b) \sin \phi_b$, is transmitted to the foundation, whereas the horizontal component must be developed by the beam. A half-plan of the ring is shown in figure 4–4(b), with a unit length segment in the inset; here, the horizontal reaction $\mathbf{H} = -N_\phi(\phi_b) \cos \phi_b$ is seen to be the radial component of the hoop tension $\mathbf{T}$.

We may evaluate $\mathbf{T}$ by summing forces in the X direction on figure 4–4(b).

$$2\mathbf{T} = \int_{-\pi/2}^{\pi/2} -N_\phi(\phi_b) \cos \phi_b \cos \theta \cdot b \cdot d\theta \tag{4.23a}$$

or

$$\mathbf{T} = -N_\phi(\phi_b)b \cos \phi_b \tag{4.23b}$$

For the spherical shell

$$b = a \sin \phi_b \tag{4.24}$$

so that

$$\mathbf{T} = -N_\phi(\phi_b)a \cos \phi_b \sin \phi_b \tag{4.25}$$

For the self-load case, we find by substituting equation (4.20) into equation (4.25)

$$\mathbf{T}_{DL} = qa^2 \frac{\cos \phi_b \sin \phi_b}{1 + \cos \phi_b} \tag{4.26}$$

It is clear from comparing equation (4.26) and the graph for N_θ in figure 4–3 that there will be a strain incompatibility at the junction between the shell and ring beam. For $\phi_b < 51°49'$, the hoop stress N_θ is compressive while the ring force $\mathbf{T}_{DL}$ is tensile for $\phi_b < 90°$. Thus the dome that has

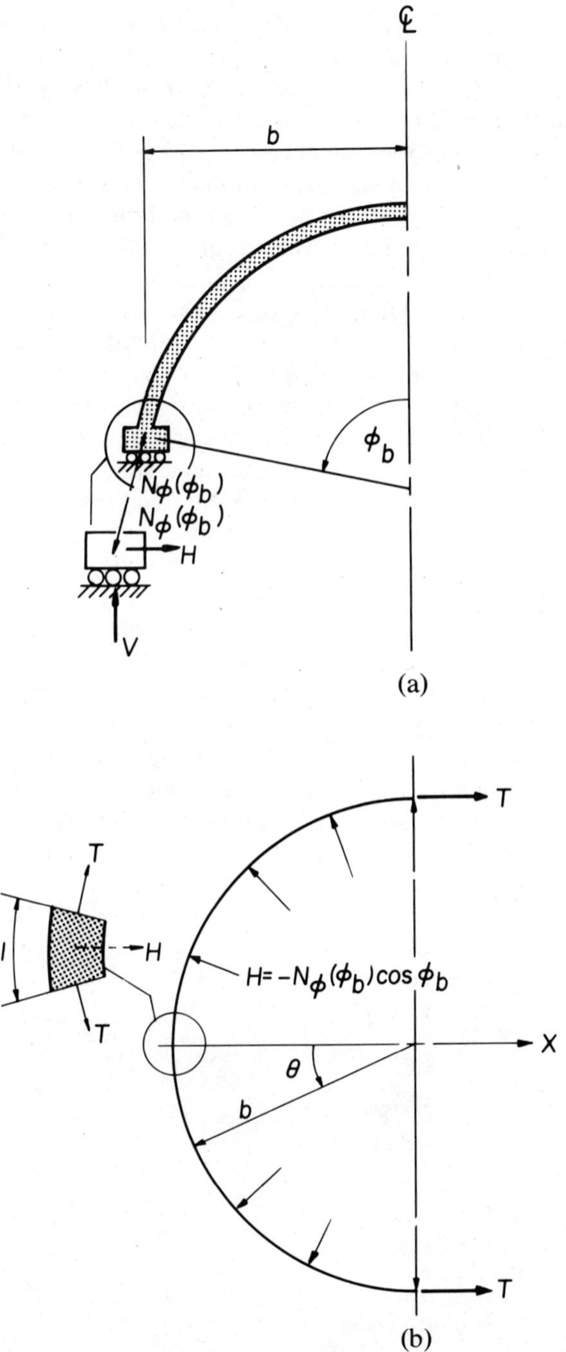

Figure 4–4. Ring Beam for a Shell of Revolution

been shown to be in a state of compression through membrane theory analysis must somehow accommodate the circumferential expansion of the base ring accompanying the tensile force T_{DL}. Clearly, this cannot be accomplished with membrane action alone, and bending forces must be considered to satisfy deformational continuity. In the terminology of section 4.2, we have violated the ideal *boundary conditions*. In some instances, such as for the dome shown in figure 2–8(j), prestressing of the base ring is effective in reducing the circumferential imcompatibility. In general, the bending effects introduced by this type of support are usually confined to a relatively small portion of the shell adjacent to the base, and membrane action will still predominate throughout most of the shell. It should also be noted that equation (4.23) is not restricted to spherical shells, but is valid for all shells of revolution supported in this fashion.

It is apparent from figure 4–3 that the stresses increase rapidly for deep spherical shells with $\phi_b > 120°$. For such shells, it is logical to support the shell at a point somewhat above the base. Such a situation is depicted in figure 2–8(t), although the description of discrete column supports requires an extension of the ideal membrane case as we will see later. If we have, for example, a complete sphere supported by an ideal membrane boundary at an angle $\phi = \phi_r$ as shown in figure 4–5, we can easily obtain the stress pattern from figure 4–3. For $\phi \leq \phi_r$ we compute the stresses as before. For $\phi > \phi_r$, the lower portion of the shell behaves just like a spherical cap with $\phi_b = \pi - \phi_r$ and a loading, as shown in the inset, which is just opposite in sense to that shown on figure 4–2. Thus if we enter $\bar{\phi} = \pi - \phi$ for ϕ on figure 4–3, the dead load stress resultants for the lower portion of the shell are identical in magnitude but opposite in sign to those read from the graph.

Another common loading condition for spherical domes is that of uniform normal pressure. Consider a positive internal pressure p on the spherical dome. We substitute $R_\phi = R_\theta = a$; $q_\phi = 0$; $q_n = p$ and $\phi_p = 0$ into equation (4.14a), whereupon

$$N_\phi(\phi) = \frac{1}{a \sin^2 \phi} \int_0^\phi p \cos \phi \, a^2 \sin \phi \, d\phi$$

$$= \frac{pa^2}{a \sin^2 \phi} \left[\frac{1}{2} \sin^2 \phi \right]_0^\phi$$

$$= \frac{pa}{2} \tag{4.27}$$

From equation (4.5)

$$N_\theta(\phi) = a\left(p - \frac{N_\phi}{a} \right)$$

$$= \frac{pa}{2} \tag{4.28}$$

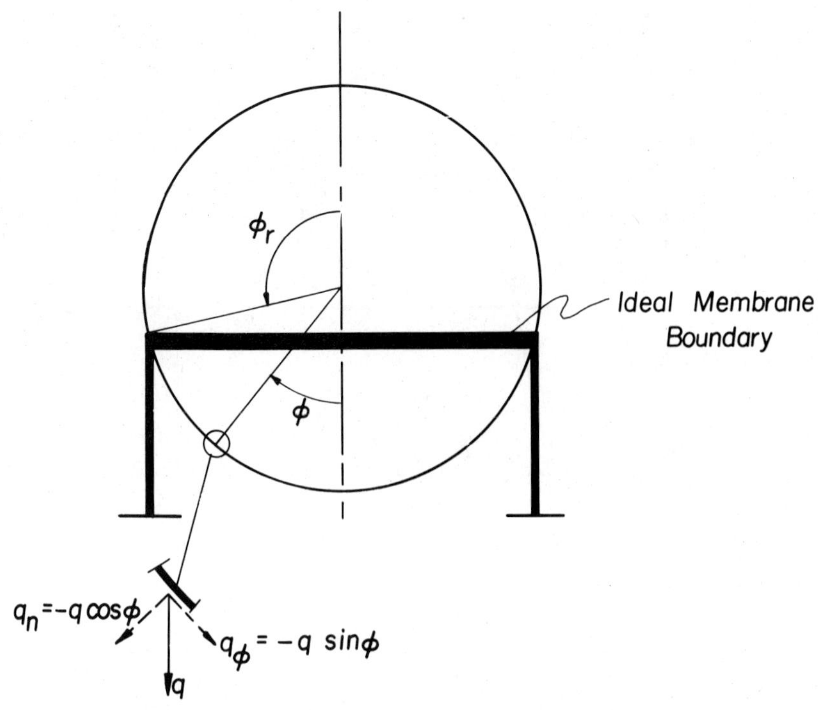

Figure 4–5. Complete Spherical Shell with an Intermediate Continuous Support

so that we have a uniform state of stress, $N_\phi = N_\theta = pa/2$, throughout the shell.

4.3.2.3 Hyperboloidal Shells. As a second illustration of a membrane theory solution for an axisymmetrically loaded shell of revolution, we consider the form illustrated in figures 4–6 and 2–8(g). This surface is generated by the rotation of a hyperbola of one sheet and is a shell with *negative* Gaussian curvature, since the centers of curvature corresponding to R_ϕ and R_θ lie on opposite sides of the meridian (see section 2.7 and figures 2–6 and 2–11). This form is of considerable practical importance, because of the wide use of such shells for reinforced concrete *hyperbolic cooling towers* [figure 2–8(o)]. These massive structures may approach a height of 600 ft (183m), span 300 ft (92m) in diameter, and yet have an average thickness of less than 9 in (23 cm), a striking testimony to the efficiency of thin shell structures.

The equation of the generating curve is

$$\frac{R_0^2}{a^2} - \frac{Z^2}{b^2} = 1 \tag{4.29}$$

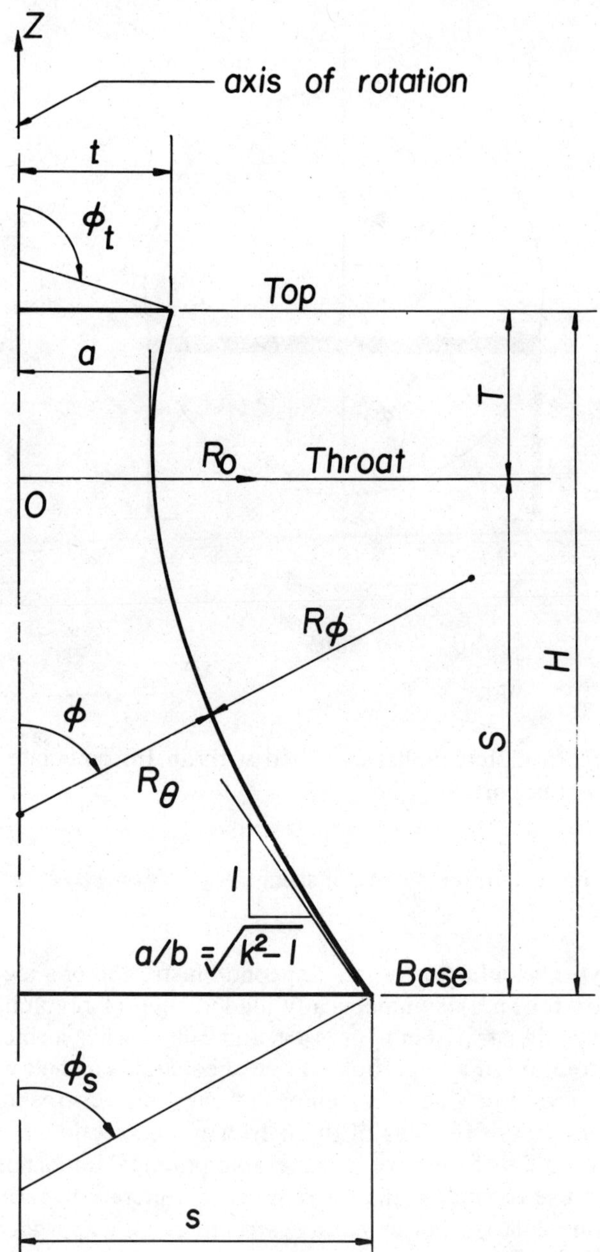

Figure 4–6. Hyperboloidal Shell Geometry

in which b is a characteristic dimension of the shell that may be evaluated by substituting the base coordinates (s, S) or the top coordinates (t, T) into equation (4.29) as

$$b = \frac{aT}{\sqrt{(t^2 - a^2)}} = \frac{aS}{\sqrt{(s^2 - a^2)}} \tag{4.30}$$

The ratio a/b is the slope of the asymptote to the generating hyperbola shown on figure 4–6, and the parameter

$$k = \sqrt{\left(1 + \frac{a^2}{b^2}\right)} \tag{4.31}$$

may be viewed as an indicator of the deviation of the profile from the degenerate case of the cylinder, $k = 1$, with a larger k corresponding to a more pronounced curvature of the meridian. As may be seen from equation (4.30), b, and hence the geometric profile can be set independently using specified top or base dimensions. For cooling tower applications, it is sometimes desirable to use different values of b for the top and bottom, and thus creates a compound shell (sometimes called an offset hyperboloid) with a smooth transition at $Z = 0$.

With the parameter k defined, we may rewrite equation (4.29) as

$$R_0^2 - (k^2 - 1)Z^2 = a^2 \tag{4.32}$$

The next step is to derive expressions for the principal radii of curvature in terms of the curvilinear coordinate ϕ. Direct substitution of the equation of the meridian into equations (2.29), as previously done for the spherical shell, leads to cumbersome expressions, since Z is not given as an explicit function of ϕ. Instead we consider equation (4.32) along with the expression for the differential arc length, equation (2.45a), which becomes for this case,

$$ds_\phi^2 = dZ^2 + dR_0^2$$

or

$$R_\phi^2 \, d\phi^2 = Z_{,\phi}^2 \, d\phi^2 + R_{0,\phi}^2 \, d\phi^2 \tag{4.33}$$

We then solve equation (4.32) for Z

$$Z = \left(\frac{R_0^2 - a^2}{k^2 - 1}\right)^{1/2} \tag{4.34}$$

and compute

$$Z_{,\phi} = \frac{R_0 R_{0,\phi}}{Z(k^2 - 1)}$$

Since

$$R_{0,\phi} = R_\phi \cos \phi \tag{4.35a}$$

$$Z_{,\phi} = \frac{R_0 R_\phi \cos \phi}{[(R_0^2 - a^2)(k^2 - 1)]^{1/2}} \tag{4.35b}$$

After substituting equations (4.35a) and (4.35b) into equation (4.33) and clearing fractions, we may cancel $R_\phi^2 \, d\phi^2$ in each term, so that

$$(R_0^2 - a^2)(k^2 - 1) = R_0^2 \cos^2 \phi + (R_0^2 - a^2)(k^2 - 1) \cos^2 \phi \tag{4.36}$$

from which, after some manipulation, we find

$$R_\theta = \frac{R_0}{\sin \phi} = \frac{a\sqrt{(k^2 - 1)}}{[k^2 \sin^2 \phi - 1]^{1/2}} \tag{4.37}$$

Finally, from equations (4.37) and (4.35a)

$$R_\phi = \frac{-a\sqrt{(k^2 - 1)}}{[k^2 \sin^2 \phi - 1]^{3/2}} \tag{4.38}$$

We now consider the elementary case of the self-weight. The principal radii of curvature are given by equations (4.37) and (4.38), and the loading components are $q_\phi = q \sin \phi$ and $q_n = -q \cos \phi$ as derived in figure 4–2. Also, we take $\phi' = \phi_t$, and equation (4.10) becomes

$$N_\phi(\phi) = N_\phi(\phi_t)\frac{\sin^2 \phi_t}{\sin^2 \phi} \cdot \frac{[k^2 \sin^2 \phi - 1]^{1/2}}{[k^2 \sin^2 \phi_t - 1]^{1/2}} - \frac{[k^2 \sin^2 \phi - 1]^{1/2}}{a \sin^2 \phi \sqrt{(k^2 - 1)}}$$

$$\cdot \int_{\phi_t}^{\phi} (-q \cos^2 \phi - q \sin^2 \phi)\frac{a^2(k^2 - 1) \sin \phi \, d\phi}{[k^2 \sin^2 \phi - 1]^2} \tag{4.39}$$

Assuming a stress-free top edge, $N_\phi(\phi_t) = 0$, and $q = $ constant (uniform thickness), equation (4.39) integrates to [7]

$$N_\phi(\phi) = \frac{qa[k^2 \sin^2 \phi - 1]^{1/2}}{\sin^2 \phi \sqrt{(k^2 - 1)}} [\zeta_1(\phi) - \zeta_1(\phi_t)] \tag{4.40a}$$

in which the integrated function

$$\zeta_1(\phi) = \frac{-\cos \phi}{2[k^2 \sin^2 \phi - 1]} + \frac{1}{4k\sqrt{(k^2 - 1)}} \, ln\left(\frac{\sqrt{(k^2 - 1)} - k \cos \phi}{\sqrt{(k^2 - 1)} + k \cos \phi}\right)$$

$$\tag{4.40b}$$

The circumferential stress N_θ is computed from equation (4.5) as

$$N_\theta(\phi) = \frac{a\sqrt{(k^2 - 1)}}{[k^2 \sin^2 \phi - 1]^{1/2}} \left[-q \cos \phi + \frac{N_\phi(\phi)[k^2 \sin^2 \phi - 1]^{3/2}}{a\sqrt{(k^2 - 1)}}\right] \tag{4.41}$$

Some further comments are in order regarding the preceding solution for the self-weight stress resultants. If the shell thickness changes with the height of the shell, the solution as derived must be generalized slightly. Assuming that the shell thickness is or may be approximated as constant within a defined subregion of the shell, the basic solution, equation (4.39), may be applied in a piecewise fashion. For any subregion r bound by ϕ_t^r and ϕ_s^r as shown in figure 4–7,

$$\phi_s^{r-1} = \phi_t^r \geq \phi^r \geq \phi_s^r = \phi_t^{r+1}$$

$$r = 1,\ldots, n$$

(4.42a)

and

$$q^r = \gamma h^r$$

(4.42b)

where h^r is the local constant thickness. We start at the top of the shell, $r = 1$, for which the solution is given by equation (4.40a). Then we evaluate equation (4.40a) at $\phi = \phi_s^1 = \phi_t^2$ to compute $N_\phi(\phi_s^1) = N_\phi(\phi_t^2)$, which is substituted into the first term of equation (4.39). Thus the solution for region $r = 2$ is given by equation (4.40a), plus the first term of equation (4.39), with ϕ_t taken as ϕ_t^2. For the general region r, we have

$$N_\phi(\phi^r) = N_\phi(\phi_t^r)\frac{\sin^2 \phi_t^r}{\sin^2 \phi^r} \cdot \frac{[k^2 \sin^2 \phi^r - 1]^{1/2}}{[k^2 \sin^2 \phi_t^r - 1]^{1/2}}$$

$$+ q^r a\frac{[k^2 \sin^2 \phi^r - 1]^{1/2}}{\sin^2 \phi^r \sqrt{(k^2 - 1)}} [\zeta_1(\phi^r) - \zeta_1(\phi_t^r)]$$

(4.43)

$N_\theta(\phi^r)$ is calculated as before from equation (4.41).

Although the coordinate ϕ has proved to be convenient for integrating the membrane theory equations, it is somewhat awkward for physically locating a particular position on the shell. It is obvious that the vertical coordinate Z would be more meaningful from the standpoint of practical construction. Corresponding to any value of ϕ, R_0 may be computed from equation (4.37), and then Z can be found from equation (4.34). Conversely, for a specified Z, R_0 is computed from equation (4.32), and ϕ is conveniently obtained by solving equation (4.37) for $\sin \phi$.

$$\phi = \sin^{-1}\left\{\frac{R_0}{(a^2 + k^2(R_0^2 - a^2))^{1/2}}\right\}$$

(4.44)

Values of ϕ in the first and second quadrant correspond to the lower and upper portions of the shell, respectively.

To compute the limits of integration, ϕ_t and ϕ_s, when the corresponding value of R_0, t, and s, are known, equation (4.44) can be used directly.

For spherical shells, we were able to deduce some general characteristics of the response to self-load by making a nondimensional plot of the

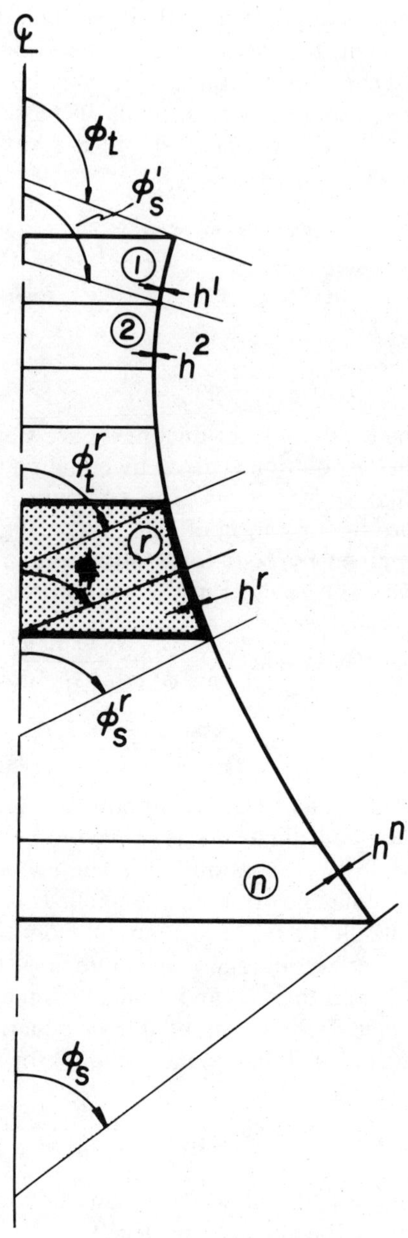

Figure 4–7. Hyperboloidal Shell Divided into Constant Thickness Regions

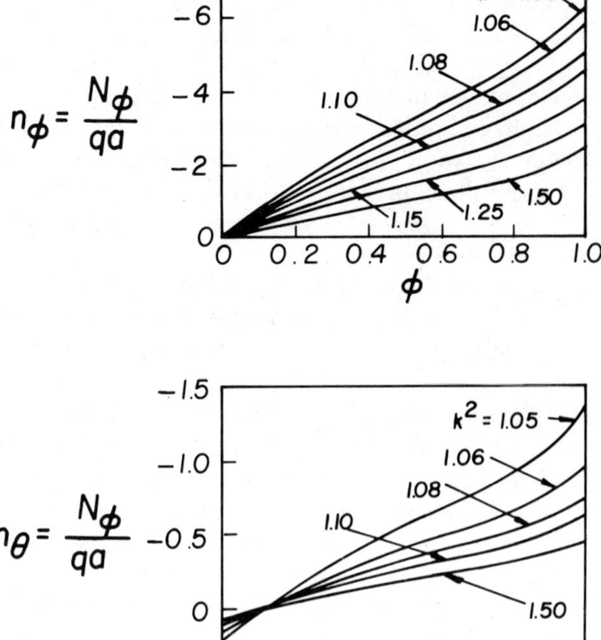

Source: P. L. Gould and S. L. Lee, ''Hyperbolic Cooling Towers under Seismic Load,'' *Journal of the Structural Division, ASCE*, 93, no. ST5 (October 1967): 95.

Figure 4–8. Nondimensional Stress Resultants for a Hyperboloidal Shell with $a/t = 0.90$ and $a/s = 0.55$ (Reprinted with permission of American Society of Civil Engineers)

membrane stress resultants. It is also useful to examine such a study for the hyperboloidal shell. If we define a nondimensional meridional coordinate

$$\Phi = \frac{\phi_t - \phi}{\phi_t - \phi_s} \tag{4.45}$$

and assume the thickness to be constant, we may plot nondimensional membrane stress resultants for various values of k^2, a/s, and a/t. Such a plot is shown in figure 4–8 for a shell of typical proportions, $a/s = 0.90$ and $a/t = 0.55$. To give some idea of the actual size of a corresponding cooling tower, with $k^2 = 1.18$ and $a = 70$ ft (23m), $s = 127$ ft (39m), $t = 78$ ft (24m), $S = 80$ ft (25m), and $T = 258$ ft (79m), so that the total height is 338 ft (104m).

From figure 4–8, we see that under self dead load the entire shell is in a state of compression, except for a portion above the throat, which has a relatively small tensile hoop stress. From the standpoint of reinforced concrete, this stress pattern is quite favorable. With respect to the base region, $\Phi = 1.0$, there are several points to be noted. First, the state of circumferential compression does not match the anticipated tension in the ring support, as previously illustrated in the discussion of spherical shells. Second, hyperboloidal shells which are used in cooling tower applications must have openings at the base to allow air to enter. This is accomplished by supporting the shell on an annular ring of closely spaced columns, as shown in figure 2–8(p), producing a so-called mixed boundary, where the shell may be presumed to have zero meridional displacement within the column width and zero meridional stress between the columns. The combination of the discrete support and the ring beam clearly indicates a boundary far beyond the idealized membrane case. This is considered further in subsequent sections. (Complete tabulations of membrane theory stress resultants for hyperboloidal shells of typical dimensions are provided in several available articles.[8])

4.3.2.4 Toroidal Shells. The toroidal shell is quite useful and efficient for pressure vessel applications. Also, segments of toroidal shells are frequently used as transitions between cylindrical tanks and shallow spherical or flat ends. This type of *compound* shell will be treated in some depth later.

Examining the toroidal geometry, figure 4–9, we see that the surface is generated by the rotation of a closed curve, usually a circle, about an axis lying outside the curve.

The definitive geometry is conveniently established from figure 4–9. The meridian is the circle $ABCD$, with radius a. If we examine a normal to the surface defined by a meridional angle ϕ, we observe that it pierces the surface at two points; that is, there is not a one-to-one correspondence between the curvilinear coordinate ϕ and a unique point on the surface. The consequence of this anomaly is that the segment of the shell within the semicircle BAD must be considered separately from that within BCD.

Considering the exterior segment BAD denoted by the superscript e,

$$R_\theta^e = b \csc \phi^e + a \tag{4.46}$$

where $b =$ the constant distance from the center of the generating circle to the axis of rotation.

The center of curvature of the meridian BAD is on the same side of the

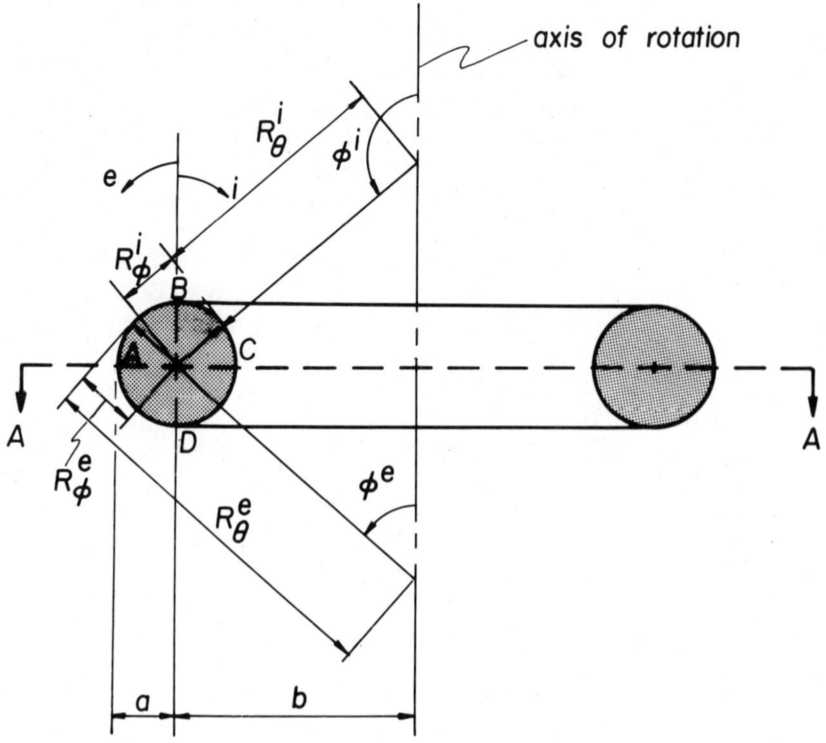

Figure 4–9. Toroidal Shell Geometry

meridian as the center of curvature of R_θ^e denoting a shell of *positive Gaussian curvature* so that

$$R_\phi^e = +a \tag{4.47}$$

For the interior segment *BCD* denoted by the superscript *i*,

$$R_\theta^i = b \csc \phi^i - a \tag{4.48}$$

Observe that, for this segment, the center of curvature of the meridian lies on the opposite side of the meridian from the center of curvature of R_θ^i, indicating that this segment has *negative Gaussian curvature*. Thus

$$R_\phi^i = -a \tag{4.49}$$

At this point, it is appropriate to comment further on the choice of the correct algebraic sign of the principal radii of curvature when the *magnitudes* are obtained from a geometric rather than a mathematical argument, as we have done here. Specifically, the condition of negative Gaussian curvature implies only that the product $1/R_\phi R_\theta$ is negative, and does not indicate which of the radii has the negative sign. Also, a shell with positive Gaussian curvature may seemingly have both principal radii with negative signs. From the fundamental geometrical definition of the shell of revolution, figure 2–11, the horizontal radius $R_0 = R_\theta \sin \phi$ is always positive. With $\sin \phi$ remaining positive for $0 \le \phi \le \pi$, R_0 and hence R_θ will always be positive, and thus the signs on equation (4.47) and (4.49) are necessary to give the correct sign for the Gaussian curvature. It is easily verified that equations (4.46) and (4.48) satisfy the Gauss–Codazzi condition. In the toroid, we have encountered a shell which has positive Gaussian curvature in one region and negative curvature in the other, with transition points at B and D. At B and D, $R_\theta \to \infty$, indicating a zero Gaussian curvature condition.

The most common loading case for toroids is a uniform normal positive internal pressure p. Referring to figure 2–11, this loading corresponds to the positive sense of $\mathbf{t}_n$ for the positive curvature segment and the negative sense of $\mathbf{t}_n$ for the negative curvature segment. Therefore, for segment BAD, $q_n = q_n^e = +p$, and for segment BCD, $q_n = q_n^i = -p$. Further, because of symmetry, we may restrict our consideration to the quarter-circles AB for e $(\pi/2 \ge \phi^e \ge 0)$ and BC for $i(\pi \ge \phi^i \ge \pi/2)$. We select the indefinite integral form, equation (4.8), for ψ, since this shell does not have an external boundary where ψ or N_ϕ is known or can be specified *a priori*. For e, $q_n = q_n^e = p$, $R_\phi = R_\phi^e = +a$ and $R_\theta = R_\theta^e = b \csc \phi^e + a$, so that

$$\psi_1 = \int p \cos \phi^e \cdot a \cdot (b \csc \phi^e + a) \sin \phi^e \, d\phi^e$$

$$= pa\left(b \sin \phi^e + \frac{a}{2} \sin^2 \phi^e + C_1\right)$$

and

$$N_\phi^e = \frac{\psi_1}{R_\theta^e \sin^2 \phi^e}$$

$$= \frac{pa}{(b \csc \phi^e + a) \sin^2 \phi^e}\left(b \sin \phi^e + \frac{a}{2} \sin^2 \phi^e + C_1\right)$$

$$= \frac{pa}{b + a \sin \phi^e}\left(b + \frac{a}{2} \sin \phi^e + \frac{C_1}{\sin \phi^e}\right)$$

For a finite meridional stress at $\phi^e = 0$ or π (points B and D), $C_1 = 0$, so that

$$N_\phi^e = \frac{pa(2b + a \sin \phi^e)}{2(b + a \sin \phi^e)} \tag{4.50}$$

To evaluate the circumferential stress resultant N_θ, we use equation (4.5)

$$N_\theta^e = (b \csc \phi^e + a)\left[p - \frac{pa}{2a} \cdot \frac{(2b + a \sin \phi^e)}{(b + a \sin \phi^e)}\right]$$

Multiplying the numerator and denominator of the r.h.s. by $\sin \phi^e$ and simplifying,

$$N_\theta^e = \frac{p}{\sin \phi^e}\left(b + a \sin \phi^e - b - \frac{a \sin \phi^e}{2}\right)$$

$$= \frac{pa}{2} \tag{4.51}$$

which of course does not vary with ϕ.

Using a parallel argument for segment BCD with $q_n = q_n^i = -p$, $R_\phi = R_\phi^i = -a$ and $R_\theta = R_\theta^i = b \csc \phi^i - a$,

$$N_\phi^i = \frac{pa(2b - a \sin \phi^i)}{2(b - a \sin \phi^i)} \tag{4.52}$$

and

$$N_\theta^i = \frac{pa}{2} \tag{4.53}$$

At the common points B and D, $\phi^e = 0$ and π and $\phi^i = \pi$ and 0, respectively, and

$$N_{\phi B}^e = N_{\phi B}^i = N_{\phi D}^e = N_{\phi D}^i = pa \tag{4.54}$$

Interestingly, we see that the circumferential stress N_θ throughout the shell as well as the meridional stress N_ϕ at the top and bottom circles are independent of the relative plan size of the torous, as represented by b, and are only dependent on the radius of the generating circle a. Also, as the mean plan radius b becomes large as compared to a, N_θ^e and $N_\phi^i \to pa$ throughout the shell. These observations have rather simple physical interpretations which give some insight into the general load resisting characteristics of membrane shells.

First, to investigate the circumferential stress resultant N_θ, we consider the half-plan free body as shown in figure 4–10. We have, acting in the Y direction, a force of magnitude $p \cdot$ [Projected area of outer circle $-$ Projected area of inner circle]. From figure 4–9 the difference in the projected areas is $[4ab + 2 \cdot \frac{1}{2}\pi a^2 - (4ab - 2 \cdot \frac{1}{2}\pi a^2)] = 2\pi a^2$, or simply twice the cross-sectional area, so that the resultant force is $2\pi pa^2$. This force must be balanced by the force in the shell wall arising from N_θ. Since the pres-

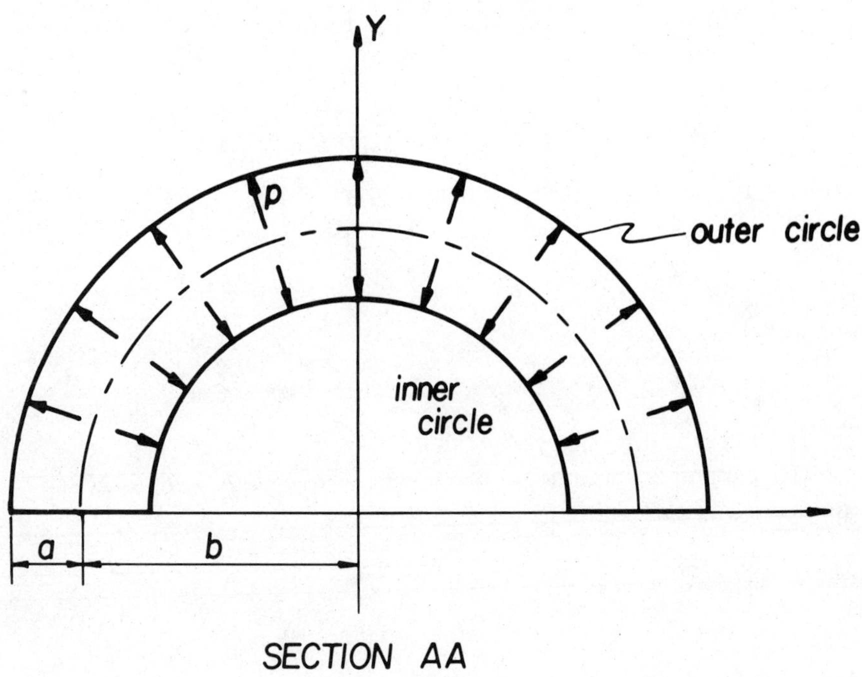

Figure 4–10. Toroidal Shell under Uniform Pressure Load

sure p is constant throughout the cross section, it is reasonable to regard N_θ as constant. Therefore,

$$2N_\theta \cdot 2\pi a = 2p\pi a^2$$

or

$$N_\theta = \frac{pa}{2} \tag{4.55}$$

which verifies equations (4.51) and (4.53). In the next section, we further illustrate that equations of overall (as opposed to differential) equilibrium on strategically chosen sections of shells can frequently lead to simple solutions for the membrane stress resultants.

Next, we seek to interpret the values of N_ϕ at the top and bottom circles. From equations (4.46) and (4.48), the circumferential radius $R_\theta \to \infty$ at B and D. Considering the so-called third equation of membrane equilibrium, equation (4.3c),

$$\frac{N_\phi}{R_\phi} + \frac{N_\theta}{R_\theta} = q_n \tag{4.3c}$$

we have

$$\frac{N_\phi}{a} + \frac{N_\theta}{\infty} = p$$

or

$$N_\phi = pa \qquad (4.56)$$

which verifies equation (4.54). Once N_θ has been derived from the consideration of overall equilibrium [i.e., figure 4–10 and equation (4.55)], N_ϕ may be computed at any point on the meridian using equation (4.3c).

In general, we make frequent recourse to equation (4.3c) to physically interpret the results of mathematical analyses. This equation, which describes the resistance of the shell to transverse loading, is perhaps the most incisive single equation in the theory of shells. Moreover, it clearly illustrates the basic difference between plate and shell action in terms of the influence of the initial curvature of the surface.

4.3.3 Alternate Formulation for Axisymmetric Loading

4.3.3.1 Overall Equilibrium. In the previous section, the circumferential stress computation for a symmetrically loaded toroidal shell was verified using an equation expressing the overall equilibrium across an internal section. This procedure is quite efficient in a variety of cases. Consider the general axisymmetrically loaded shell of revolution shown in figure 4–11 and pass a reference section normal to the axis of rotation. The trace of this section on the shell is the horizontal circle defined by the meridional angle ϕ.

To maintain equilibrium in the Z direction, the axial component of the meridional stress N_ϕ multiplied by the circumference of the section must balance the resultant axial load above the section, $Q(\phi)$, taken positive in the *negative Z* direction to ensure that a compressive force produces a negative meridional stress.

The axial component of $N_\phi(\phi)$ is given by $N_\phi(\phi) \cdot \sin \phi$ so that

$$Q(\phi) = N_\phi(\phi) \sin \phi \cdot 2\pi R_0(\phi) \qquad (4.57)$$

from which

$$N_\phi(\phi) = \frac{Q(\phi)}{2\pi R_0(\phi) \sin \phi} \qquad (4.58)$$

Then, $N_\theta(\phi)$ is computed, as before, from equation (4.3c).

To compute the resultant vertical load $Q(\phi)$ when the shell is subject

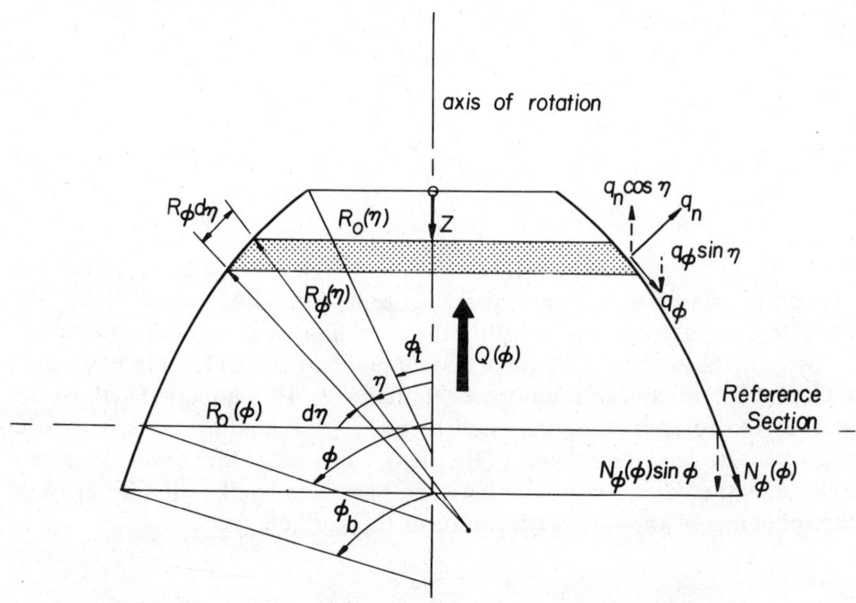

Figure 4–11. Symmetrically Loaded Shell of Revolution

to a distributed surface loading, we examine a differential ring element above the section. This ring element is defined by the auxiliary meridional angle η and has a meridional length equal to $R_\phi(\eta)\,d\eta$. The distributed surface loading, represented by q_n and q_ϕ, contributes axial components $+q_n(\eta)\cos\eta$ and $-q_\phi(\eta)\sin\eta$ per unit area of middle surface, respectively, as shown in figure 4–11. Then

$$Q(\phi) = \int_{\phi_t}^{\phi} [q_n(\eta)\cos\eta - q_\phi(\eta)\sin\eta][2\pi R_0(\eta)][R_\phi(\eta)\,d\eta] \quad (4.59)$$

which may be directly substituted into equation (4.58) to find $N_\phi(\phi)$.

As an elementary example, we will again solve the spherical dome under self-dead load, previously considered in section 4.3.2.2 and illustrated in figure 4–2. From equation (4.59) and figure 4–2, we have $q_n = -q\cos\eta$, $q_\phi = q\sin\eta$, $R_0 = a\sin\eta$, and $R_\phi = a$, so that

$$Q(\phi) = \int_{\phi_t}^{\phi} (-q\cos^2\eta - q\sin^2\eta)(2\pi a\sin\eta)a\,d\eta$$

$$= 2\pi qa^2(\cos\phi - \cos\phi_t) \quad (4.60a)$$

From equation (4.58),

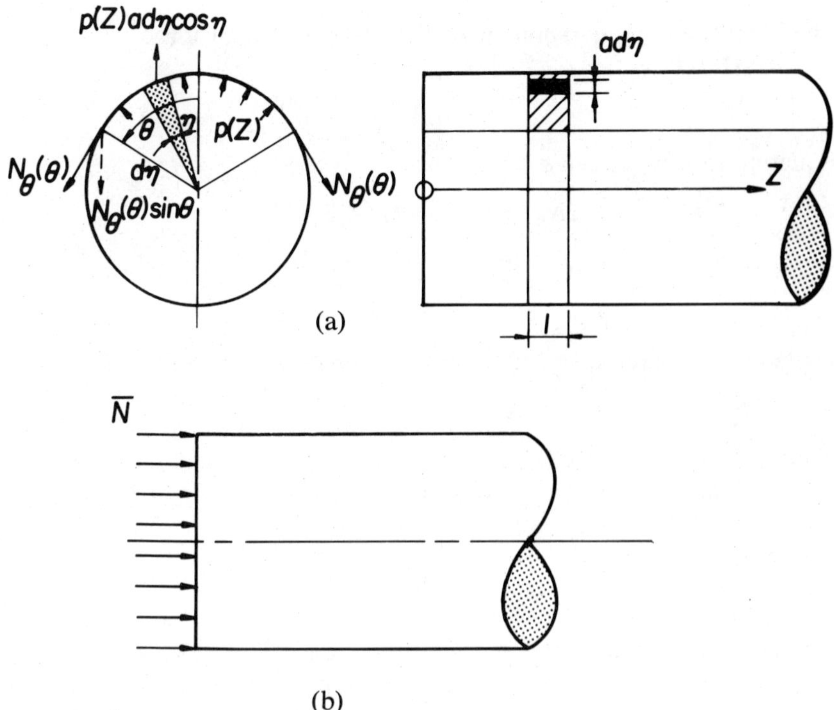

Figure 4–12. Symmetrically Loaded Cylindrical Shell

$$N_\phi = \frac{Q(\phi)}{2\pi a \, \sin^2 \phi}$$

$$= \frac{qa}{\sin^2 \phi} (\cos \phi - \cos \phi_t) \tag{4.60b}$$

which is identical to equation (4.18).

4.3.3.2 Circular Cylindrical Shells. A similar approach may be applied to a circular cylindrical shell subject to an internal pressure $p(Z)$. Consider a unit length axial slice of such a cylinder with radius a as shown in figure 4–12(a). In this case, the total resultant force along the $\theta = 0$ axis is

$$Q(\theta) = 2\int_0^\theta p(Z) \cos \eta \cdot 1 \cdot a \, d\eta$$

$$= 2p(Z)a \sin \theta \tag{4.61a}$$

98

The corresponding component of the circumferential force which must balance $Q(\theta)$ is

$$2N_\theta(\theta) \sin \theta \cdot 1 \qquad (4.61b)$$

Equating forces, we have

$$2N_\theta(\theta) \sin \theta = 2p(Z)a \, \sin \theta$$

or

$$N_\theta = p(Z)a \qquad (4.62a)$$

For a uniform axial distribution of pressure, $p(Z) = p$ and

$$N_\theta = pa \qquad (4.62b)$$

which is, of course, well known from elementary strength of materials. Obviously, equations (4.62) follow directly from equation (4.3c), since $R_\phi = \infty$. This solution also holds for incomplete or open cylindrical shells bound by a maximum $\theta = \theta''$. This type of shell, illustrated in figure 2–8(h), is examined in detail in a later section.

We now turn to the computation of the meridional stress resultant. Here, ϕ is not suitable for the meridional coordinate, since $\phi = \pi/2$ all along the axis. Instead, the axial coordinate Z is appropriate. We return to equations (4.2) with $\alpha = Z$, $\beta = \theta$, $A = 1$, and $B = a$. For axisymmetrical loading, equation (4.2b) vanishes and equation (4.2c) confirms the result given in equation (4.62). This leaves equation (4.2a), which becomes

$$\frac{1}{a}(aN_Z),_z + q_z = 0$$

or

$$N_{Z,z} = -q_z$$

so that

$$N_Z = -\int q_z \, dZ \qquad (4.63a)$$

We may also write the integral in the alternate form introduced in section 4.3.2.1.

$$N_Z(Z) = N_Z(0) - \int_0^Z q_z \, dZ \qquad (4.63b)$$

Equations (4.63a) and (4.63b) indicate that, for symmetrically loaded cylindrical shells, the meridional stress resultant is uncoupled from the normal loading and the circumferential stress resultant, and is a function only of the axial loading and boundary condition. For example, if the shell shown in figure 4–12(a) is subjected to a uniform axial edge load $\bar{N}$ per unit length of circumference as shown in figure 4–12(b), and $q_z = 0$, equation (4.63) gives

$$N_z(Z) = \bar{N} = \text{constant} \tag{4.64}$$

so that the effect of the edge loading penetrates along the axis of the cylinder and must be developed on the opposite boundary. This unattenuated propagation of edge loads is quite plausible if one recalls that the cylindrical shells have meridians composed of straight lines parallel to such axial loads. A similar situation arises in the membrane behavior of shells of negative Gaussian curvature, where tangential edge loads tend to propagate along the straight lines on the surface.[9] This is discussed further in later sections.

It is interesting to compare the solutions for the toroidal shell and the cylindrical shell under uniform normal pressure loading. For this purpose, the coordinates ϕ and θ on the toroid should be regarded as equivalent to θ and Z, respectively, in the cylinder. Now, if $b >> a$, we find [from equation (4.50) or (4.52) for the toroid and equation (4.62) for the cylinder] that $(N_\phi)_{\text{toroid}} = (N_\theta)_{\text{cylinder}} = pa$. Furthermore, if the cylindrical shell is closed by a flat circular plate, $N_z = p \cdot (\text{Area/Circumference}) = pa/2$, which would match N_θ as computed for the toroid in equations (4.51) and (4.53). Thus the membrane resistance mechanism of the two geometries is remarkably similar, with the torous representing, in effect, a self-closing cylinder.

4.3.3.3 Conical Shells. We now investigate conical shells. Initially, examine the frustum as shown in figure 4–13(a). The shell is conveniently described in terms of the vertical coordinate Z, since the meridional angle $\phi = \pi/2 - \alpha = \text{constant}$. At any section,

$$R_0 = t + Z \tan \alpha \tag{4.65}$$

and

$$R_\theta = \frac{R_0}{\sin \phi} = \frac{t + Z \tan \alpha}{\cos \alpha} \tag{4.66}$$

Since the meridian is straight, $R_\phi = \infty$, and from equation (4.3c) we immediately obtain

$$N_\theta = q_n \frac{t + Z \tan \alpha}{\cos \alpha} \tag{4.67}$$

Then, we concentrate on the meridional stress resultant N_ϕ.

Assume that the shell is subjected to an internal suction q, and that the

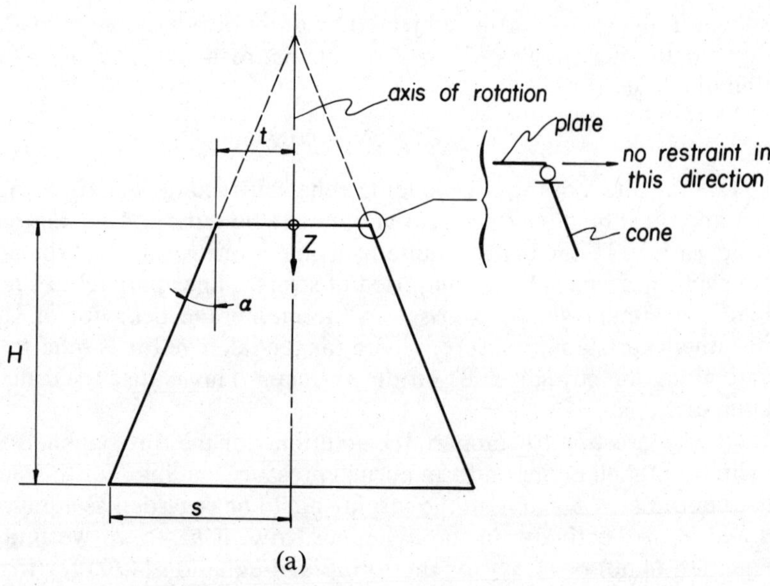

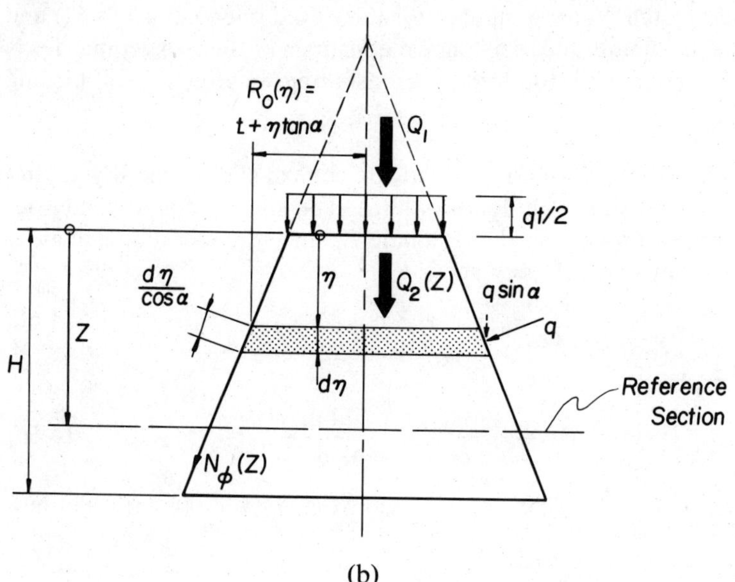

Figure 4–13. Conical Shell

top is closed by a flat plate which is free to move radially and to rotate freely at the junction with the conical shell. The idealized junction detail is shown in the inset. Under these conditions, the plate transmits a total force of only $q\pi t^2$ in the $+Z$ direction to the shell. Therefore, the resultant axial force at the top is

$$\mathbf{Q}_1 = -q\pi t^2 \tag{4.68}$$

This force is distributed over the circumference of the top circle, $2\pi t$, providing an edge load of $qt/2$ on the shell as shown in figure 4–13(b).

We now want to derive the resultant force $\mathbf{Q}_2(Z)$ due to the uniform suction. We define an auxiliary vertical coordinate η and resolve the load accordingly. Then we have

$$\mathbf{Q}_2(Z) = \int_0^Z [-q \sin \alpha][2\pi(t + \eta \tan \alpha)]\frac{d\eta}{\cos \alpha}$$

$$= -2\pi q \tan \alpha \int_0^Z (t + \eta \tan \alpha)\,d\eta$$

$$= -2\pi q \tan \alpha \, Z\left(t + \frac{Z}{2} \tan \alpha\right) \tag{4.69}$$

Since

$$N_\phi(Z) \sin \phi = \frac{Q(Z)}{2\pi R_0(Z)} \tag{4.70a}$$

and

$$Q(Z) = \mathbf{Q}_1 + \mathbf{Q}_2(Z) \tag{4.70b}$$

then,

$$N_\phi(Z) = \frac{\mathbf{Q}_1 + \mathbf{Q}_2(Z)}{2\pi \cos \alpha(t + Z \tan \alpha)}$$

$$= \frac{-q\{t^2 + 2 \tan \alpha Z[t + (Z/2) \tan \alpha]\}}{2 \cos \alpha(t + Z \tan \alpha)} \tag{4.71}$$

From equation (4.67), with $q_n = -q$

$$N_\theta(Z) = -q\,\frac{t + Z \tan \alpha}{\cos \alpha} \tag{4.72}$$

We note in passing that if α is set equal to 0, we have a cylinder with radius t. Equations (4.71) and (4.72) would reduce to

$$N_\phi = -\frac{qt}{2} \tag{4.73}$$

and

$$N_\theta = -qt \tag{4.74}$$

Since the N_ϕ comes entirely from the top plate reaction in the axial direction, these equations agree with the solution for cylindrical shells found in the previous section.

We may now investigate a complete cone by letting $t \to 0$, whereby

$$N_\phi = -\frac{q}{2} Z \cdot \frac{\tan \alpha}{\cos \alpha} \tag{4.75}$$

and

$$N_\theta = -qZ \cdot \frac{\tan \alpha}{\cos \alpha} \tag{4.76}$$

The complete cone is an illustration of a pointed closed shell, shown in figure 4–1(b). In contrast to the dome, which was examined in detail in sections 4.3.2.1 and 4.3.2.2, N_ϕ and $N_\theta \to 0$ at the pole. Also, recall that for a spherical shell under constant pressure, we found $N_\phi = N_\theta$ throughout, whereas for the conical shell, the circumferential stress is double the meridional stress.

4.3.4 Compound Axisymmetrically Loaded Shells

We now examine shells which have a meridian defined by more than one geometric curve; such combinations are called *compound shells*. A common usage of this form is for pressure vessels, in which a basic cylindrical tube may be capped by a so-called head and/or bottom which are often shallow spherical shells. Examples of such shells are illustrated by figures 2–8(r, s, and u), and an idealized case is shown in figure 4–14. One's first impulse might be to make the ends hemispheres to provide a smooth transition; however, the spherical shell, being a shell with positive Gaussian curvature, is nondevelopable and is not readily formed by bending flat sheet. Rather, it is considerably easier to form a relatively shallow spherical cap by rolling and dishing operations guided by suitable templates.

Turning to the shell shown in figure 4–14 under a positive internal pressure p, we find from equations (4.27) and (4.28) that for the spherical segment

$$N_\phi(\phi) = N_\theta(\phi) = \frac{pa_1}{2} \tag{4.77}$$

For the cylindrical shells, from equation (4.62b),

$$N_\theta = pa_2 \tag{4.78}$$

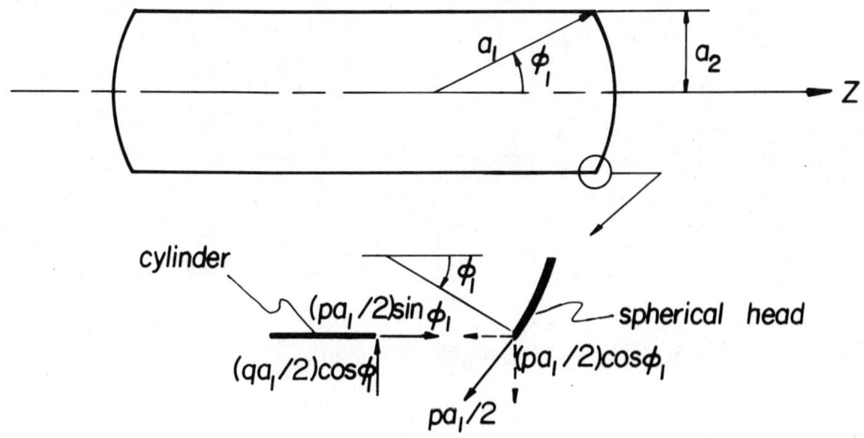

Figure 4–14. Cylindrical Shell with a Spherical Head

As far as the meridional stress resultant in the cylindrical shell is concerned, we have shown in equation (4.64) that N_Z = a constant dependent on the boundary value. As shown in the inset, the meridional stress in the sphere imparts an edge force per unit length of $(pa_1/2)\sin\phi_1$ to the cylinder so that

$$N_Z(Z) = \left(\frac{pa_1}{2}\right)\sin\phi_1 \tag{4.79}$$

We now attempt to ascertain the admissibility of the membrane solution for the compound shell. First, note that the sudden transition between the cylindrical and spherical shell is a violation of the *geometrical* guideline for the existence of a membrane state of stress as discussed in section 4.2. It is rather obvious that the radial component of $N_\phi(\phi_1)$, $pa_1/2$ $\cos\phi_1$, cannot be balanced by a membrane force in the cylinder, but will introduce transverse shears and, subsequently, bending which will affect both shells. Another result apparent from equations (4.77) and (4.78) is that the circumferential or hoop stress N_θ will likely be different for both shells at the junction. This leads to discontinuities in the circumferential strain and the radial deformation which we cannot quantify at this stage of the development, but which serve as further evidence of the inadequacy of the membrane theory in the junction region. Note that even if a complete hemisphere is used for a cap, with $a_1 = a_2 = a$,

$$N_\theta \text{ (hemisphere)} = \frac{pa}{2} \tag{4.80a}$$

$$N_\theta \text{ (cylinder)} = pa \tag{4.80b}$$

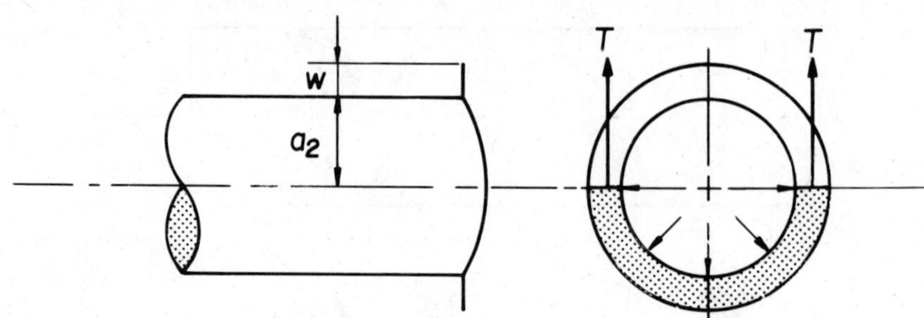

Figure 4–15. Ring Stiffener for a Shell Junction

so that the circumferential strain discrepancy will still remain, even though the unbalanced radial force $(pa/2)\cos\phi_1$ vanishes.

An attractive method of balancing the radial component of the thrust in the cap is the provision of a ring stiffener or collar at the junction. An example is provided by figure 2–8(r), and an idealization is shown in figure 4–15. The ring tension **T** in the stiffener can balance the radial thrust in the same manner as a cylindrical shell resists normal pressure, so that $2T = (pa_1/2)\cos\phi\,(2a_2)$, with the assumption that $w << a_2$. The ring, however, will restrain the radial expansion of the cylinder, and thus violate the ideal membrane *boundary* requirements.

We now consider two additional possibilities of providing a smooth transition. In figure 4–16(a), we show an ellipsoidal cap, whereas in figure 4–16(b), we provide a segment of a toroidal shell between the cylinder and the spherical cap. The latter form is sometimes called a *torospherical head*.

Considering the ellipsoidal head first, we note that a smooth transition is provided between the cylinder and the cap along the meridian, so that transverse shear in the cylinder is not required for radial equilibrium. However, the hoop stresses computed for an ellipsoid again do not match those for a cylinder as calculated from equation (4.80b);[10] thus, an incompatibility in the circumferential strain and subsequent radial deformation occurs which must, as before, be corrected by transverse shears acting on the cylinder and on the ellipsoid. Also, the problem of forming a deep curved cap, as mentioned previously, remains.

The torospherical head shown in figure 4–16(b) avoids a portion of the deep cap, but it turns out that the hoop stresses and the corresponding radial deformations, as computed from membrane theory, result in incompatibilities at both the cylindrical–toroidal and toroidal–spherical junctions. A satisfactory solution requires the bending theory, and the ultimate efficiency of the compound shells illustrated will depend significant-

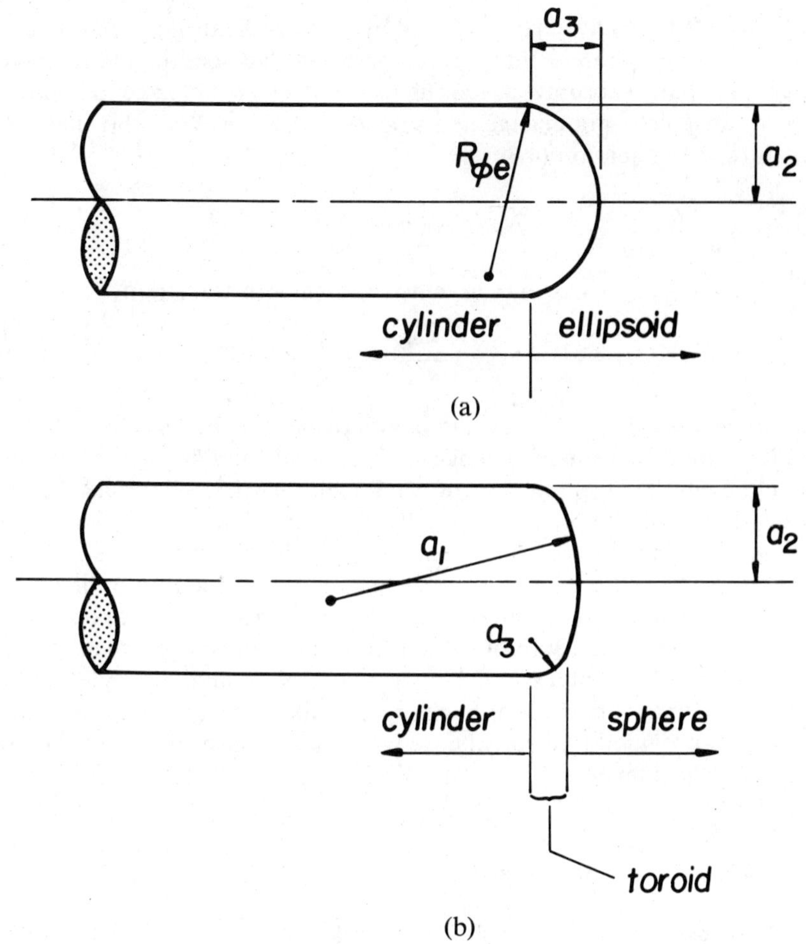

Figure 4–16. Compound Shells

ly on the penetration of the bending effects from the junction region. This is investigated quantitatively in chapter 9.

Another interesting aspect of the analysis of compound pressure vessels emerges from considering the normal equilibrium equation

$$\frac{N_\phi}{R_\phi} + \frac{N_\theta}{R_\theta} = q_n \tag{4.3c}$$

which was earlier termed the single most incisive equation in thin shell theory. We first apply the overall equilibrium approach developed in section 4.3.3 to the shells shown in figure 4–16, and particularly consider equation (4.59). At any section defined by the coordinate ϕ, $N_\phi(\phi)$ is basically a function of the resultant force between that section and the pole and cannot change abruptly as long as $R_0 \sin \phi$ does not change. Similarly, $R_\theta = R_0/\sin \phi$ does not change at a smooth transition. With this in mind, we can rewrite equation (4.3c) as

$$N_\theta = R_\theta\left(p - \frac{N_\phi}{R_\phi}\right) \tag{4.81}$$

With the foregoing arguments in mind, at a smooth transition,

$$N_\theta : \left[\frac{-N_\phi}{R_\phi} + p\right] \tag{4.82}$$

Thus, even if the transition is smooth with respect to the meridian and the loading is smooth, a sudden change in R_ϕ can radically change the magnitude and even the algebraic sign of N_θ. Recall that the cylindrical shell in figure 4–16 resists the pressure p by N_θ/R_θ, since R_ϕ is ∞. This occurs regardless of the magnitude of N_ϕ. Suddenly, at the transition, we encounter a finite value of R_ϕ, $R_{\phi e}$ for the ellipsoid and a_3 for the toroid. Since N_ϕ is positive, equation (4.82) reveals that N_θ decreases in proportion to the now finite value of R_ϕ. Both of these transitions can be presumed to produce large negative values of the hoop stress N_θ,[11] and such shells may fail because of circumferential buckling if proper strengthening and stiffening is not provided. An actual case where this occurred is described in Fino and Schneider.[12]

4.3.5 Nonsymmetric Loading

4.3.5.1 Transformation of Equilibrium Equations. We have thus far considered the membrane theory analysis of shells of revolution subject to loading that is nonvarying with respect to the circumferential coordinte θ, so-called symmetric loading. The solutions obtained in the previous section are based principally on equation (4.7a), which is written in terms of the auxiliary variable ψ. This equation was obtained by suppressing the θ dependence. We now return to equations (4.6) and eliminate ξ by taking $\partial/\partial\phi$ of equation (4.6a), $\partial/\partial\theta$ of equation (4.6b), and by dividing both sides by $R_\phi R_\theta \sin \phi$ which gives[13]

$$\frac{1}{R_\phi R_\theta \sin \phi}\left[\frac{R_\theta^2 \sin \phi}{R_\phi}\,\psi,_\phi\right],_\phi + \frac{1}{R_\phi \sin^2 \phi}\,\psi,_{\theta\theta} =$$

$$\frac{1}{R_\phi R_\theta \sin \phi}\left[(q_n \cos \phi - q_\phi \sin \phi)R_\theta^3 \sin^2 \phi\right],_\phi$$

$$+ R_\theta(q_{n,\theta\theta} + q_{\theta,\theta} \sin \phi) \tag{4.83}$$

Equation (4.83) is a second order partial differential equation in ψ. Once ψ is determined, ξ may be found by solving equation (4.6a) or (4.6b).

4.3.5.2 Fourier Series Representation. Since we have a partial differential equation to solve, we try the standard method of separation of variables. Specifically, we apply the Fourier series technique, whereby all loadings and dependent variables are taken in the form of Fourier series

$$\{\mathbf{q}\} = \begin{Bmatrix} q_\phi \\ q_\theta \\ q_n \end{Bmatrix} = \sum_{j=0}^{\infty} \begin{Bmatrix} q_\phi^j \cos j\theta \\ q_\theta^j \sin j\theta \\ q_n^j \cos j\theta \end{Bmatrix} \tag{4.84a}$$

$$\begin{Bmatrix} N_\phi \\ N_\theta \\ S \end{Bmatrix} = \sum_{j=0}^{\infty} \begin{Bmatrix} N_\phi^j \cos j\theta \\ N_\theta^j \cos j\theta \\ S^j \sin j\theta \end{Bmatrix} \tag{4.84b}$$

$$\begin{Bmatrix} \psi \\ \xi \end{Bmatrix} = \sum_{j=0}^{\infty} \begin{Bmatrix} \psi^j \cos j\theta \\ \xi^j \sin j\theta \end{Bmatrix} \tag{4.84c}$$

Substituting equations (4.84) into equations (4.6a) and (4.6b) and equation (4.83), we have

$$\sum_{j=0}^{\infty} \left\{ \left[\frac{R_\theta^2 \sin \phi}{R_\phi} \psi^j{,}_\phi + j\xi^j \right] \cos j\theta = [(q_n^j \cos \phi - q_\phi^j \sin \phi) \right.$$
$$\left. \cdot R_\theta^3 \sin^2 \phi] \cos j\theta \right\} \tag{4.85a}$$

$$\sum_{j=0}^{\infty} \left\{ \left[\xi^j{,}_\phi + j \frac{R_\theta}{\sin \phi} \psi^j \right] \sin j\theta = -[(q_\theta^j \sin \phi - jq_n^j) \right.$$
$$\left. \cdot R_\phi R_\theta^2 \sin \phi] \sin j\theta \right\} \tag{4.85b}$$

$$\sum_{j=0}^{\infty} \left\{ \left[\frac{1}{R_\phi R_\theta \sin \phi} \left[\frac{R_\theta^2 \sin \phi}{R_\phi} \psi{,}_\phi^j \right]{,}_\phi - \frac{j^2}{R_\phi \sin^2 \phi} \psi^j \right] \cos j\theta = \right.$$
$$\left[\frac{1}{R_\phi R_\theta \sin \phi} [(q_n^j \cos \phi - q_\phi^j \sin \phi) R_\theta^3 \sin^2 \phi]{,}_\phi \right.$$
$$\left. - jR_\theta(jq_n^j - q_\theta^j \sin \phi) \right] \cos j\theta \right\} \tag{4.85c}$$

We may treat each equation of the series separately, and then obtain the total solution by summing to a suitable truncation limit on j. We recognize that the individual equations for each harmonic are ordinary, rather than partial, differential equations, since the θ-dependent terms may be cancelled. This is, of course, a formidable mathematical simplification. Also, we see that equations (4.85a) and (4.85b), with $j = 0$, reduce to equations (4.7a) and (4.7b), respectively, which have previously been investigated with respect to axisymmetric loading. Thus, the previous solution for axisymmetric loading, as given by equation (4.8) and (4.9), serves also as the solution of the $j = 0$ harmonic for the nonsymmetric loading situation, with the loading taken as q_ϕ^0 and q_n^0, the $j = 0$ components of the surface loading. These solutions do not include the case of a $j = 0$ component of q_θ which, as mentioned in section 4.3.2.1, is quite rare. This case can be accommodated, however, by simply interchanging the $\sin j\theta$ and $\cos j\theta$ terms in equations (4.84a–c), obtaining an analogous set of equations to equations (4.85a–c), and combining the results of the two solutions.

Since the $j = 0$ case has already been solved, we now seek a solution to equation (4.85c) for a general harmonic $j > 0$. After cancelling the $\cos j\theta$ terms, we have for harmonic j

$$\frac{1}{R_\phi R_\theta \sin \phi} \left[\frac{R_\theta^2 \sin \phi}{R_\phi} \, \psi^j{}_{,\phi} \right]_{,\phi} - \frac{j^2}{R_\phi \sin^2 \phi} \, \psi^j =$$

$$\frac{1}{R_\phi R_\theta \sin \phi} \left[(q_n^j \cos \phi - q_\phi^j \sin \phi) R_\theta^3 \sin^2 \phi \right]_{,\phi}$$

$$- jR_\theta \, (jq_n^j - q_\theta^j \sin \phi) \quad (4.86)$$

We postpone consideration of the general solution of equation (4.86) until further specializations are introduced. For now, note that once ψ^j has been derived, we may solve for ξ^j from either equation (4.85a) or (4.85b). Since ξ^j occurs as an undifferentiated term in equation (4.85a), this equation is commonly selected. For each harmonic, the stress resultants at any circumferential angle may then be computed by evaluating N_ϕ^j and S^j, using equations (4.4), and N_θ^j, using equation (4.3c), and then substituting the appropriate value of θ into equation (4.84b). In passing, note that the "price paid" for the mathematical simplification of reducing the partial differential equation to an uncoupled series of ordinary differential equations is reflected in the total number of harmonic solutions that must be summed, in a given case, to adequately represent the effect of the particular surface loading. Each harmonic solution entails a complete analysis of the shell, whereas the partial differential equation, (4.83), need only be solved once. In spite of this, the Fourier series approach has been widely used to solve shells of revolution subject to unsymmetric loading, because it is fundamentally simpler from a mathematical viewpoint.

The eventual solution of equation (4.86) is strongly dependent on the harmonic number j. The case $j = 1$ is called *antisymmetric* loading and is somewhat more tractable than the case $j > 1$, which will be called *asymmetric* loading.

4.3.6 Antisymmetrical Loading

4.3.6.1 Integration of Equilibrium Equations and Boundary Requirements. Consider equation (4.86), with $j = 1$. Introduce a further auxiliary variable[13]

$$Y(\phi) = \psi^1(\phi)R_\theta \sin \phi \tag{4.87}$$

and substitute into equation (4.86), with $j = 1$, to get

$$\left[\frac{1}{R_\phi \sin \phi} Y_{,\phi} \right]_{,\phi} = \frac{1}{R_\theta \sin \phi} [(q_n^1 \cos \phi - q_\phi^1 \sin \phi)$$

$$\cdot R_\theta^3 \sin^2 \phi]_{,\phi} - R_\theta R_\phi(q_n^1 - q_\theta^1 \sin \phi) \tag{4.88}$$

The transformation between equations (4.86) and (4.88) is most easily verified by substituting equation (4.87) into equation (4.88) and performing the indicated differentiation with the aid of the Gauss–Codazzi relation, equation (2.47).

Equation (4.88) is in a form which permits Y to be evaluated by successive integrations with respect to ϕ. Then ψ is found from equation (4.87), and the remaining auxiliary variables and stress resultants are determined as detailed in section 4.3.5.2. The procedure is straightforward and, since it is discussed in detail in Novozhilov,[14] it is not repeated here. Rather, we present the results which may be easily verified as a solution by direct substitution into the membrane theory equilibrium equations, (4.3a–c). The stress resultants corresponding to $j = 1$ are

$$N_\phi = N_\phi^1 \cos \theta; \quad N_\theta = N_\theta^1 \cos \theta; \quad S = S^1 \sin \theta \tag{4.89}$$

where the Fourier coefficients are given by

$$N_\phi^1 = \frac{1}{R_0^2 \sin \phi} \left\{ C_2 + C_1 \int_{\phi'}^{\phi} R_\phi \sin \phi \, d\phi + \int_{\phi'}^{\phi} \Phi(\phi)R_\phi \sin \phi \, d\phi \right\} \tag{4.90a}$$

$$N_\theta^1 = q_n^1 R_\theta - \frac{R_\theta}{R_\phi} N_\phi^1 \tag{4.90b}$$

$$S^1 = N_\phi^1 \cos \phi - \frac{C_1}{R_0} + \chi(\phi) \tag{4.90c}$$

and the functions $\Phi(\phi)$ and $\chi(\phi)$ are

110

$$\Phi(\phi) = (q_n^1 \cos \phi - q_\phi^1 \sin \phi)R_0 R_\theta - \int_{\phi'}^{\phi} (q_n^1 \sin \phi$$

$$+ q_\phi^1 \cos \phi - q_\theta^1)R_\phi R_0 \, d\phi \tag{4.91a}$$

$$\chi(\phi) = \frac{1}{R_0} \int_{\phi'}^{\phi} (q_n^1 \sin \phi + q_\phi^1 \cos \phi - q_\theta^1)R_\phi R_0 \, d\phi \tag{4.91b}$$

Here, we assume that both N_ϕ and S are specified on the same boundary, $\phi = \phi'$, and that C_1 and C_2 are integration constants. From equation (4.90a),

$$C_2 = N_\phi^1(\phi')R_\theta^2(\phi') \sin \phi' \tag{4.92a}$$

and from equation (4.90c),

$$C_1 = R_0(\phi')[N_\phi^1(\phi') \cos \phi' - S^1(\phi')] \tag{4.92b}$$

so that a stress-free top edge gives $C_1 = C_2 = 0$. Also, for a dome we take $\phi' = 0$. Then, $R_0 = 0$ and $C_1 = C_2 = 0$. The more general case, where N_ϕ and S are specified at different boundaries, is treated in an analogous manner.

To evaluate the stress resultants at the pole of a dome for $j = 1$, we consider first equation (4.90a), with $\phi' = 0$, and $C_1 = C_2 = 0$, and recognize that we have a 0/0 indeterminate form. We apply L'Hospital's rule by differentiating the numerator and denominator of equation (4.90a). Noting that

$$(R_0^2 \sin \phi)_{,\phi} = \sin \phi \cdot 2R_0 R_\phi \cos \phi + R_0^2 \cos \phi$$

$$= R_0 \cos \phi \, (2R_\phi \sin \phi + R_0) \tag{4.93}$$

we get

$$\lim_{\phi \to 0} N_\phi^1(\phi) = \lim_{\phi \to 0} \frac{\Phi R_\phi \sin \phi}{R_0 \cos \phi \, (2R_\phi \sin \phi + R_0)} \tag{4.94}$$

which remains an indeterminate form. Therefore, we must apply L'Hospital's rule once more to equation (4.94). Considering the numerator of equation (4.94)

$$(\Phi R_\phi \sin \phi)_{,\phi} = R_\phi \sin \phi \, \Phi_{,\phi} + \Phi \cdot (R_\phi \sin \phi)_{,\phi}$$

$$\lim_{\phi \to 0} (\Phi R_\phi \sin \phi)_{,\phi} = 0 + 0 = 0 \tag{4.95}$$

Differentiating the denominator gives

$$2(R_0 \sin \phi \cdot R_\phi \cos \phi)_{,\phi} + (R_0^2 \cos \phi)_{,\phi} \tag{4.96}$$

The second term in equation (4.96) will go to 0 in the limit by analogy with equation (4.93). Considering the first term, we have

$$2(R_0 \sin \phi \cdot R_\phi \cos \phi)_{,\phi} =$$

$$2[R_\phi \cos \phi \cdot R_\phi \cos \phi + R_0 \sin \phi \cdot (R_\phi \cos \phi)_{,\phi}] \qquad (4.97)$$

As $\phi \to 0$, the second term of equation (4.97) will also go to 0, while the first term becomes $2R_\phi^2(0)$. Thus, we finally conclude that

$$\lim_{\phi \to 0} N_\phi^1(\phi) = \frac{0}{2R_\phi^2(0)} = 0 \qquad (4.98a)$$

or

$$N_\phi^1(0) = 0 \qquad (4.98b)$$

We may then establish from equation (4.90c) that

$$S^1(0) = 0 \qquad (4.99)$$

since $C_1 = 0$ and $\chi(0) = 0$ by L'Hospital's rule applied to equation (4.91b).

Examining equation (4.90b) for N_θ^1 as $\phi \to 0$, the second term goes to 0, while the first term remains. Thus we find

$$N_\theta^1(0) = q_n^1(0)R_\theta(0) \qquad (4.100)$$

We have established in section 4.3.2.1 that $N_\phi(0) = N_\theta(0)$ for a dome. Equation (4.98b) indicates that $N_\theta^1(0) = 0$, which in view of equation (4.100), implies that $q_n^1(0)$ must be equal to 0 for a properly defined antisymmetric loading. An example is presented in the following section.

Finally, with respect to the general analysis of antisymmetrically loaded shells of revolution, note that we have evaluated the constants of integration through the consideration of (a) specified values of N_ϕ^1 and S^1 at given boundaries for open shells; or (b) pole conditions for domes. Now it is required that the remaining boundary develop the calculated values of N_ϕ and S to ensure membrane action. As illustrated in figure 4–17(a), the displacements corresponding to these stress resultants, in the meridional direction for N_ϕ and in the tangential direction for S, must be 0.

Such a set of boundary conditions (meridional and tangential displacements equal to 0 in one boundary) apparently are sufficient to ensure membrane behavior, provided that the corresponding geometrical and loading restrictions are observed. The question arises as to the possibility of specifying an alternate set of conditions for an open shell; e.g., meridional displacement equal to 0 on one boundary, and tangential displacement equal to 0 on the other, as shown in figure 4–17(b). The study of this possibility requires the consideration of so-called *pure bending deforma-*

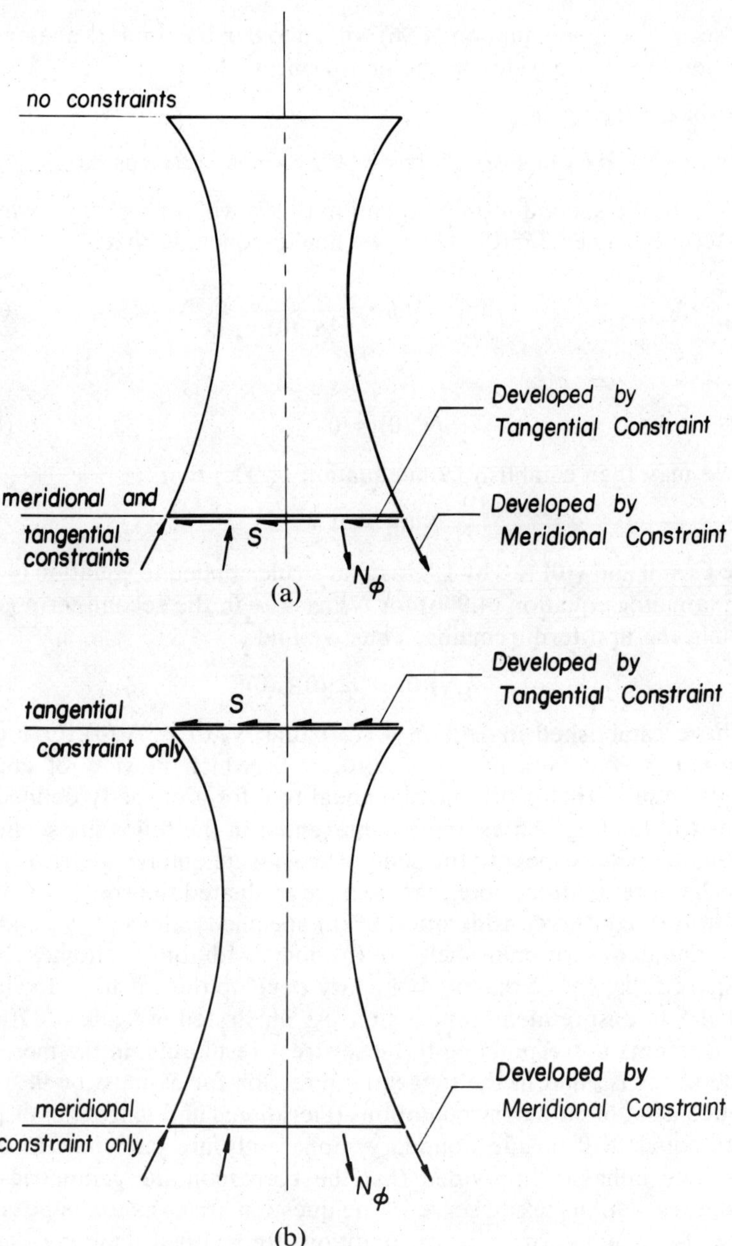

Figure 4–17. Boundary Requirements for Shells of Negative Gaussian Curvature

tions, which are deformations without any extension or in-plane shearing of the middle surface.

Although a formal study of pure bending deformations is beyond the scope of this text, one can show that shells for which this behavior is possible are inherently unstable. The circumstances for which these deformations are possible are not primarily a function of the *applied loading*, but are related to the *geometrical properties* and the *boundary conditions*. Mathematically, one can show that such deformations constitute a nontrivial solution to the equations describing the middle surface displacement of the shell when the loading terms q_ϕ, q_θ, and q_n are equal to 0; e.g., a nontrivial homogeneous solution to the equations describing the displacement of the middle surface. A closely related configuration, *inextensional deformation* originally conceived by Lord Rayleigh in the study of the vibrations of cylindrical shells,[15] has been used to facilitate instability analysis of shells.

Obviously, the possibility of pure bending deformations in actual structures is to be avoided. One can show that a support for which both the meridional and circumferential displacements are constrained on one boundary, figure 4–17(a), is *sufficient* to prevent pure bending deformations for all shells of revolution, and is also *necessary* to remove this possibility for shells with *negative* Gaussian curvature.[16] In other words, there are some geometries for which a shell with negative Gaussian curvature might undergo pure bending deformation if both the meridional and tangential displacements are not prevented at one boundary, as in figure 4–17(b).

An interesting parable based on an awareness of the possibility of these deformations drastically reducing the stable configuration of prefabricated hyperboloidal cooling towers is found in the treatise of V. Z. Vlasov.[17] He noted

The lattice designs of Engineer Shukov which are well known in construction practice, outlined by the surface of a hyperboloid of one sheet and consisting of rods arranged in straight lines, forming and tied together in horizontal rings, may constitute instantaneously varying unstable systems under certain combinations of geometric sizes and types of asymmetrical loads (wind loads, for example). Unless such structures are properly reinforced by additional structural elements (for example, tension members located in the planes of the rings), they possess very little load-bearing capacity.[a]

4.3.6.2 Spherical Shell under Wind Loading. It is common practice to represent dynamic loadings such as wind and earthquake by statically equivalent or pseudostatic loading. While this procedure is not universally applicable and can often lead to erroneous conclusions, it is often expe-

---

[a] Reprinted with permission of NASA.

114

dient, and, if due care is exercised, it can provide a representation that may be adequate for design. With this preface, consider a pseudostatic wind loading on a spherical dome as shown in figure 4–18.[18] The wind pressure is considered to be normal to the surface, and the components are

$$q_\phi = 0 \tag{4.101a}$$

$$q_\theta = 0 \tag{4.101b}$$

$$q_n = -p \sin \phi \cos \theta \tag{4.101c}$$

where p = the static wind pressure intensity.

Therefore, from equation (4.84a),

$$q_n^1 = -p \sin \phi \tag{4.102}$$

At this point, a word of warning is in order. It is fairly obvious that the pseudostatic wind loading given on figure 4–18 and in equation (4.101) has been constructed as a reasonable facsimile of a possible pressure distribution to fit the antisymmetrical load form, $j = 1$. Extreme caution should be exercised in using this form of loading in actual design. When asymmetric loading is discussed later, we show that the stress resultants in shells of revolution are very sensitive to the *circumferential* distribution of the forcing function, and that to represent a loading which has significant $j > 1$ content by only a $j = 1$ component similar to equation (4.101) can be quite unconservative.

Proceeding with the solution for the spherical dome, we have $R_\phi = R_\theta = a$; $C_1 = C_2 = 0$; $\phi' = 0$; and $q_\phi^1 = q_\theta^1 = 0$, $q_n^1 = -p \sin \phi$ substituted into equations (4.90) and (4.91). After integration, the solution is[19]

$$N_\phi^1 = -pa \frac{\cos \phi}{3 \sin^3 \phi} (2 + \cos \phi)(1 - \cos \phi)^2 \tag{4.103a}$$

$$N_\theta^1 = -(pa \sin \phi + N_\phi^1) \tag{4.103b}$$

$$S^1 = -pa \frac{(2 + \cos \phi)(1 - \cos \phi)^2}{3 \sin^3 \phi} \tag{4.103c}$$

The solution is readily verified by direct substitution in the equilibrium equations, (4.3a–c).

Equations (4.102) and (4.103) are graphed in nondimensional form on figure 4–19 for a hemisphere. As discussed earlier with respect to figure 4–3, the graph holds as well for spherical domes with $\phi_b < \pi/2$ by considering the portion above that ordinate. Two points of special interest on this graph are: a), S^1 is maximum at $\phi = \pi/2$; and b), $N_\phi^1 = 0$ at $\phi = \pi/2$. To provide a physical explanation of these results, we again find the over-

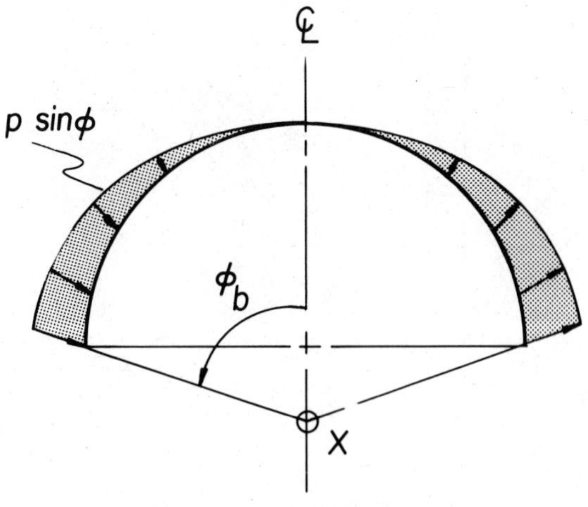

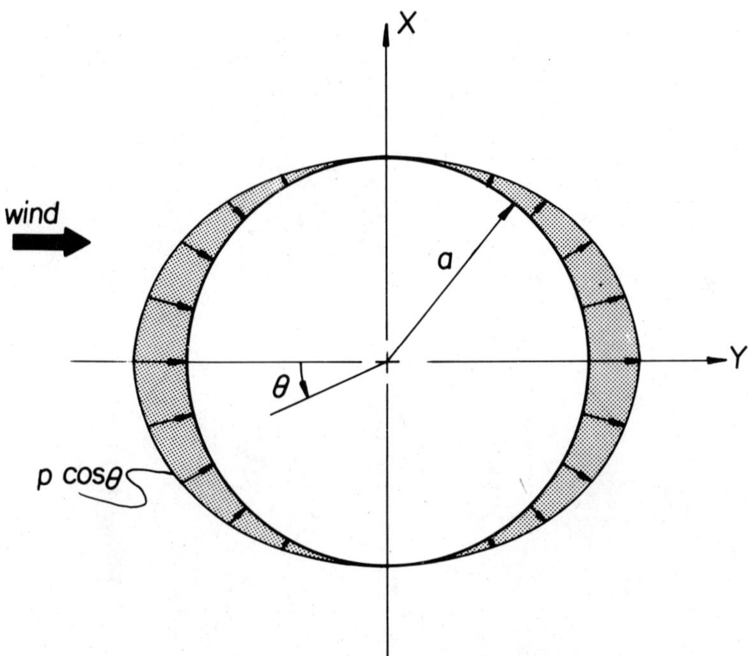

Figure 4–18. Static Design Wind Load on a Spherical Shell

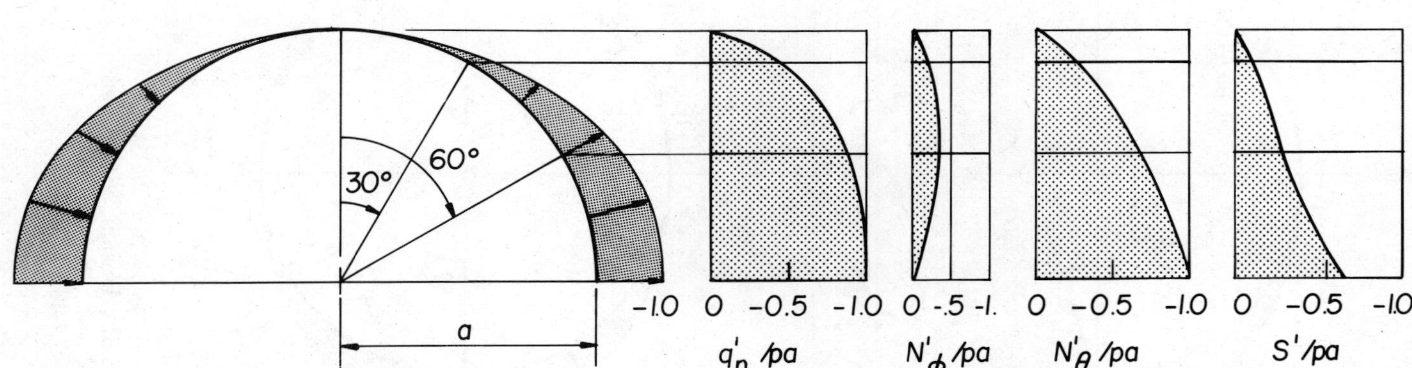

Figure 4–19. Stress Resultants Due to Design Wind Load on a Spherical Dome

all equilibrium approach to be helpful. The appropriate equilibrium equations, written with respect to the X–Y coordinates shown in figure 4–18, are $\Sigma M_X = 0$ and $\Sigma F_Y = 0$. Here the X axis is drawn through the center of curvature and Y is parallel to the assumed direction of the wind. We show the appropriate free body diagrams in figure 4–20. The positive signs of N_ϕ and S are established from figure 3–6.

First, considering $\Sigma M_X = 0$, we have

$$\int_{-\pi}^{\pi} N_\phi\left(\frac{\pi}{2}, \theta\right) \cdot a\, d\theta \cdot a \cos\theta + \int_{0}^{\pi/2} \int_{-\pi}^{\pi} q_n(\phi, \theta)\, a\, d\theta \cdot a\, d\phi \cdot 0$$

$$+ \int_{-\pi}^{\pi} S\left(\frac{\pi}{2}, \theta\right) \cdot a\, d\theta \cdot 0 = 0 \qquad (4.104)$$

The last two integrals vanish because $q_n(\phi, \theta)$ and $S(\pi/2, \theta)$ pass through the X axis, so that we have remaining

$$\int_{-\pi}^{\pi} N_\phi\left(\frac{\pi}{2}, \theta\right) a^2 \cos\theta\, d\theta = N_\phi^1\left(\frac{\pi}{2}\right) a^2 \int_{-\pi}^{\pi} \cos^2\theta\, d\theta$$

$$= \pi a^2 N_\phi^1\left(\frac{\pi}{2}\right)$$

$$= 0$$

We thus confirm that

$$N_\phi^1\left(\frac{\pi}{2}\right) = 0 \qquad (4.105)$$

as shown on figure 4–19.

Then, we evaluate $\Sigma F_Y = 0$

$$\int_{\theta=-\pi}^{\pi} \int_{\phi=0}^{\pi/2} -q_n(\phi, \theta) \sin\phi \cos\theta \cdot a\, d\phi \cdot a \sin\phi\, d\theta$$

$$+ \int_{-\pi}^{\pi} S\left(\frac{\pi}{2}, \theta\right) \cdot \sin\theta \cdot a\, d\theta$$

$$= pa^2 \int_{-\pi}^{\pi} \int_{0}^{\pi/2} \sin^3\phi \cos^2\theta\, d\phi\, d\theta + aS^1\left(\frac{\pi}{2}\right) \int_{-\pi}^{\pi} \sin^2\theta\, d\theta$$

$$= pa^2\pi \left[-\frac{1}{3} \cos\phi(\sin^2\phi + 2) \right]_0^{\pi/2} + aS^1\left(\frac{\pi}{2}\right)\pi$$

$$= \frac{2}{3}pa + S^1\left(\frac{\pi}{2}\right)$$

$$= 0 \qquad (4.106)$$

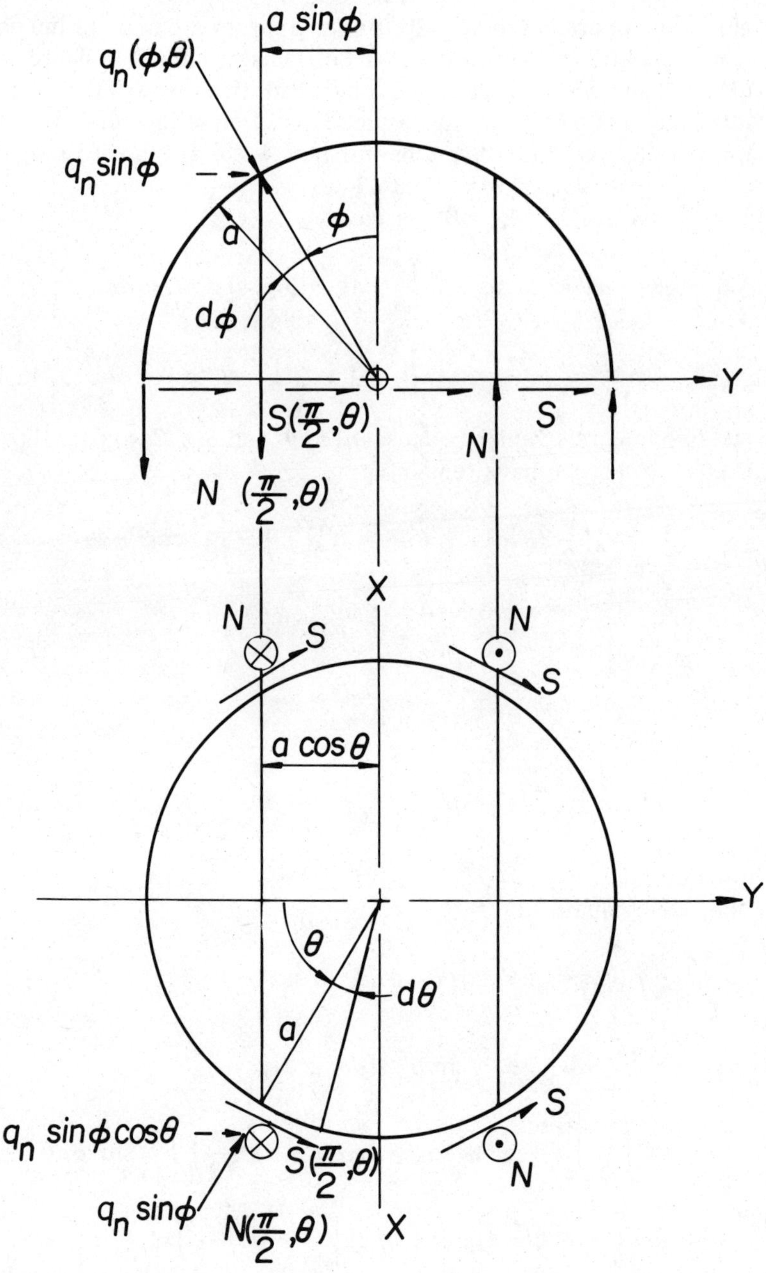

Figure 4–20. Equilibrium of a Spherical Shell under Design Wind Load

so that

$$\frac{S^1(\pi/2)}{pa} = -\frac{2}{3} \qquad (4.107)$$

as shown in figure 4–19.

The verification of the stress resultants using the overall equilibrium equations suggests that these two equations, written with reference to a general coordinate ϕ, should provide an independent derivation for N_ϕ^1 and S^1, whereupon N_θ^1 would be computed, as usual, from equation (4.3c). Note that in the general case $\phi_b \neq \pi/2$, so that N_ϕ and S would participate in $\Sigma M_X = 0$, and N_ϕ in $\Sigma F_Y = 0$. The procedure of summing forces and moments across a section to compute bending and shear stresses is strikingly similar to the basic approach of simple beam theory. In fact, the $j = 1$ case is indeed equivalent to beam behavior except, of course, beam theory alone does not admit the computation of the circumferential stress N_θ. The use of beam theory to compute the stresses in antisymmetrically loaded shells of revolution is developed in some detail in Popov.[20]

4.3.6.3 Hyperboloidal Shells under Seismic Loading. Another dynamic effect frequently simulated by a static representation is an earthquake or seismic load. Records of actual or artificial earthquakes are generally available in component form, two horizontal components (N–S) and (E–W), and a vertical component. The vertical component is an axisymmetric loading which fits the solution developed in section 4.3.2.1. Regarding the horizontal components, it is frequently assumed that they are not in phase; i.e., the peak effects from the two directions do not occur at the same time on the structure. Since shells of revolution are axisymmetric in construction, it is here regarded as sufficient to consider only the stronger of the two components. The X axis in figure 4–21 is oriented with the direction of this component.

From the results of a linear dynamic analysis or from some approximations thereto, a total design base shear **V** may be computed corresponding to the strong horizontal motion earthquake component. The base shear is frequently written as a percentage C of the dead weight of the structure (e.g., 15% g). Although details of the computation of **V** are beyond the scope of this book, note that the dynamic characteristics of the structure must enter into any rational treatment. An excellent comprehensive discussion of this subject is given in Clough,[21] and an application to shells of revolution is contained in Gould, Sen, and Suryoutomo.[22] The remainder of this section is concerned with the determination of the distribution of surface loading that produces a specified value of the base shear **V** and the evaulation of the corresponding membrane stress resultants.

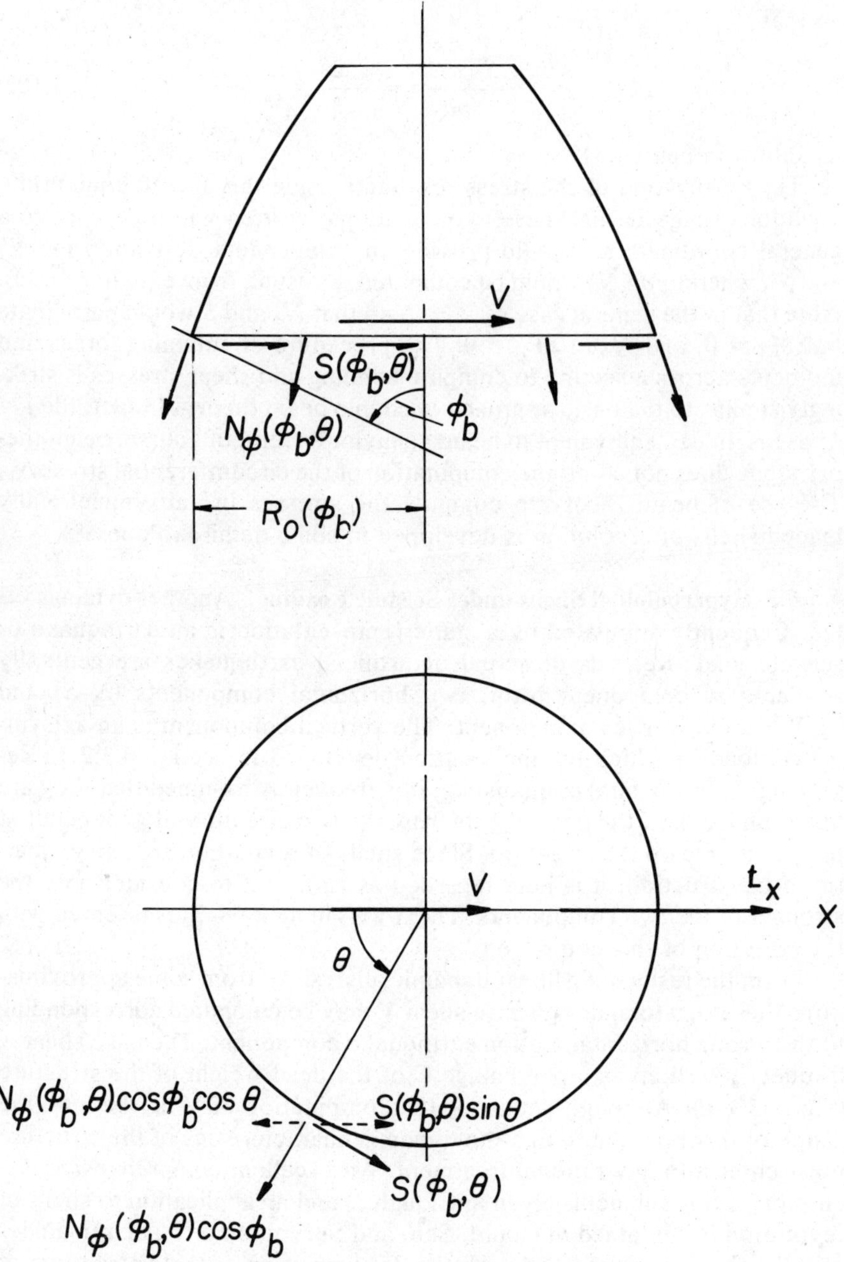

Figure 4–21. Base Shear Due to Horizontal Seismic Loading

As depicted in figure 4–21, the design base shear $\mathbf{V}$ is equal to the total horizontal component of N_ϕ and S, or

$$\mathbf{V} = \int_{-\pi}^{\pi} [S(\phi_b, \theta) \sin \theta - N_\phi(\phi_b, \theta) \cos \phi_b \cos \theta] R_0(\phi_b)\, d\theta \qquad (4.108a)$$

We assume that the base moves uniformly without circumferential distortion, so that only the $j = 1$ harmonic participates. Then, we have

$$\mathbf{V} = R_0(\phi_b) \left[S^1(\phi_b) \int_{-\pi}^{\pi} \sin^2 \theta\, d\theta - N_\phi^1(\phi_b) \cos \phi_b \int_{-\pi}^{\pi} \cos^2 \theta\, d\theta \right]$$

$$= \pi R_0(\phi_b)[S^1(\phi_b) - N_\phi^1(\phi_b) \cos \phi_b] \qquad (4.108b)$$

The subsequent analysis for the membrane stress resultants is dependent on the surface loading which produces $\mathbf{V}$. We know that the equivalent surface loading should have a resultant in the direction of $\mathbf{V}$; i.e.

$$\iint_{\text{Surface}} \mathbf{q}\, ds_\alpha ds_\beta = \mathbf{V} = V\mathbf{t}_x \qquad (4.109)$$

where $\mathbf{q} = q\mathbf{t}_x$ is defined in figure 3–4. Since $\mathbf{V}$ is assumed to act in the horizontal plane, $\mathbf{q}$ at any level acts in the horizontal plane parallel to the assumed strong direction of the earthquake, indicated here by the X axis. Referring to figure 4–22(b), we see that $\mathbf{q}(\phi)$ has a circumferential component of $q(\phi) \sin \theta$, and a radial component of $q(\phi) \cos \theta$. Considering the meridional view, figure 4–22(a), the radial component, $q(\phi) \cos \theta$, is projected from the lower figure and resolved into a meridional component, $q_\phi \cos \theta \cdot \cos \phi$, and a normal component, $q(\phi) \cos \theta \sin \phi$. Thus, referring to figure 3–6 for the correct signs, we have

$$q_\phi = -q(\phi) \cos \phi \cos \theta \qquad (4.110a)$$

$$q_\theta = q(\phi) \sin \theta \qquad (4.110b)$$

$$q_n = -q(\phi) \sin \phi \cos \theta \qquad (4.110c)$$

We compare equations (4.110a–c) with equation (4.84a) and conclude that

$$q_\phi^1 = -q(\phi) \cos \phi \qquad (4.110d)$$

$$q_\theta^1 = q(\phi) \qquad (4.110e)$$

$$q_n^1 = -q(\phi) \sin \phi \qquad (4.110f)$$

It remains to determine the meridional variation of q, $q(\phi)$, so that equation (4.109) is satisfied.

Consider a cooling tower structure in the form of a hyperboloid of revolution as illustrated in figure 4–23. An actual cooling tower is shown in

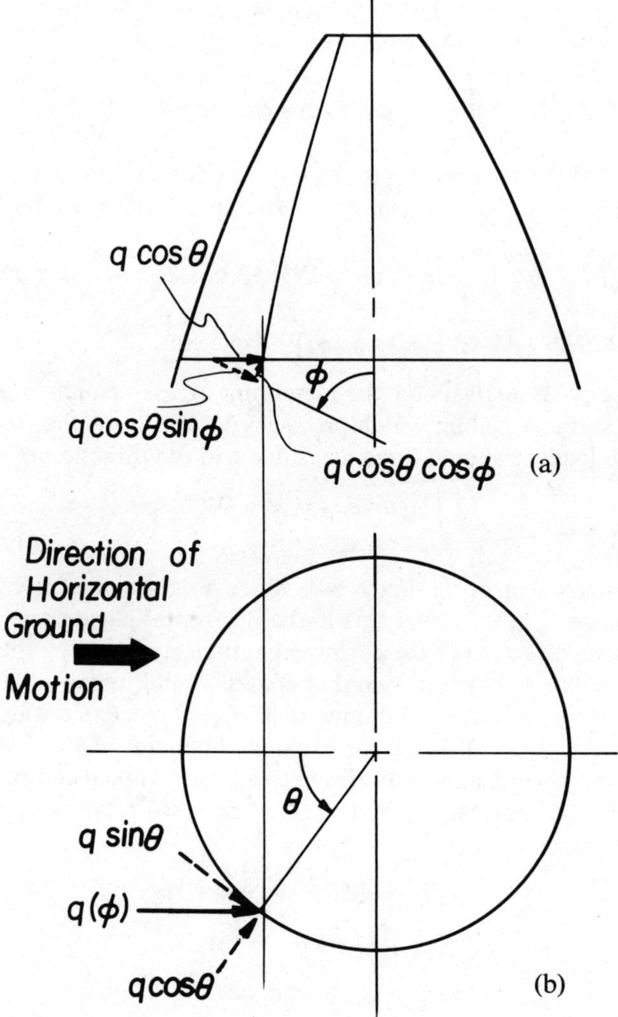

Figure 4–22. Static Horizontal Seismic Loading on a Shell of Revolution

figure 2–8(o). First, we relate the total base shear magnitude V to the total shell weight W. Following the overall equilibrium method,

$$W = N_\phi(\phi_b) \sin \phi_b \cdot 2\pi R_0(\phi_b) \qquad (4.111a)$$

where $N_\phi(\phi_b)$ is due to self-dead load. Substituting equation (4.40a) into equation (4.110)

$$W = 2\pi q_{DL} a^2 [\zeta_1(\phi_b) - \zeta_1(\phi_t)] \qquad (4.111b)$$

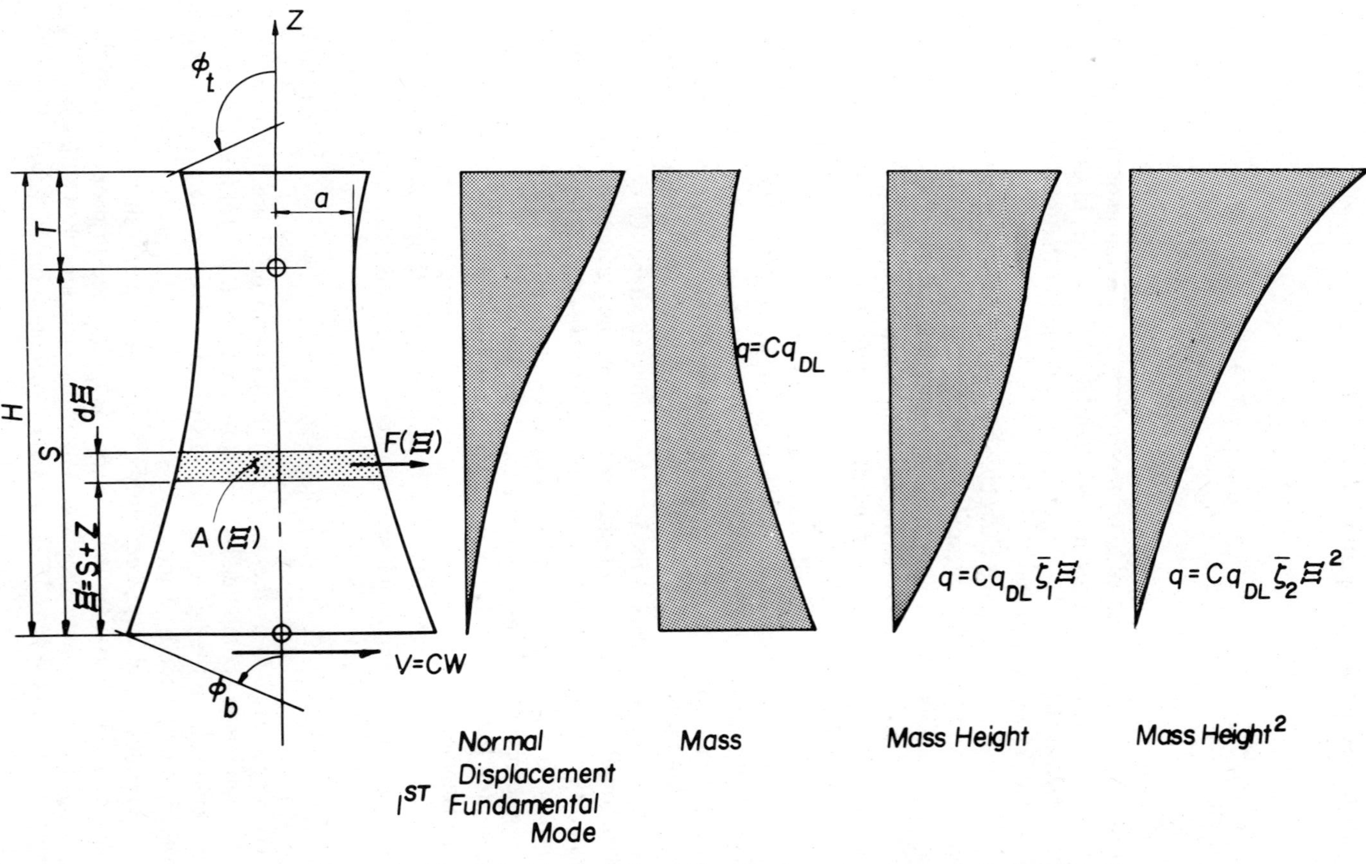

Figure 4–23. Seismic Loading Distributions for Hyperboloidal Shells

where ζ_1 is defined in equation (4.40b). Then

$$C = \frac{V}{W} \tag{4.112a}$$

or

$$V = CW \tag{4.112b}$$

Either C or V are assumed to be specified by consideration of the dynamic characteristics of the system, as described, for example, in Clough. Note that equation (4.111) may be useful for calculating the total weight of a tower of given dimensions.

Now we refer to some characteristics of the linear dynamic response of a tower in the $j = 1$ harmonic. We can show that this response can be represented as a linear combination of the longitudinal modes of free vibration,[23] and that the first mode is the most important for the seismic analysis. The first mode normal displacement for a tower of typical dimensions is shown in figure 4–23. Keeping in mind that the total base shear V is presumed to be known, the actual magnitude of the normal displacement is not needed for this discussion; what is important here is the shape. The inertial forces produced by a harmonic response are, in fact, proportional to the mode shape. This follows from D'Alembert's principle. Thus, it is reasonable to try and arrive at a distribution of the base shear V which resembles the dynamic deflected shape. Of course, the actual longitudinal response is a combination of the participating modes and is an ever-changing function of time; the best we can attempt with a static approach is a one-term approximation at a given time, and this is reasonably selected as the first mode shape. Thus we shall try to arrive at a distribution $q(\phi)$, which has a shape similar to the first mode of the normal displacement.

The most common distribution used for shell design has been to assume that the seismic load is everywhere proportional to the gravity load, or

$$q(\phi) = Cq_{DL} \tag{4.113}$$

This is referred to as the *Mass (M)* distribution on figure 4–23 and the components are computed by substituting equation (4.113) into (4.110). It is obvious from the figure that this distribution does not realistically represent the first mode shape.

Another possibility, frequently adopted in the analysis of tall building structures, is to assume that the load is distributed in proportion to the weight and the height of each element of the structure above the base. We refer to this as the *Mass-Height (MH)* distribution on the figure. This distribution has been adapted to cooling towers.[24] To derive this shape, we refer to a differential element $d\Xi$, which is located at a distance $\Xi = S + Z$ above the base as shown on figure 4–23. The coordinate Ξ is thus an axial coordinate measured from the base. The surface loading acting on the differential element $d\Xi$ (which represents the appropriate proportion of the base shear V as defined in equation (4.109)) is given by

$$F(\Xi) = V \frac{\Xi \, dW}{\sum \Xi \, dW} = V \frac{W,_\Xi \, \Xi \, d\Xi}{\int_0^H W,_\Xi \Xi \, d\Xi} \qquad (4.114)$$

where W is given by equation (4.111). Then $F(\Xi)$ is divided by the differential area

$$A(\Xi) = 2\pi R_0(\Xi) \, d\Xi \qquad (4.115)$$

to get $q(\Xi) = F(\Xi)/A(\Xi)$. The resulting expression is easily transformed into a function of ϕ

$$q(\phi) = Cq_{DL}\bar{\zeta}_1\Xi(\phi) \qquad (4.116)$$

through equation (4.34). $\bar{\zeta}_1$ is a constant defined in Appendix 4A. From the graph of the MH distribution, it is obvious that this appears to be far more realistic than the M distribution when compared to the mode shape.

A third possibility may be included, whereby the shear is distributed according to the weight and the *square* of the height of each element above the base. This is termed the *mass-height*[2] (MH^2) distribution. Following the same procedure, we get[25]

$$q(\phi) = Cq_{DL}\bar{\zeta}_2\Xi^2(\phi) \qquad (4.117)$$

where $\bar{\zeta}_2$ is a constant defined in the Appendix 4A. From the graph, we see that perhaps the MH^2 distribution is even a better approximation than the MH.

The components of surface load for each of the assumed distributions may be substituted into equations (4.90), (4.91), and (4.92) to evaluate the stress resultants, which has been done in various publications. Since the results are not readily available in collected form, the pertinent formulae are summarized in Appendix 4A. All quantities are nondimensionalized and are tersely presented in a form suitable for computer input preparation. Tabulated results for a wide range of tower dimensions are also available in Closure to Reference[26] for the M distribution.

Although the detailed derivation of the various solutions are probably of interest only to the specialist, it is instructive to look at a comparative

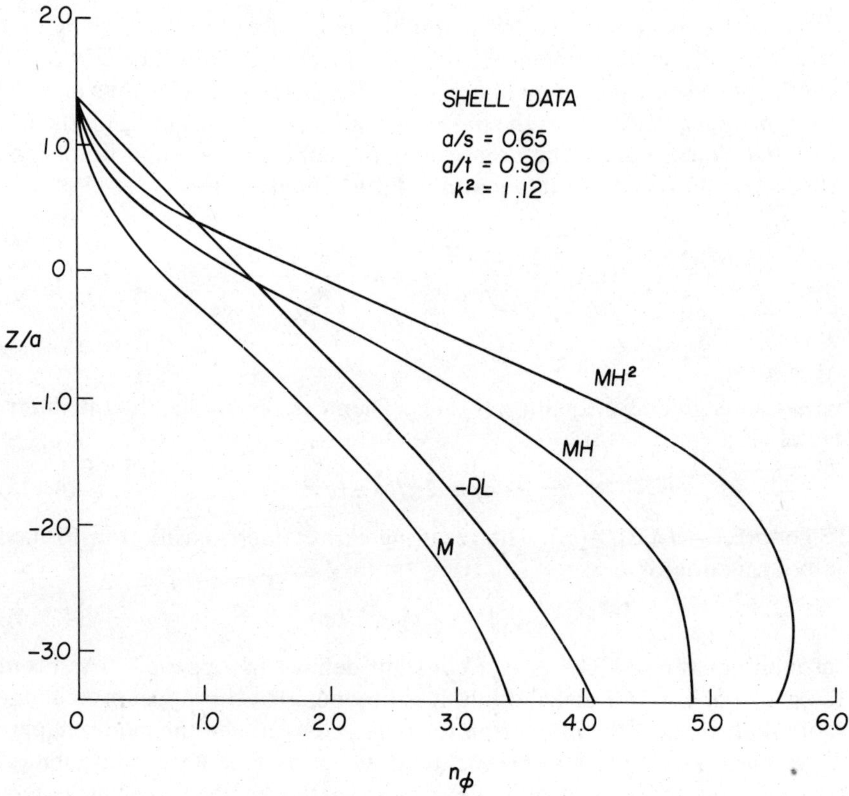

Figure 4–24. Effect of Base Shear Distribution on the Meridional Stress Resultant for a Hyperboloidal Shell

study of the stress resultants obtained using the various distributions. In figure 4–24, we show the nondimensional meridional stress resultant $n_{\phi 1}$ = $N_{\phi 1}/(q_{DL}a)$ for a shell of typical proportions using the three proposed distributions. Also shown is the negative of the dead-load stress resultant, equation (4.40a), suitably nondimensionalized. When the seismic load graph falls to the right of the dead load graph, this indicates a net tension at $\theta = 0$ due to the combined loading. It is for this net tension that a concrete shell must be reinforced. Also it is appropriate to note here that the designer should be very careful when considering combined lateral and gravity loading, which produces a net tension situation. Each loading should be factored by an appropriate amount prior to combining effects with opposite signs. For example, consider figure 4–24 at $Z/a = -1.0$, where $N_{\phi}^{(DL)} = -2.5$ and $N_{\phi}^{1(MH)} = 3.3$. Using load factors from a standard code,[27]

$$N_{\phi}^{(NET)} = 1.3 \cdot 3.3 - 0.9(2.5) = 2.04$$

An alternate (not recommended) would be to apply a factor of safety to the net stress. For example, using a nominal factor of safety of 1.5,

$$N_\phi^{(\text{NET})} = 1.5(3.3 - 2.5) = 1.20$$

It is apparent that the latter procedure can lead to unconservative results when the lateral and dead load stresses are close in magnitude but opposite in sign.

4.3.7 Asymmetrical Loading

4.3.7.1 Integration of Equations. The analysis of shells of revolution for harmonics $j > 1$ is quite different from the $j = 0$ and $j = 1$ cases. While there are a relative abundance of quadrature solutions available for the symmetric and antisymmetric loading cases, this is not the case for asymmetric loading; the governing equation, equation (4.86), is not solvable in general terms but, rather, whatever solutions that exist are dependent on some special form of the variable coefficients.

At this point, recall that the $j = 0$ and $j = 1$ solutions were directly related to and derivable from static equations of overall equilibrium. The $j = 0$ solution represents force equilibrium parallel to the axis of rotation, whereas the $j = 1$ case reflects equilibrium of forces normal to the axis and equilibrium of moments about the axis of overturning. In contrast, $j > 1$ does not correspond to any equation of overall equilibrium. All loading cases for $j > 1$ are said to be *self-equilibrated* with respect to the overall equilibrium of the shell. Thus there is no hope of computing the stress resultants using equations of overall equilibrium on a cleverly chosen free-body diagram; rather, the differential equations offer the only approach. One should note at the outset (and it is demonstrated in subsequent sections) that the membrane stress resultants in shells of revolution may be very sensitive to the influence of the higher harmonic components of the surface loading, and that shells which are subject to this form of loading must be investigated very thoroughly.

4.3.7.2 Spherical Shell with No Surface Loading. One case of a specific geometry that can be solved for $j > 1$ is the spherical shell. We limit our investigation to the case of no surface loading, which corresponds to a homogeneous set of equations. This form is sufficient to describe an important practical situation, the edge-loaded shell of revolution. Once a homogeneous solution is available, a particular solution for a given surface loading is easily obtained using the standard method of variation of parameters.

It is expedient to return to the original untransformed membrane equations, equation (4.3). We introduce the separated form of the stress resultants, equation (4.84b), for harmonic j into equation (4.3). Noting that equation (4.3c) gives $N_\theta = -N_\phi$, since $q_n = 0$, we have

$$N^j_{\phi,\phi} + 2N^j_\phi \cot \phi + \frac{j}{\sin \phi} S^j = 0 \qquad (4.118a)$$

$$S^j_{;\phi} + 2S^j \cot \phi + \frac{j}{\sin \phi} N^j_\phi = 0 \qquad (4.118b)$$

These equations may be separated by introducing the auxiliary variables[28]

$$\Omega^j_1 = N^j_\phi + S^j \qquad (4.119a)$$

$$\Omega^j_2 = N^j_\phi - S^j \qquad (4.119b)$$

giving

$$\Omega^j_{1,\phi} + \left(2 \cot \phi + \frac{j}{\sin \phi}\right)\Omega^j_1 = 0 \qquad (4.120a)$$

$$\Omega^j_{2,\phi} + \left(2 \cot \phi - \frac{j}{\sin \phi}\right)\Omega^j_2 = 0 \qquad (4.120b)$$

which have solutions[28]

$$\Omega^j_1 = C^j_1 \frac{\cot^j (\phi/2)}{\sin^2 \phi} \qquad (4.121a)$$

$$\Omega^j_2 = C^j_2 \frac{\tan^j (\phi/2)}{\sin^2 \phi} \qquad (4.121b)$$

From equation (4.119),

$$N^j_\phi = \frac{1}{2 \sin^2 \phi} \left[C^j_1 \cot^j \left(\frac{\phi}{2}\right) + C^j_2 \tan^j \left(\frac{\phi}{2}\right) \right] \qquad (4.122a)$$

$$S^j = \frac{1}{2 \sin^2 \phi} \left[C^j_1 \cot^j \left(\frac{\phi}{2}\right) - C^j_2 \tan^j \left(\frac{\phi}{2}\right) \right] \qquad (4.122b)$$

The complete homogeneous solution is given in the Fourier series form of equation (4.84b) by

$$N_\phi(\phi,\theta) = -N_\theta(\phi, \theta) = \frac{1}{2 \sin^2 \phi} \sum_{j=0}^{\infty}$$

$$\left[C^j_1 \cot^j \left(\frac{\phi}{2}\right) + C^j_2 \tan^j \left(\frac{\phi}{2}\right) \right] \cos j\theta \qquad (4.123a)$$

$$S(\phi, \theta) = \frac{1}{2 \sin^2 \phi} \sum_{j=0}^{\infty}$$

$$\left[C_1^j \cot^j \left(\frac{\phi}{2}\right) - C_2^j \tan^j \left(\frac{\phi}{2}\right) \right] \sin j\theta \quad (4.123b)$$

The constants for each harmonic, C_1^j and C_2^j, must be evaluated from a set of nonzero force boundary conditions which are also expanded in Fourier series. One possibility is

$$N_\phi(\phi', \theta) = \sum_{j=0}^{\infty} N_\phi^j(\phi') \cos j\theta \qquad (4.124a)$$

$$S(\phi'' \theta) = \sum_{j=0}^{\infty} S^j(\phi'') \sin j\theta \qquad (4.124b)$$

In these equations, ϕ' and ϕ'' denote the boundaries of the spherical shell where N_ϕ and S are specified. Also, the form of equation (4.123) indicates that N_ϕ and S may both be designated at the same boundary—e.g., $\phi' = \phi''$ in equation (4.124)—or that N_ϕ or S, alone, may be chosen at both boundaries. In any case, only *two* boundary conditions are available, and the computed values of N_ϕ and/or S at the other boundaries must be developed by the support in order for the solution to remain valid.

For a spherical dome, $\cot \phi/2 \to 0$ at the pole, so that $C_1^j = 0$ and only one condition, N_ϕ or S, may be specified on the lower boundary.

4.3.7.3 Shell of Revolution Supported on Columns. It is frequently necessary to provide interruptions at the base of a shell above the ground for functional reasons. One obvious possibility is to support the shell on a series of concentrically placed columns, figures 2–8(p,t,u). This obviously introduces an interruption in the continuous, nondeforming boundary required to maintain membrane action and suggests that the stress pattern, particularly in the vicinity of the support, may be severely altered.

A fairly common approach is to analyze the shell by a superposition technique as shown in figure 4–25. In what follows, the discussion is restricted to axisymmetric loading and membrane theory analysis to expediently introduce the concept, but this type of model has been generalized for nonsymmetric loading and bending effects as well.[29] Also, here the columns are assumed to be equispaced around the circumference, and the axis of the column is taken as tangent to the meridian at the boundary $\phi = \phi_b$. For columns oriented vertically, the same theory can be applied with some minor modifications.

We now consider the superposition in figure 4–25 in detail. We assume that the axisymmetrical surface loading q is carried equally by the n_c equispaced columns. As shown in figure 4–25(a), the column reaction R_c is

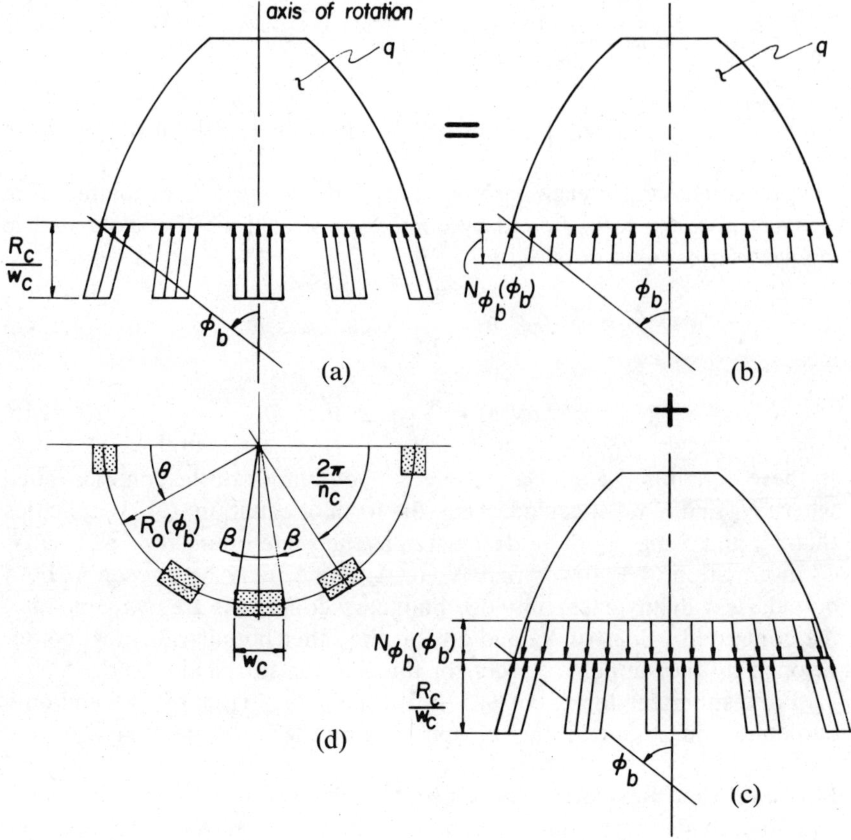

Figure 4–25. Superposition Representations of a Column-Supported Shell of Revolution

taken to be uniformly distributed over the column width w_c, so that the intensity of the reaction per unit length of circumference R_{cl} is R_c/w_c. Loading case (a) is represented by the superposition of cases (b) and (c). Case (b) is the familiar axisymmetric loading case previously studied in great detail in sections 4.3.2 and 4.3.3. The continuous meridional stress at the boundary $\phi = \phi_b$ is denoted by $N_{\phi b}(\phi_b)$. Case (c) is an edge loading applied along $\phi = \phi_b$. It consists of (i) the negative of the continuous boundary reaction in the region between the column and (ii) the difference between the column reaction and the continuous boundary reaction within the column width. Since case (b) is readily solvable, the remaining problem is to address case (c).

First, we evaluate the column reaction in terms of the known meridional stress from the continuous boundary analysis, case (b), as

$$R_c = N_{\phi b}(\phi_b) \cdot \frac{2\pi R_0(\phi_b)}{n_c} \tag{4.125}$$

The intensity is given by

$$R_{cl} = \frac{R_c}{w_c} \tag{4.126}$$

Referring to figure 4–25(d), we write w_c in terms of the subtended angle 2β measured in the horizontal plane

$$w_c = 2R_0(\phi_b) \cdot \beta \tag{4.127}$$

Then, substituting equations (4.127) and (4.125) into equation (4.126), we have

$$R_{cl} = \frac{\pi N_{\phi b}(\phi_b)}{\beta n_c} \tag{4.128}$$

R_{cl} has units of force/length.

Now we wish to express the edge loading shown in figure 4–25(c) as a function of the circumferential coordinate θ. Referring to figure 4–25(d), we have

$$N_{\phi c}(\phi_b, \theta) = [-N_{\phi b}(\phi_b) + \delta R_{cl}] \tag{4.129}$$

where $\quad \delta = 1$ if $m(2\pi/n_c) - \beta \le \theta \le m(2\pi/n_c) + \beta$

$\quad (m = -n_c/2, \ldots, -1, 0, 1, \ldots, n_c/2)$

$\quad \delta = 0$ for all other θ.

The limits on δ are set presuming an even number of columns and can be adjusted accordingly for an odd number, and $N_{\phi c}$ is taken as positive in the positive sense of $N_{\phi b}$.

We expand $N_{\phi c}$ in a Fourier series similar to equation (4.124a),

$$N_{\phi c}(\phi_b, \theta) = \sum_{j=0}^{\infty} N_{\phi c}^j(\phi_b) \cos j\theta \tag{4.130}$$

First, the Fourier coefficient $N_{\phi c}^0$ is given by

$$N_{\phi c}^0(\phi_b) = \frac{1}{2\pi} \int_{-\pi}^{\pi} N_{\phi c}(\phi_b, \theta) \, d\theta \tag{4.131a}$$

From equations (4.129) and (4.128),

132

$$N^{\circ}_{\phi c}(\phi_b) = \frac{1}{2\pi} \int_{-\pi}^{\pi} \left[-N_{\phi b}(\phi_b) + \delta \cdot \frac{N_{\phi b}(\phi_b)}{\beta n_c} \right] d\theta$$

$$= \frac{N_{\phi b}(\phi_b)}{2\pi} \left[-2\pi + \frac{n_c \cdot \pi}{\beta n_c} \cdot 2\beta \right]$$

$$= 0 \tag{4.131b}$$

This confirms the physical observation that the *net* meridional resultant in figure 4–25(c) is 0, since the total of the column reactions is equal and opposite to the uniform boundary reaction. This edge loading is then said to be *self-equilibrated*.

Next, for $j > 0$, we have

$$N^{j}_{\phi c}(\phi_b) = \frac{1}{\pi} \int_{-\pi}^{\pi} N_{\phi c}(\phi_b, \theta) \, d\theta \tag{4.132}$$

Referring to figure 4–25(d), where the circumferential spacing of the columns is given as $2\pi/n_c$, we have

$$N^{j}_{\phi c}(\phi_b) = \frac{1}{\pi} \left\{ \int_{-\pi}^{\pi} -N_{\phi b}(\phi_b) \cos j\theta \, d\theta + R_{cl} \left[\int_{0}^{\beta} \cos j\theta \, d\theta \right.\right.$$

$$+ \int_{(2\pi/n_c)-\beta}^{(2\pi/n_c)+\beta} \cos j\theta \, d\theta + \int_{2(2\pi/n_c)-\beta}^{2(2\pi/n_c)+\beta} \cos j\theta \, d\theta + \ldots$$

$$\ldots + \int_{(n_c-1)(2\pi/n_c)-\beta}^{(n_c-1)(2\pi/n_c)+\beta} \cos j\theta \, d\theta + \left.\left. \int_{2\pi-\beta}^{2\pi} \cos j\theta \, d\theta \right] \right\} \tag{4.133}$$

$$N^{j}_{\phi c}(\phi_b) = \frac{N_{\phi b}(\phi_b)}{\pi} \left\{ -\frac{2}{j} \sin j\pi + \frac{\pi}{j\beta n_c} \left(\sin j\beta \right.\right.$$

$$+ \sum_{m=1}^{n_c-1} \left[\sin j\left(m\frac{2\pi}{n_c} + \beta \right) - \sin j\left(m\frac{2\pi}{n_c} - \beta \right) \right]$$

$$\left.\left. + \sin j2\pi - \sin j\left(2\pi - \beta \right) \right) \right\} \tag{4.134}$$

For an integer value of j, the first and next to last terms drop out. Furthermore, we can show that for $j \neq in_c$, where i is an integer, the remaining terms sum to 0, so that $N^{j}_{\phi c}(\phi_b) = 0$ if $j \neq in_c$. When $j = in_c$, the complete symmetry results in the total integral being n_c times the integral over one column, or

$$N_{\phi c}^{j}(\phi_b) = n_c \frac{N_{\phi b}(\phi_b)}{\pi} \cdot 2 \cdot \frac{\pi \sin j\beta}{j\beta n_c}$$

$$= \frac{2}{j\beta} N_{\phi b}(\phi_b) \sin j\beta \quad (j = in_c) \qquad (4.135)$$

The above assertions are easily verified for a specified numerical value of n_c.

We thus conclude that only those harmonics j which are integer multiples of the total number of columns n_c participate in the solution, and we may rewrite equation (4.130) as

$$N_{\phi c}(\phi_b, \theta) = \sum_{j=n_c,2n_c,\ldots}^{\infty} N_{\phi c}^{j}(\phi_b) \cos j\theta \qquad (4.136)$$

We now turn to the spherical shell treated in section 4.3.7.2 as an example of a column supported shell. An actual case is shown in figure 2–8(t). Note that if the shell is not a dome, there are two constants of integration to be evaluated. We have already specified $N_{\phi}(\phi_b)$, and since we cannot provide an ideal condition because of the columns, the lower boundary must develop the corresponding value of $S(\phi_b)$. This leaves us the option of specifying $N_{\phi}(\phi_t)$ or $S(\phi_t)$ for an open shell, and $N_{\phi}(\phi_t) = 0$ is probably the most realistic choice. Thus we must provide suitable resistance to S at both boundaries in the case of the open shell, or at the base in the case of a dome. This can be accomplished with a circumferentially stiff edge member.

For our example, we chose a hemispherical dome supported on four columns as shown in figure 4–26 subject to dead load q. From equation (4.20), the solution for case (b) of the superposition is

$$N_{\phi b} = \frac{-qa}{1 + \cos \phi} \qquad (4.137)$$

At $\phi = \phi_b = \pi/2$, $N_{\phi b} = N_{\phi b}(\phi_b) = -qa$, and equation (4.135) becomes

$$N_{\phi c}^{j}(\phi_b) = -\frac{2qa}{j\beta} \sin j\beta \qquad (4.138a)$$

From equation (4.136), taking $n_c = 4$,

$$N_{\phi c}\left(\frac{\pi}{2}, \theta\right) = \sum_{j=4,8,12}^{\infty} -\frac{2qa}{j\beta} \sin j\beta \cos j\theta \qquad (4.138b)$$

Recalling the Fourier series expansion for the stresses in the spherical

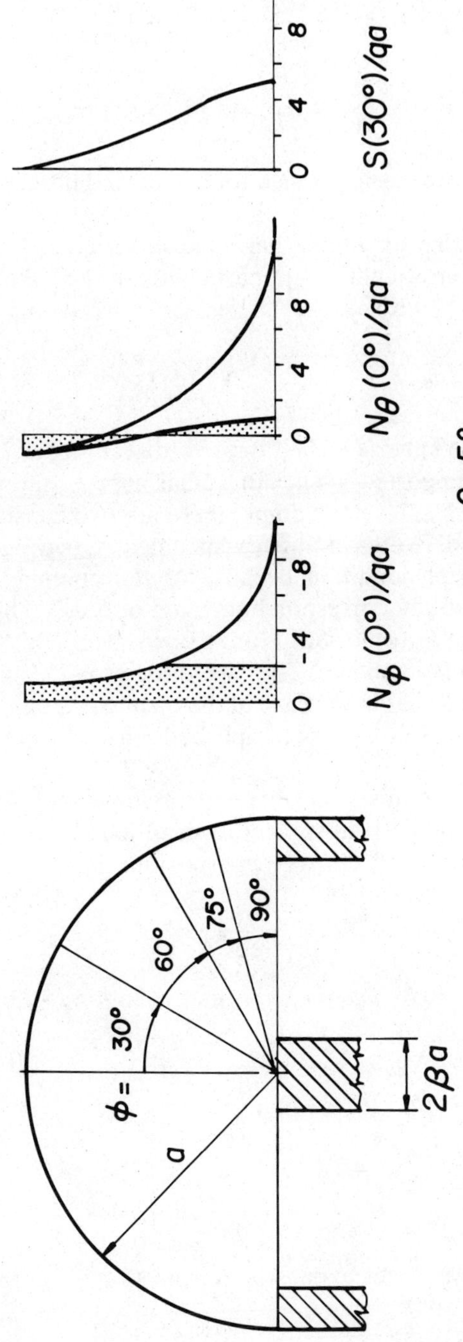

Figure 4–26. Stress Resultants for a Column-Supported Hemispherical Dome under Dead Load

shell, equations (4.123a) and (4.123b), taking $C_1^j = 0$ because we are considering a dome, and evaluating equation (4.123a) at $\phi_b = \phi/2$,

$$N_\phi\left(\frac{\pi}{2}, \theta\right) = \frac{1}{2} \sum_{j=0}^{\infty} C_2^j \tan^j\left(\frac{\pi}{4}\right) \cos j\theta \qquad (4.139)$$

Equating the coefficients of $\cos j\theta$ for each term in the series in equations (4.138b) and (4.139), we have

$$C_2^j = \frac{-4qa}{j\beta} \sin j\beta \quad (j = 4, 8, 12) \qquad (4.140)$$

since $\tan^j(\pi/4) = 1$, and $C_2^j = 0$ for $j \neq 4, 8, 12$. Now we may write the solution to case (c) of the superposition by substituting C_2^j into equations (4.123)

$$N_{\phi c}(\phi, \theta) = -N_{\theta c}(\phi, \theta)$$

$$= \frac{-2qa}{\beta \sin^2 \phi} \sum_{j=4,8,12}^{\infty} \frac{\sin j\beta}{j} \cdot \tan^j\left(\frac{\phi}{2}\right) \cos j\theta \qquad (4.141a)$$

$$S_c(\phi, \theta) = \frac{2qa}{\beta \sin^2 \phi} \sum_{j=4,8,12}^{\infty} \frac{\sin j\beta}{j} \tan^j\left(\frac{\phi}{2}\right) \sin j\theta \qquad (4.141b)$$

and the complete solution as case (b) plus case (c), which gives, in view of equations (4.20) and (4.21),

$$N_\phi(\phi, \theta) = \frac{-qa}{1 + \cos \phi} + N_{\phi c}(\phi, \theta) \qquad (4.142a)$$

$$N_\theta(\phi, \theta) = qa\left(\frac{1}{1 + \cos \phi} - \cos \phi\right) - N_{\phi c}(\phi, \theta) \qquad (4.142b)$$

$$S(\phi, \theta) = S_c(\phi, \theta) \qquad (4.142c)$$

Again we note that the value of $S = S_c$ computed from equation (4.142c) must be developed at the lower boundary.

To study the characteristics of this solution, we show the nondimensionalized stress resultants for the hemisphere in figure 4–26. The half angle β is taken as $5°$, N_ϕ and N_θ are graphed at $\theta = 0°$, the centerline of the column, and S is shown at $\theta = 30^\theta$. The series converges rather slowly with about 14 terms ($n = 56$) required to compute, within 1%, the known value of N_ϕ at the base, $360°/(4 \cdot 2 \cdot 5°) \cdot N_{\phi b}(\phi_b)$, or 9 times the continuous boundary reaction. The continuous boundary stress resultants are superimposed and shaded to indicate the penetration of the discontinuous boundary reaction amplification into the shell.

Although we have obtained a convergent membrane theory solution for the spherical dome supported on columns, all shells with concentrated edge effects are not able to sustain such loading through membrane action. Only a limited number of specific geometries have been investigated, and no general proof is offered; however, it appears that the attenuation of concentrated meridional edge effects through principally membrane action occurs only in positive curvature shells. Physically, it can be surmised that the straight characteristic lines present on zero- and negative-curvature shells offer a path for the propagation of the edge effect to the opposite boundaries, unless bending distortions are present.[30]

4.3.7.4 Cylindrical Shell under Circumferentially Varying Normal Pressure. Consider a circumferentially varying normal pressure $p(\theta)$ on a cylindrical shell. This could simulate a pseudostatic wind loading on a cylindrical tower, as shown in figure 4–27. The circumferential variation of the pressure would generally be established through field measurement or wind tunnel tests. For this example, p can be taken as constant with respect to Z, but a variation with Z can be treated by a simple extension of the following development.

We follow a procedure suggested in Rish and Steel.[31] It is convenient to return to the membrane theory equations in terms of curvilinear coordinates α and β, equations (4.2). We take $\alpha = Z$, $\beta = \theta$; $A = 1$, $B = a$, $R_\alpha = \infty$, $R_\beta = a$; $q_\alpha = q_\beta = 0$, and $q_n = -p(\theta)$. Then we have

$$N_{Z,Z} + \frac{1}{a} S_{,\theta} = 0 \qquad (4.143a)$$

$$S_{,Z} + \frac{1}{a} N_{\theta,\theta} = 0 \qquad (4.143b)$$

$$\frac{N_\theta}{a} = -p(\theta) \qquad (4.143c)$$

Because these equations are relatively simple, we are able to treat the partial differential equation form directly rather than to introduce separated variables. However, later we find it convenient to represent the integrated solution in terms of the separated variables.

Proceeding from equation (4.143c), we have

$$N_\theta(Z, \theta) = -p(\theta)a \qquad (4.144)$$

Next we take $\partial/\partial\theta$ of equation (4.144), substitute into equation (4.143b) and integrate to get

$$S(Z, \theta) = p(\theta)_{,\theta} Z + f_1(\theta)$$

With a stress-free top edge, $S(0, \theta) = 0$; therefore $f_1(\theta) = 0$, and

$$S(Z, \theta) = p(\theta)_{,\theta} Z \qquad (4.145)$$

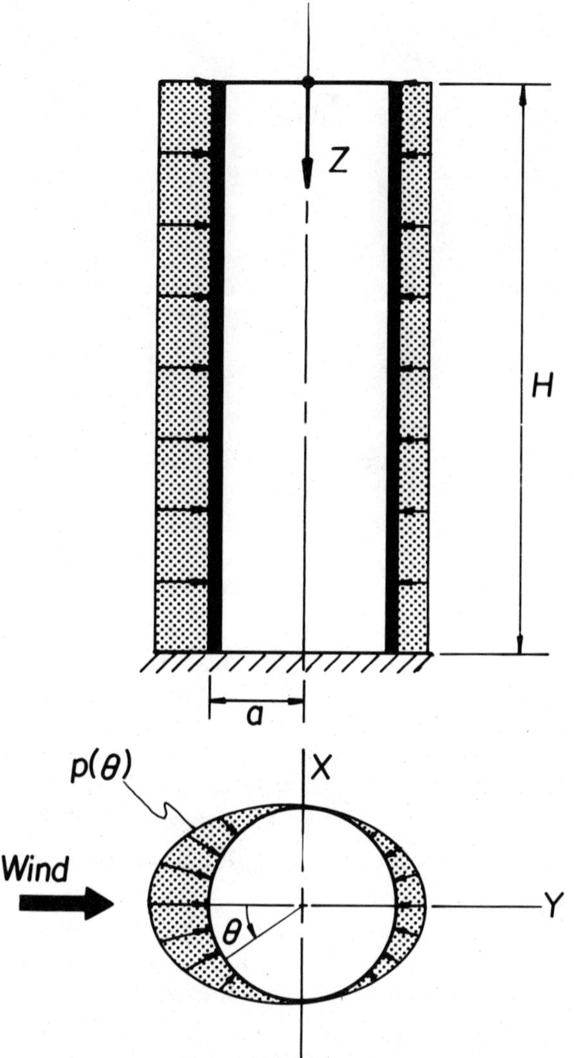

Figure 4–27. Normal Loading on a Cylindrical Shell

Finally we take $\partial/\partial\theta$ of equation (4.145), substitute into equation (4.143a), and integrate once more to obtain

$$N_Z(Z,\theta) = -p(\theta),_{\theta\theta}\frac{Z^2}{2a} + f_2(\theta)$$

Again invoking the stress-free condition at $Z = 0$, $N_Z(0,\theta) = 0$, $f_2(\theta) = 0$, and

138

$$N_Z(Z, \theta) = -p(\theta),_{\theta\theta}\frac{Z^2}{2a} \qquad (4.146)$$

We now represent $p(\theta)$ by the Fourier series

$$p(\theta) = \sum_{j=0}^{\infty} p^j \cos j\theta \qquad (4.147)$$

and take the stress resultants in the separated form, similar to equation (4.84b), with Z replacing ϕ:

$$\begin{Bmatrix} N_Z \\ N_\theta \\ S \end{Bmatrix} = \sum_{j=0}^{\infty} \begin{Bmatrix} N_Z^j(Z) \cos j\theta \\ N_\theta^j(Z) \cos j\theta \\ S^j(Z) \sin j\theta \end{Bmatrix} \qquad (4.148)$$

where

$$N_Z^j(Z) = j^2 p^j \frac{Z^2}{2a} \qquad (4.149a)$$

$$N_\theta^j(Z) = -p^j a \qquad (4.149b)$$

$$S^j(Z) = jp^j Z \qquad (4.149c)$$

These very simple expressions for the stress resultants for a general harmonic offer a rare insight into the interesting properties of the higher harmonic components. Consider, for example, the solution expressed by equations (4.148) and (4.149) contrasted with a solution for the resultant lateral force and overturning moment by ordinary beam theory:

1. The circumferential stress resultant N_θ is not computable by beam theory.

2. The shearing stress resultant is linearly proportional to the axial coordinate Z just as in beam theory. However, the shell solution shows that S is also proportional to the harmonic number j. This discrepancy is, of course, increased as the contribution of the higher harmonics increases.

3. The meridional stress resultant is proportional to the square of the axial coordinate by either theory; however, the shell solution reveals that N_Z is also proportional to j^2, indicating that the meridional stress resultant is highly sensitive to the higher harmonic components.

To help put these observations in perspective, note two characteristics of the circumferential normal pressure distribution $p(\theta)$:

1. High harmonic number components of $p(\theta)$ result from distributions of the pressure that vary rapidly around the circumference. In general, the more rapid the fluctuations, the greater relative participation of the higher harmonics.

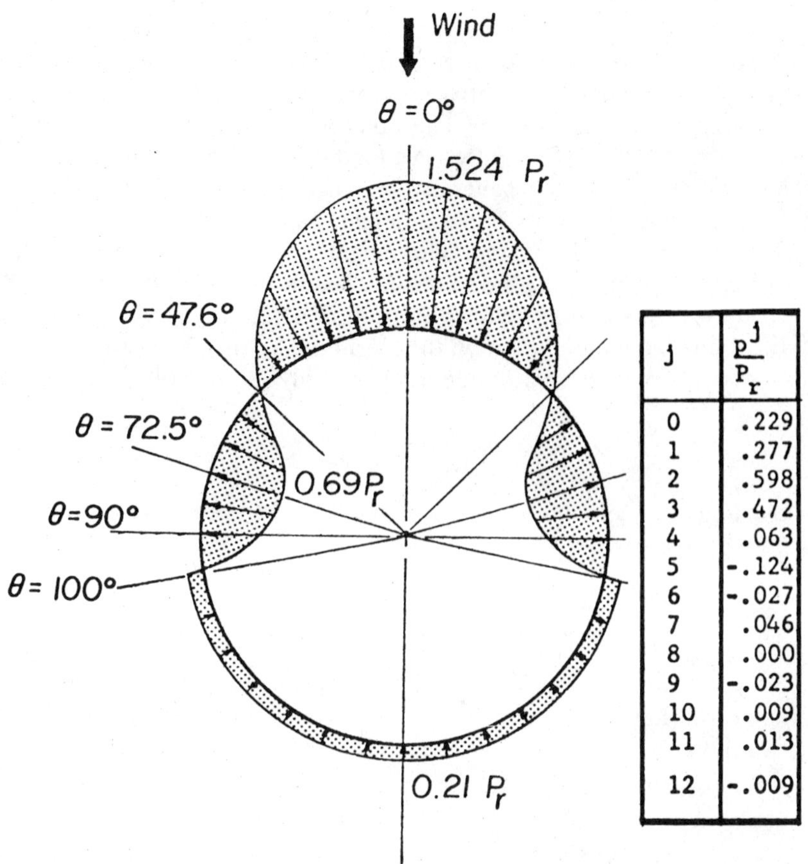

j	$\dfrac{p^j}{P_r}$
0	.229
1	.277
2	.598
3	.472
4	.063
5	-.124
6	-.027
7	.046
8	.000
9	-.023
10	.009
11	.013
12	-.009

Figure 4–28. Design Wind Pressure for Circular Towers (Reprinted with permission of American Society of Civil Engineers)[8]

2. Harmonic components $j = 0$ and $j > 1$ do not contribute to the resultant lateral force or overturning moment on the shell. Only $j = 1$ affects these resultants. If the meridian was curved, the $j = 0$ component would influence the axial force equilibrium, but the $j > 1$ components are always self-equilibrated with respect to the overall equilibrium of the shell.

As an illustration, consider the cylinder shown in figure 4–27 subject to the pressure distribution shown in figure 4–28. This loading was derived from measurements taken from a wind tunnel experiment and is representative of the circumferential pressure distribution on large cylindrical and hyperboloidal towers.[32] The Fourier coefficients for this distribution are also shown on the figure. The loading is given in terms of a reference pressure P_r, which is generally specified as a function of geographical location, terrain, exposure, and height. For simplicity, P_r is taken as unity here, and also as constant with height. A stepwise variation in P_r is easily handled by applying the solution, equations (4.143) through (4.149), in a piecewise fashion as detailed in section 4.3.2.3.

We focus the computation on the meridional stress N_Z at $\theta = 0$ for the base, $Z = H$. With $P_r = 1.0$, equations (4.148) and (4.149) give

$$N_Z(H, 0) = \frac{H^2}{2a} \sum_{j=0}^{\infty} j^2 p^j \tag{4.150}$$

for the Fourier coefficients shown on figure 4–28. Summing for harmonics 0 to 12, we get

$$N_Z(H, 0) = 5.42 \frac{H^2}{2a} \tag{4.151}$$

Now, for comparison, we evaluate $N_Z(H, 0)$ from beam theory. Since only the first harmonic contributes to the resultant force per unit height R_Y, we have

$$R_Y = \int_{-\pi}^{\pi} p^1 \cos \theta \cdot a \, d\theta \cdot \cos \theta$$

$$= p^1 a \pi$$

and the bending moment is

$$M_X = R_Y \frac{H^2}{2} = \frac{p^1 a \pi H^2}{2}$$

Since we are considering the stress resultant that acts on the circular middle surface line, we compute N_Z using the familiar linear stress formula $N_Z = M_Z a / I_X$, where I_X is the moment of inertia of a circle of radius a about a diameter, πa^3.[33] Thus we have

$$N_Z^1(H, 0) = p^1 \frac{H^2}{2a} \tag{4.152}$$

For the tabulated value of p^1 found in figure 4–28, equation (4.152) becomes

$$N_Z^1(H, 0) = 0.277 \frac{H^2}{2a} \tag{4.153}$$

Comparing equations (4.153) and (4.151) clearly illustrates the sensitivity of N_z to the higher harmonic components. This emphasizes the practical importance of obtaining realistic circumferential pressure distributions on tower type structures.

4.3.7.5 Hyperboloidal Shells under Wind Loading. As mentioned previously, analytical solutions to the membrane equations for $j > 1$ are only available for isolated cases. As a means of demonstrating a possible approach to the general solution of equation (4.86), and also of describing a problem with considerable practical utility, consider the hyperboloidal shell of revolution, figure 4–6, subject to the normal wind pressure shown in figure 4–28.

In this case, the loading takes the form

$$q_\phi = 0 \qquad (4.154a)$$

$$q_\theta = 0 \qquad (4.154b)$$

$$q_n = \sum_{j=0}^{\infty} p^j \cos j\theta \qquad (4.154c)$$

where the Fourier coefficients are tabulated in figure 4–28.

The solutions for harmonics $j = 0$ and $j = 1$ are obtained in identical fashion to the dead-load case, section 4.3.2.3, and the seismic load case, section 4.3.6.3, respectively. They are available in Lee and Gould[34] and Gould.[35]

For the case $j > 1$, a different approach is required. We begin with equation (4.86) and differentiate the first term on each side using the product rule. The resulting equation is

$$\psi_{,\phi\phi} + \chi_1(\phi)\psi_{,\phi} + \chi_2(\phi)\psi = \chi_3(\phi) \qquad (4.155)$$

after dividing through by the leading coefficient. In equation (4.155),

$$\chi_1(\phi) = \left(2\frac{R_\phi}{R_\theta} - 1\right)\cot\phi - \frac{1}{R_\phi}R_{\phi,\phi} \qquad (4.156a)$$

$$\chi_2(\phi) = -\frac{j^2 R_\phi}{R_0 \sin\phi} \qquad (4.156b)$$

$$\chi_3(\phi) = \frac{R_\phi}{R_0 R_\theta}\left[(q_n^j \cos\phi - q_\phi^j \sin\phi)R_0^2 R_\theta\right]_{,\phi}$$

$$- jR_\phi^2(jq_n^j - q_\theta^j \sin\phi) \qquad (4.156c)$$

The expansion in equations (4.155) and (4.156) has been retained in a general form to point out that we are confronted with a second order linear differential equation with *variable* coefficients. It is obvious from equations (4.156a) and (4.156b) that the variable coefficients may be very complicated expressions for many surfaces.

A strategy for dealing with such an equation is suggested in Novozhilov.[36] The technique involves a straightforward elimination of the first order term $\psi_{,\phi}$ through the introduction of an additional auxiliary variable, and also another less obvious transformation on the independent variable ϕ. The resulting equation has the form

$$\bar{\psi}_{,\bar{\beta}\bar{\beta}} + \rho(\bar{\beta})^2\bar{\psi} = \chi_4(\phi) \tag{4.157}$$

in which $\bar{\psi}$ and $\bar{\beta}$ are the transformed variables, and $\chi_4(\phi)$ is the appropriately modified r.h.s. The details of the transformation are left to the cited references; here we examine the remaining variable coefficient ρ in more detail. The coefficient is a very involved function of the geometry and harmonic number j and is given in Lee and Gould[37] in terms of a new function, χ_5, as

$$\rho(\bar{\beta})^2 = \rho(\phi)^2 = 1 + \chi_5(\phi) \tag{4.158}$$

The remaining solution procedure is outlined as follows:

1. Expand $\rho(\bar{\beta})$ in a Fourier cosine series in $\bar{\beta}$

$$\rho(\bar{\beta})^2 = \sum_{k=1}^{\infty} \alpha^k \cos 2k\bar{\beta} \tag{4.159}$$

   and substitute into equation (4.157).

2. The equation is now in the form of Hill's equation, which is solved using techniques described in Hill, and also in Whittaker and Watson.[38] The solution is in the form of a rapidly converging infinite series. From a historical viewpoint, it is interesting to note that this equation was first studied by the noted astronomer, G. W. Hill, with reference to the motion of the lunar perigee, which is the point of the orbit of the moon that is nearest to the earth.

3. Using the solution derived in (2), the original stress resultant variables are recovered through the various transformations.

4. Through some numerical studies using the rather involved series solution obtained from Hill's technique, a very close approximation was found by considering only one term of the Fourier series expansion for $\rho(\bar{\beta})^2$, α^o.[39] In effect, this reduces equation (4.157) to a second order linear differential equation with a *constant* coefficient. With the coefficient $\rho(\bar{\beta})^2$ replaced by the constant α^o in equation (4.157), the equation is easily solvable in terms of elementary trigonometric functions.

A comprehensive set of tables for hyperboloidal shells with the general properties of hyperbolic cooling towers that are subject to the wind loading shown in figure 4–28 are given in Lee and Gould.[40]

The solution for the cooling tower exhibits similar characteristics to those previously observed for the cylindrical shell under wind loading. In particular, two points should be reemphasized:

1. The solution is very sensitive to the harmonic number j.
2. The wind load stresses are opposite in sign to the gravity load stresses, and the net stresses, critical for design, should be calculated using proper load factors. The sensitivity of the cooling tower to the net tension is similar to the combined seismic–gravity loadings discussed in section 4.3.6.3.

This example indicates that the solution to the $j > 1$ case may be quite complicated for other than cylindrical or spherical geometries. The transformation to the Hill's equation form has apparently only been investigated for the hyperboloidal geometry and may be useful for other geometries as well. Also the hyperboloidal shell under unsymmetric loading has been solved in the partial differential equation form, similar to that used for the cylindrical shell in section 4.7.3.4, rather than from the harmonic decomposition standpoint.[41]

This example of a solution for asymmetric loading probably approaches the limit to which analytical solutions are efficiently employed. Computer-based numerical techniques have been extensively used for the hyperboloidal problem,[42] but certainly the independent check of selected cases with such analytical solutions that are available serve to establish confidence in the numerical procedures. Also, note that a comprehensive collection of analytical solutions to rotational shell problems have been compiled in a semitabular form by Baker et al.[43] and provide a valuable resource for the design of such shells.

4.4 Shells of Translation

A surface of translation may be generated by passing one plane curve over another. The translational shell is often treated separately from the rotational shell, especially in research; however, there may be considerable overlap. For example, zero-curvature shells of revolution, such as cylinders and cones, may be formed by the translation of a straight line over a circle, or, in the case of the cylinder, by the translation of the circle along the straight line. Even the negative curvature hyperboloid of revolution may be produced by the trace of a skewed line around a circle. In the context of shell analysis, it is generally simpler to regard shells as rotational rather than translational surfaces when there is a choice. Therefore, in our subsequent treatment, we are only concerned with shells that do not possess rotational symmetry.

A popular use of the translational shell is to provide a roof without interior supports over what can be a quite large plan area. Some notable examples are shown in figures 2–8. As can be seen from figures 2–8(a) and (b), such shells are often supported only at a few points to provide an un-interrupted open space. Because of the wide variety of boundary conditions encountered, we may find that the membrane theory solution, by itself, is a poor approximation to the stress distribution in such a shell, and that bending effects will play an important or even dominant role in the design. The bending analysis of translational shells is discussed in chapter 9.

Further, we find it convenient to separate the membrane analysis of translational shells into two parts: (a) circular cylindrical shells; and (b) shells of double curvature.

4.4.1 Circular Cylindrical Shells

Single or multiple circular cylindrical shells have been widely employed in so-called *barrel shell* roofs, such as that illustrated in figure 2–8(h). Extensive discussions of this application are available.[44] Here we concentrate on the basic features of such shells.

The geometry is defined in figure 4–29. We have attempted to remain consistent with section 4.3.3.2 and figure 4–12. An exception is the adoption of X rather than Z as the axial coordinate to conform to the prevalent practice in the literature. Thus we have $\alpha = X, \beta = 0; A = 1, B = a$, and $R_\alpha = \infty, R_\beta = a$. The shell is bounded by the two arcs at $X = 0$ and $X = L$, and the two straight edges at $\theta = \pm\theta_k$.

We start with the membrane theory equilibrium equation, (4.2), with the appropriate geometric parameters inserted:

$$N_{X,X} + \frac{1}{a} S_{,\theta} + q_X(X,\theta) = 0 \tag{4.160a}$$

$$S_{,X} + \frac{1}{a} N_{\theta,\theta} + q_\theta(X,\theta) = 0 \tag{4.160b}$$

$$\frac{N_\theta}{a} = q_n(X,\theta) \tag{4.160c}$$

If we compare equations (4.160a–c) with equation (4.143), we find that they are identical except for the loading terms.

As such, with reference to section 4.3.7.4, the equations are readily solved to give

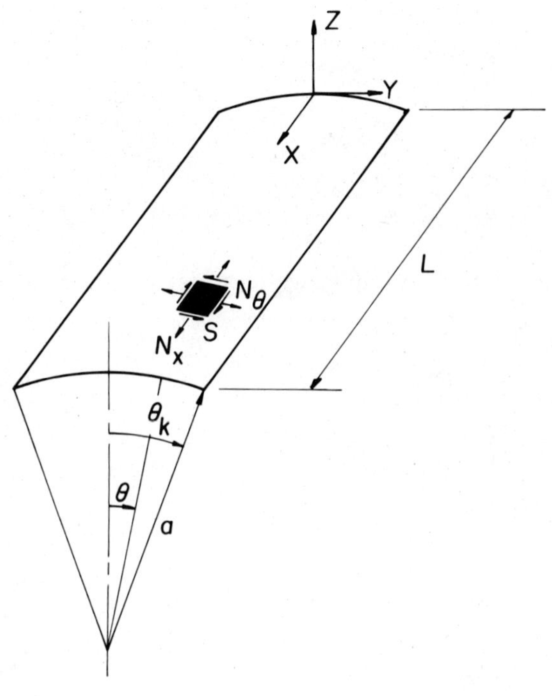

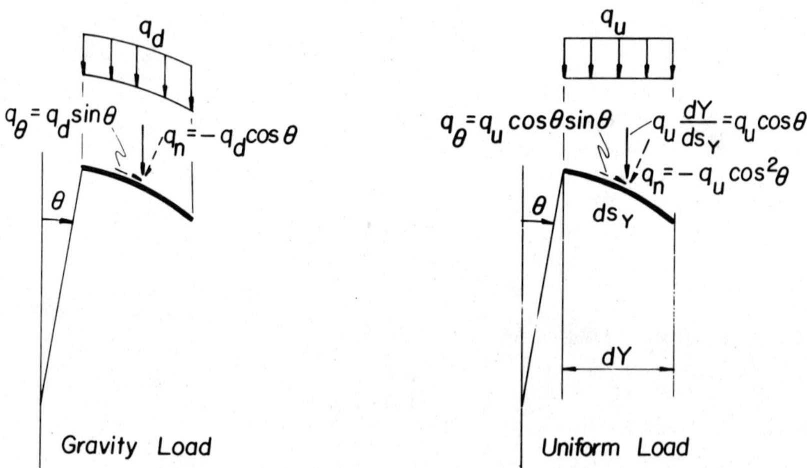

Figure 4–29. Open Cylindrical Shell

146

$$N_\theta(X,\theta) = q_n a \tag{4.161a}$$

$$S(X,\theta) = -\int \frac{1}{a} N_{\theta,\theta} \, dX - \int q_\theta \, dX + f_1(\theta) \tag{4.161b}$$

$$N_X(X,\theta) = -\frac{1}{a}\int S_{,\theta} \, dX - \int q_X \, dX + f_2(\theta) \tag{4.161c}$$

As an elementary example, we consider a circumferentially uniform gravity load $q_d(X)$ shown in figure 4–29.

$$q_X(X,\theta) = 0 \tag{4.162a}$$

$$q_\theta(X,\theta) = q_d(X) \sin\theta \tag{4.162b}$$

$$q_n(X,\theta) = -q_d(X) \cos\theta \tag{4.162c}$$

Proceeding, we find

$$\left.\begin{aligned}
N_\theta &= -a q_d(X) \cos\theta
\end{aligned}\right\} \tag{4.163a}$$

$$\left.\begin{aligned}
N_{\theta,\theta} &= a q_d(X) \sin\theta \\[4pt]
S &= -2 \sin\theta \int q_d(X) \, dX + f_1(\theta) \\[4pt]
S_{,\theta} &= -2 \cos\theta \int q_d(X) \, dX + f_1(\theta)_{,\theta}
\end{aligned}\right\} \tag{4.163b}$$

$$N_X = -\frac{1}{a}\left\{ -2 \cos\theta \int\left[\int q_d(X)\,dX \right] dX + f_1(\theta)_{,\theta} \int dX \right\}$$

$$+ f_2(\theta) \tag{4.163c}$$

We must now inquire further into the longitudinal distribution of the dead load $q_d(X)$. There are two cases of practical interest: a) uniform distribution,

$$q_d(X) = q = \text{constant}$$

and b) periodic distribution

$$q_d(X) = \sum_{j=0}^{\infty} q_d^j \sin\frac{j\pi X}{L}$$

For the latter case, the function $q_d(X)$ may be considered as an odd function on the interval $-L \le X \le L$, as shown in figure 4–30, whereupon

$$q_d(X) = \sum_{j=1,3,\text{odd}}^{\infty} q_d^j \sin\frac{j\pi X}{L} \tag{4.164a}$$

where the Fourier coefficients are

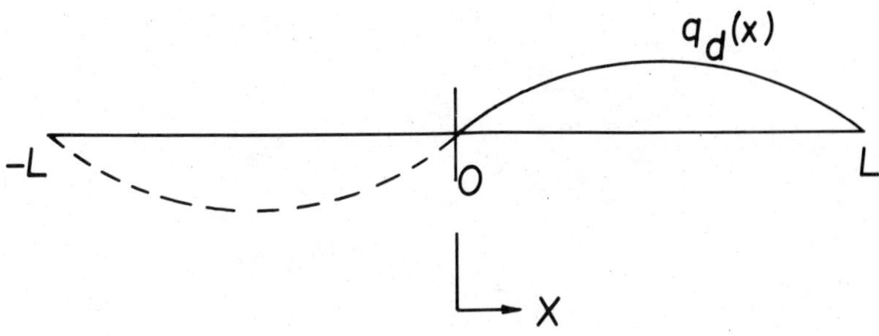

Figure 4–30. Odd Function

$$q_d^j = \frac{2}{L} \int_0^L q_d(X) \sin\frac{j\pi X}{L} \qquad (j=1,3,\ldots\text{odd}) \qquad (4.164b)$$

As an example we may approximate the uniform distribution q_d by the first term ($j = 1$) of the Fourier series, for which

$$q_d^1 = \frac{2q_d}{L} \int_0^L \sin\frac{\pi X}{L}$$

$$= \frac{4}{\pi} q_d \qquad (4.165a)$$

and

$$q_d(X) \simeq \frac{4}{\pi} q_d \sin\frac{\pi X}{L} \qquad (4.165b)$$

Finally, we evaluate the stress resultants, equations (4.161a–c), for both a uniform dead load $q_d = $ constant, and for the first term approximation of q_d as given by equation (4.165b). We proceed with the integration from the boundary $X = 0$, and assume that the shell is simply supported; e.g., $N_X(0) = N_X(L) = 0$.

For $q_d(X) = q_d = $ constant, we have from equations (4.163b) and (4.163c)

$$S = -2q_d X \sin\theta + f_1(\theta) \qquad (4.166a)$$

$$N_X = -\frac{1}{a}[-q_d X^2 \cos\theta + Xf_1(\theta)_{,\theta}] + f_2(\theta) \qquad (4.166b)$$

Applying the boundary conditions to equation (4.166b),

$$f_2(\theta) = 0$$

$$f_1(\theta)_{,\theta} = q_d L \cos\theta$$

from which

$$f_1(\theta) = q_d L \sin \theta$$

and

$$S(X,\theta) = q_d L \left(1 - \frac{2X}{L} \right) \sin \theta \qquad (4.167a)$$

$$N_X(X,\theta) = -q_d \frac{L}{a} X \left(1 - \frac{X}{L} \right) \cos \theta \qquad (4.167b)$$

Also, equation (4.163a) becomes

$$N_\theta(X,\theta) = -q_d a \cos \theta \qquad (4.167c)$$

For the single term Fourier series approximation, we substitute equation (4.165b) into equations (4.163b) and (4.163c) and carry out an analogous integration, which gives

$$S = \frac{8q_d}{\pi} \left(\frac{L}{\pi} \right) \cos \frac{\pi X}{L} \sin \theta + f_1(\theta) \qquad (4.168a)$$

$$N_X = -\frac{1}{a} \left[\frac{8q_d}{\pi} \left(\frac{L}{\pi} \right)^2 \sin \frac{\pi X}{L} \cos \theta + X f_1(\theta),_\theta \right] + f_2(\theta) \qquad (4.168b)$$

We first apply the boundary condition $N_X(0) = 0$, from which $f_2(\theta) = 0$. Also, we note that the loading $q_d(X)$ is symmetrical about $X = L/2$, so that $S(L/2) = 0$. Then, from equation (4.168a), $f_1(\theta) = 0$, and the stress resultants S and N_X are

$$S(X,\theta) = \frac{8q_d L}{\pi^2} \cos \frac{\pi X}{L} \sin \theta \qquad (4.169a)$$

$$N_X(X,\theta) = -\frac{8q_d L^2}{\pi^3 a} \sin \frac{\pi X}{L} \cos \theta \qquad (4.169b)$$

Also, equation (4.163a) becomes

$$N_\theta(X,\theta) = -\frac{4q_d a}{\pi} \sin \frac{\pi X}{L} \cos \theta \qquad (4.169c)$$

It is interesting to compare the stress resultants computed from the actual loading q_d with those evaluated from the single term Fourier approximation. Note the maximum stress resultants $S(0)$, $N_X(L/2)$, and $N_\theta(L/2)$, as shown in table 4–1.

From this comparison we see that except for $S(0)$, the maximum stress resultants are conservatively computed from the one term solution. If further refinement is desired, additional terms of the series may be added.[45]

Table 4–1
Comparison of Maximum Stress Resultants

Maximum Stress Resultant	Common Factor	Coefficient	
		Exact	Approximate
$S(0)$	$q_d L \sin \theta$	1	$8/\pi^2 = 0.81$
$N_x(L/2)$	$-q_d(L/a) \cos \theta$	$L/4 = 0.25L$	$8L/\pi^3 = 0.258L$
$N_\theta(L/2)$	$-q_d a \cos \theta$	1	$4/\pi = 1.27$

Another case of practical interest is that of a load acting in the Z direction and uniformly distributed in the Y direction, q_u. The resolution of this load is shown in figure 4–29.

$$q_X(X,\theta) = 0 \tag{4.170a}$$

$$q_\theta(X,\theta) = q_u \cos \theta \sin \theta \tag{4.170b}$$

$$q_n(X,\theta) = -q_u \cos^2 \theta \tag{4.170c}$$

A parallel set of solutions for the stress resultants, using the exact loading and the Fourier series approximation, can be derived.

We now inquire into the accuracy of the membrane theory stress analysis for a barrel shell. Obviously, the cases we have discussed thus far satisfy the requirements of smooth *loading* and regular *geometry*. When it comes to the adequacy of the physical *boundaries* to meet the requirements of the membrane theory, however, we must make a closer examination.

The transverse boundaries at $X = 0$ and $X = L$ have been assumed to be simply supported, implying that N_X should vanish. Moreover, our analysis indicates that the maximum value of S must be developed, as shown by equations (4.167a) and (4.169a). These conditions suggest the use of a diaphragm that is perfectly flexible in the X, or longitudinal direction, but infinitely rigid in the θ, or circumferential direction. A typical practical solution is shown in figure 2–8(h), where a slight overhang beyond the diaphragm is also provided. In general, the membrane condition is considered to be fairly well approximated by such a support.

Now we examine the longitudinal boundaries at $\theta = \pm\theta_k$. The membrane condition requires that $N_\theta(\theta_k)$ and $S(\theta_k)$, as evaluated from equations (4.167) or (4.169), be developed on the boundary. Furthermore, the boundary must be free to displace in the normal direction so that no transverse shear will develop. In practice, several conditions are found.[46] Two are illustrated in figure 2–8(h), where the outer edge is free, while the interior edges are continuous with the adjacent barrel but unrestrained in the vertical direction. Also, supplementary stiffening beams have been employed in some structures. However, unless a continuous support, such as a bearing wall or a stiff frame, is provided beneath all longitudinal edges, there will be a serious violation of the membrane conditions and a resulting drastic alteration of the load-resisting mechanism.

This may be explained in very simplified terms by reference to figure 4–31, where we can see the membrane condition implies that the shell resists transverse loading primarily as a series of circular arches. In contrast, where the supports are not provided along the edge, the shell must resist transverse loading primarily through flexural beam action in the longitudinal direction. Since the stress resultant N_x must develop the resisting couple at any section, the value computed from membrane theory will be considerably altered by bending action, as we will see in chapter 9. Further, the contribution of $N_\theta(\pm\theta_k)$ in resisting the vertical load will be lost. The vertical equilibrium will be provided in a large part by $S(0)$ and $S(L)$. From these heuristic observations, we may anticipate a significant readjustment of all in-plane stress resultants, as well as the possible addition of significant bending moments and transverse shear forces in the solution of the general equations for open cylindrical shells.

4.4.2 Shells of Double Curvature

If we refer to the shell element shown in the center of figure 4–32, the membrane equilibrium equations may be directly written from equations (4.2a–c) by selecting the arc lengths as the curvilinear coordinates. Therefore, we have $\alpha = s_x$ and $\beta = s_y$. However, a somewhat different approach is widely used because it simplifies the subsequent mathematics.[47,48] First, we choose the Cartesian coordinate system X–Y–Z as shown on the figure. Then, the differential area element $ds_x\,ds_y$ located at $Z(X,Y)$ is mapped onto the lower element $dX\,dY$ in the X–Y plane. Also, a shorthand notation for the incremental change of stress resultants is introduced on the figure: $N_x^+ = N_x + N_{x,x}\,dx$, etc.

Projected or pseudostress resultants N_X, N_Y, and S_{XY} are defined, such that the X–Y components of the resulting *forces* on the edges ds_x or ds_y are equal to the total *forces* on the corresponding mapped edges dX or dY. On the shell element, we have noted the local tangents

$$\tan \gamma_x (X,Y) = Z_{,X}$$

$$\tan \gamma_y (X,Y) = Z_{,Y} \tag{4.171}$$

which are useful in what follows. Proceeding, we set

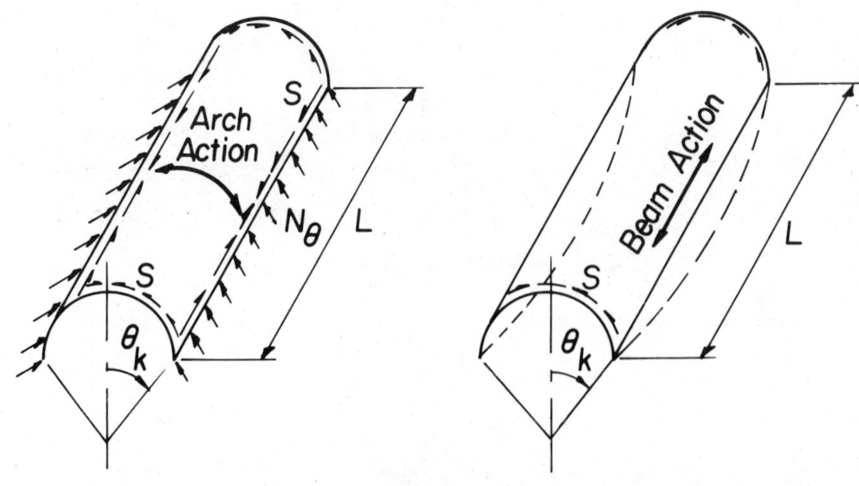

Supported Longitudinal
Boundary

Unsupported Longitudinal
Boundary

Figure 4–31. Longitudinal Boundary Conditions for Cylindrical Shells

$$
\left.\begin{aligned}
N_X\, dY &= N_x\, ds_y \cos \gamma_x \\
S_{XY}\, dY &= S\, ds_y \cos \gamma_y \\
N_Y\, dX &= N_y\, ds_x \cos \gamma_y \\
S_{XY}\, dX &= S\, ds_x \cos \gamma_x
\end{aligned}\right\} \qquad (4.172)
$$

Since

$$
\left.\begin{aligned}
\frac{dY}{ds_y} &= \cos \gamma_y \\[2ex]
\frac{dX}{ds_x} &= \cos \gamma_x
\end{aligned}\right\} \qquad (4.173)
$$

then

$$
N_x = N_X \frac{\cos \gamma_y}{\cos \gamma_x} \qquad (4.174a)
$$

$$
N_y = N_Y \frac{\cos \gamma_x}{\cos \gamma_y} \qquad (4.174b)
$$

$$
S = S_{XY} \qquad (4.174c)
$$

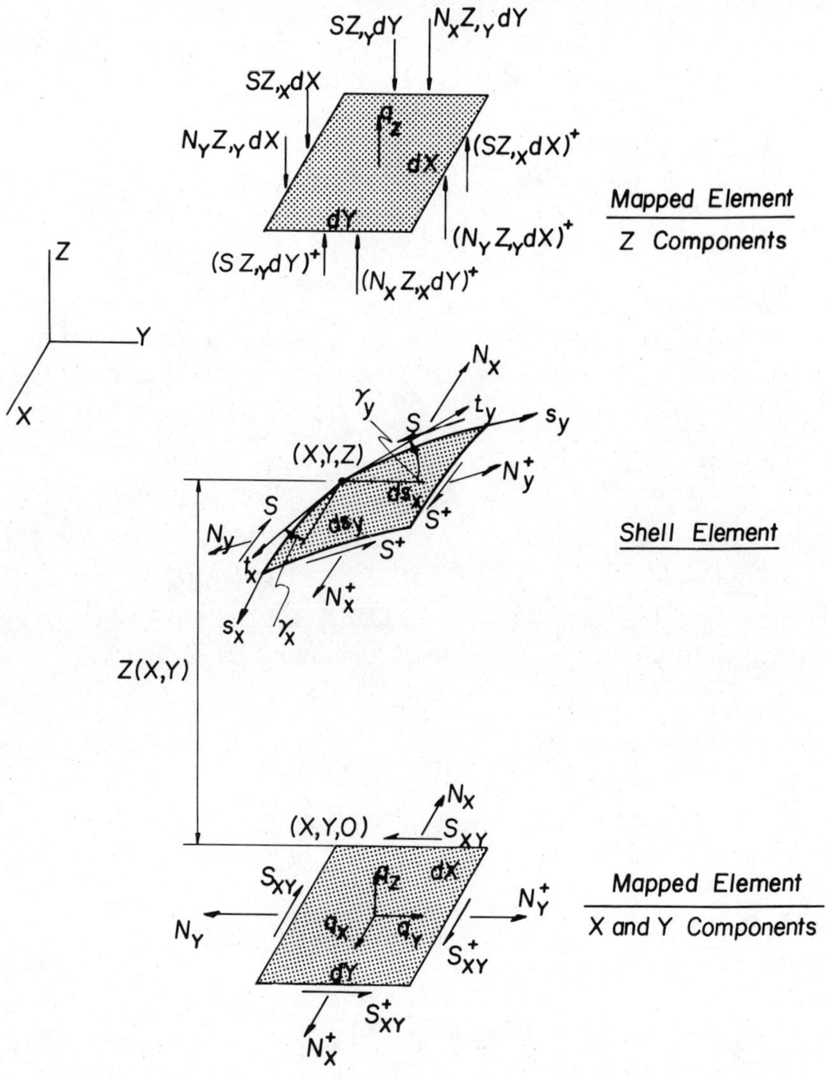

Figure 4–32. Shell Element and Mapping into Cartesian Coordinates

Now we derive the equilibrium equations for the mapped element. After solving for the projected stress resultants, the actual membrane resultants are recovered from equations (4.174a–c). Since $S_{XY} = S$ from equation (4.174c), we drop the subscript on this term.

Summing forces in the X and Y directions, we have

$$(N_X + N_{X,X}\,dX)\,dY - N_X\,dY + (S + S_{,Y}\,dY)\,dX - S\,dX + q_X\,dX\,dY = 0$$

or

$$N_{X,X} + S_{,Y} + q_X = 0 \qquad (4.175a)$$

and

$$N_{Y,Y} + S_{,X} + q_Y = 0 \qquad (4.175b)$$

To form the third equilibrium equation, we project the Z components of the forces corresponding to the actual stress resultants onto the plane, as shown on the top element in figure 4–32.

We have from N_x

$$\begin{aligned}
N_x\,ds_y\,\sin\gamma_x &= N_X \frac{\cos\gamma_y}{\cos\gamma_x} \cdot \frac{dY}{\cos\gamma_y}\,\sin\gamma_x \\[2mm]
&= N_X\,\tan\gamma_x\,dY \\[2mm]
&= N_X Z_{,X}\,dY \qquad (4.176a)
\end{aligned}$$

and from N_y

$$\begin{aligned}
N_y\,ds_x\,\sin\gamma_y &= N_y \frac{\cos\gamma_x}{\cos\gamma_y} \cdot \frac{dX}{\cos\gamma_x}\,\sin\gamma_y \\[2mm]
&= N_Y\,\tan\gamma_y\,dX \\[2mm]
&= N_Y Z_{,Y}\,dX \qquad (4.176b)
\end{aligned}$$

From the shearing forces, we find

$$\begin{aligned}
S\,ds_y\,\sin\gamma_y &= S \frac{dY}{\cos\gamma_y}\,\sin\gamma_y \\[2mm]
&= S\,\tan\gamma_y\,dY \\[2mm]
&= S Z_{,Y}\,dY \qquad (4.176c)
\end{aligned}$$

$$\begin{aligned}
S\,ds_x\,\sin\gamma_x &= S \frac{dX}{\cos\gamma_x}\,\sin\gamma_x \\[2mm]
&= S\,\tan\gamma_x\,dX \\[2mm]
&= S Z_{,X}\,dX \qquad (4.176d)
\end{aligned}$$

Summing, the projected forces in the Z direction

$$(N_X Z_{,X})_{,X} + (N_Y Z_{,Y})_{,Y} + (S Z_{,Y})_{,X} + (S Z_{,X})_{,Y} + q_Z = 0 \qquad (4.177)$$

It is convenient to expand and rearrange equation (4.177) as

$$Z_{,XX} N_X + 2Z_{,XY} S + Z_{,YY} N_Y + Z_{,X}(N_{X,X} + S_{,Y})$$

$$+ Z_{,Y}(N_{Y,Y} + S_{,X}) + q_Z = 0 \qquad (4.178)$$

We then replace the terms in parentheses in equation (4.178) by q_X and q_Y from equations (4.175a) and (4.175b), leaving

$$Z_{,XX} N_X + 2Z_{,XY} S + Z_{,YY} N_Y - Z_{,Y} q_X - Z_{,X} q_Y + q_Z = 0 \qquad (4.179)$$

as the third equilibrium equation.

We may now introduce the Pucher stress function F,[49] which is defined such that

$$F_{,YY} = N_X + \int q_X \, dX \qquad (4.180a)$$

$$F_{,XX} = N_Y + \int q_Y \, dY \qquad (4.180b)$$

$$-F_{,XY} = S \qquad (4.180c)$$

It is easily verified by substitution that if F is a continuous function with continuous partial derivatives, equations (4.175a) and (4.175b) are identically satisfied. Therefore, the membrane theory stress problem for shells of double curvature reduces to the single equation

$$Z_{,XX} F_{,YY} - 2Z_{,XY} F_{,XY} + Z_{,YY} F_{,XX} =$$

$$-q_Z + Z_{,X} q_X + Z_{,Y} q_Y + Z_{,XX} \int q_X \, dX + Z_{,YY} \int q_Y \, dY \qquad (4.181)$$

The mapping onto the X–Y plane, and the subsequent introduction of the Pucher stress function, has produced a single equation which encompasses a large class of shells. Once the stress function F is evaluated, the projected stress resultants are computed from equation (4.180), and the actual resultants from equation (4.174). The mathematical complications in solving equation (4.181) for a given geometry, and the resulting properties of the solution, are dependent on the type of the surface equation $Z(X,Y)$—e.g., elliptic, hyperbolic, or parabolic—and the complexity of the applied loading, q_X, q_Y, and q_Z, as specified in the Cartesian coordinate system.

For the purposes of illustrating the Pucher stress function technique, we consider three representative surfaces encountered quite often in shell roof structures: (a) hyperbolic paraboloid; (b) elliptic paraboloid; and (c) conoid.

4.4.3 Hyperbolic Paraboloid

4.4.3.1 Surface Properties. The hyperbolic paraboloid (HP) is perhaps the most fascinating of the translational shells. The basic form may be

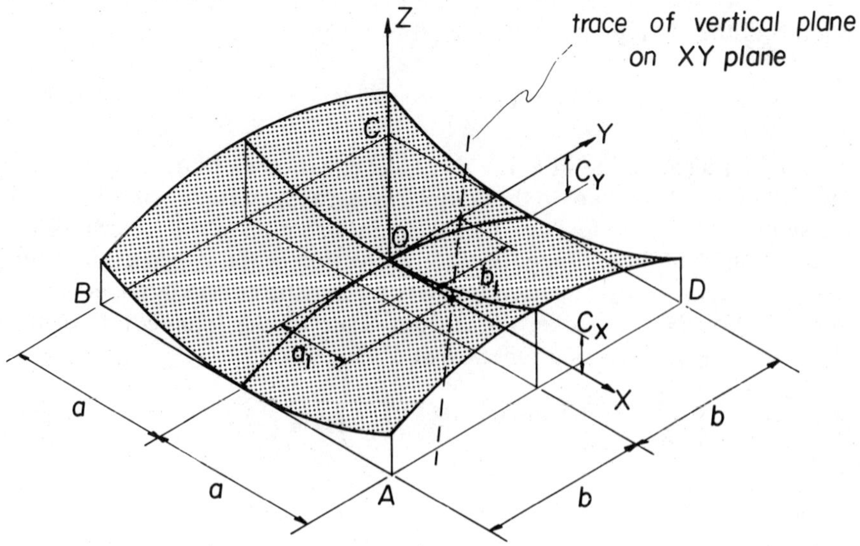

Figure 4–33. Hyperbolic Paraboloid Saddle Shell

generated by passing a convex parabola over a concave parabola, producing an anticlastic surface as shown in figure 4–33. For obvious reasons, such a hyperbolic paraboloid is commonly referred to as the "saddle shell."

The equation of the surface shown in figure 4–33 is

$$Z = \left(\frac{c_X}{a^2}\right)X^2 - \left(\frac{c_Y}{b^2}\right)Y^2 \tag{4.182}$$

We observe that the hyperbolic paraboloid is obviously a surface of negative Gaussian curvature. This is easily shown from equation (2.38), since $Z_{,XX} = 2(c_X/a^2)$, $Z_{,YY} = -2(c_Y/b^2)$ and $Z_{,XY} = 0$, from which the discriminant $\delta = Z_{,XX}Z_{,YY} - (Z_{,XY})^2 < 0$.

The stress analysis of the basic form of the hyperbolic paraboloid shown in figure 4–33 is postponed, so that we can first consider a more common configuration. We observe from figure 4–33 that the surface has negative Gaussian curvature. Referring to the classification in figure 2–7, we expect that there should be two sets of straight lines on the surface. Mathematically, these families of straight lines are called the *real characteristics*. In order to arrive at the equations for the characteristics, we use a construction suggested by Billington.[50]

First, we pass an arbitrary vertical plane through the shell. This plane intersects the $X-Y$ plane along the straight line

$$X = a_1 - \left(\frac{a_1}{b_1}\right) Y \tag{4.183}$$

shown as a dashed line in figure 4–33. In equation (4.183), a_1 and b_1 are the intercepts of the line on the X and Y axes, respectively, and $-a_1/b_1$ is the slope of the line. Next, we locate the intersection of the vertical plane with the shell surface. Since the $X-Y$ projection of the intersection is the straight line given by equation (4.183), we substitute the latter equation into equation (4.182) to get the Z coordinate of the intersection of the plane and the surface as a function of Y:

$$Z = \left[\left(\frac{c_X}{a^2}\right)\left(\frac{a_1}{b_1}\right)^2 - \left(\frac{c_Y}{b^2}\right)\right] Y^2 - 2\left(\frac{c_X}{a^2}\right)\left(\frac{a_1^2}{b_1}\right) Y + \left(\frac{c_X}{a^2}\right) a_1^2 \tag{4.184}$$

Equation (4.184) describes a straight line if the coefficient of Y^2 vanishes, which occurs when

$$\frac{a_1}{b_1} = \pm\left[\frac{(c_Y/b^2)}{(c_X/a^2)}\right] \tag{4.185}$$

Recalling that $(-a_1/b_1)$ is the slope of the projection of the intersection onto the $X-Y$ plane, the families of straight lines on the surface are defined by those vertical planes intersecting the $X-Y$ plane which produce traces with slopes given by the r.h.s. of equation (4.185). Finally, we substitute the computed (a_1/b_1), equation (4.185), into the equation of the intersection with the surface, equation (4.184), to get the equations of the families of straight lines, or characteristics

$$Z = \pm 2\frac{\sqrt{(c_X c_Y)}}{ab} a_1 Y + \frac{c_X}{a^2} a_1^2 \tag{4.186}$$

Through any point on the surface pass two characteristics. To identify the straight lines passing through a given point $(\bar{X}, \bar{Y})$, we substitute $\bar{X}$, $\bar{Y}$ and equation (4.185) into equation (4.183) and find

$$\bar{a}_1 = \bar{X} \mp \left[\frac{(c_Y/b^2)}{(c_X/a^{2)}}\right]^{1/2} \bar{Y} \tag{4.187}$$

Then, each of the computed values of $\bar{a}_1$ are substituted into equation (4.186) to obtain the equations of the intersecting straight lines.

Now that we have identified two families of intersecting straight lines on the surface, a new possibility arises. Instead of visualizing the shell as produced by the translation of one parabola over another, why not generate the shell by the translation of one or the other of the sets of straight lines? Correspondingly, we could choose our reference coordinates to coincide with a pair of the straight lines. This is especially attractive if the boundaries of the shell coincide with the straight lines.

To explore this possibility in terms of our previous equations, we have restricted our discussion to a hyperbolic paraboloid with a square planform, for which $b = a$ and $c_X = c_Y = c$. Then the slopes of the straight lines on the surface in the $X-Y$ plane are computed from equation (4.185) as

$$\left(\frac{a_1}{b_1}\right) = \pm 1 \tag{4.188}$$

Therefore, the families of straight lines are orthogonal, and we can define a new coordinate system $\tilde{X}-\tilde{Y}$ at $45°$ from the $X-Y$ system, as shown in figure 4–34. The more general transformation is treated in Billington.[51]

The relationship between the original $X-Y$ coordinates and a new pair rotated $45°$ is easily derived from figure 4–34.

$$\left.\begin{aligned} X = \tilde{X} \cos 45 - \tilde{Y} \sin 45 \\[2ex] Y = \tilde{X} \sin 45 + \tilde{Y} \cos 45 \end{aligned}\right\} \tag{4.189}$$

which, upon substitution into equation (4.182), gives

$$\begin{aligned} Z &= \frac{c}{a^2}[\tilde{X}^2 \cos^2 45 + \tilde{Y}^2 \sin^2 45 - 2\tilde{X}\tilde{Y} \cos 45 \sin 45 \\[1ex] &\qquad - \tilde{X}^2 \sin^2 45 - \tilde{Y}^2 \cos^2 45 - 2\tilde{X}\tilde{Y} \sin 45 \cos 45] \\[2ex] &= -\frac{2c}{a^2} \tilde{X}\tilde{Y} \end{aligned} \tag{4.190}$$

Now we choose new boundaries for the surface that are parallel to the straight lines instead of to the parabolas. These are indicated by dashed lines on figure 4–34 as EF, FG, GH, and HE. The plan dimension referred to in the rotated coordinate system is $d = (\sqrt{2}/2) \cdot a$, so that we may rewrite equation (4.190) as

$$\begin{aligned} Z &= \frac{-2c}{(2d/\sqrt{2})^2} \tilde{X}\tilde{Y} \\[2ex] &= \frac{-c}{d^2} \tilde{X}\tilde{Y} \\[2ex] &= k\tilde{X}\tilde{Y} \end{aligned} \tag{4.191}$$

where $k = -c/d^2$.

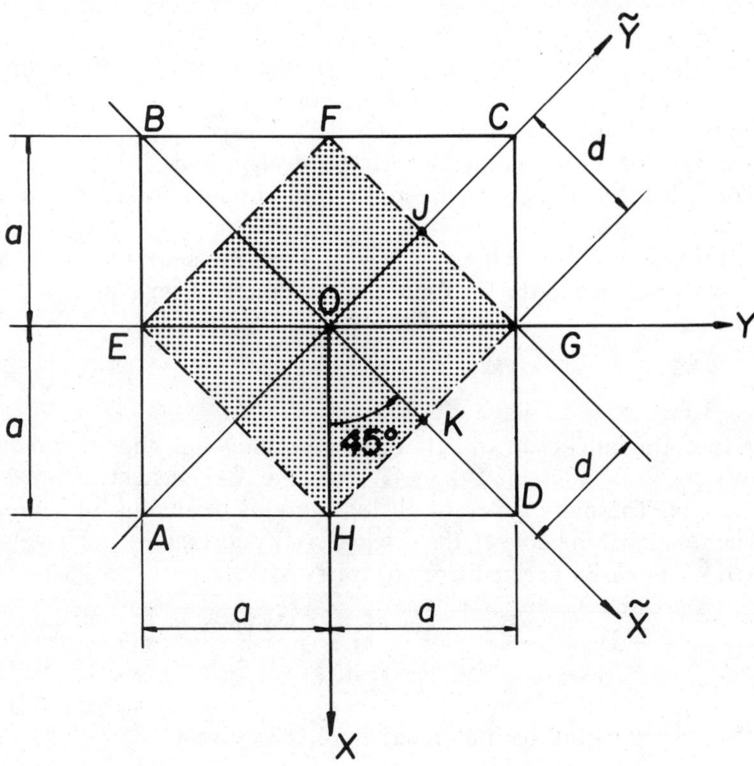

Figure 4–34. Coordinate Rotation for Hyperbolic Paraboloid

We show a quadrant of the hyperbolic paraboloid defined by equation (4.191) in figure 4–35. We may visualize this surface as being formed by a straight line parallel to the $\tilde{X}$ axis, called a *generator*, translated between the two straight lines OJ and KG, known as *directrices*, where point G has been displaced a distance $-c$ from K in the Z direction. This construction produces a warped hyperbolic paraboloidal surface with straight boundaries.

Hyperbolic paraboloids of this general form have become a symbol of grace and elegance in the architectural applications of thin shell roofs. Popularized in particular by F. Candela in Mexico,[52] several examples of actual shells are shown in figures 2–8(a–e). A noteworthy construction feature of this surface is that the formwork for reinforced concrete shells may be fabricated entirely from straight pieces. Various assemblages of four quadrants, similar to those shown in figure 4–35, are depicted in figure 4–36. These forms are referred to freely in the ensuing discussion.

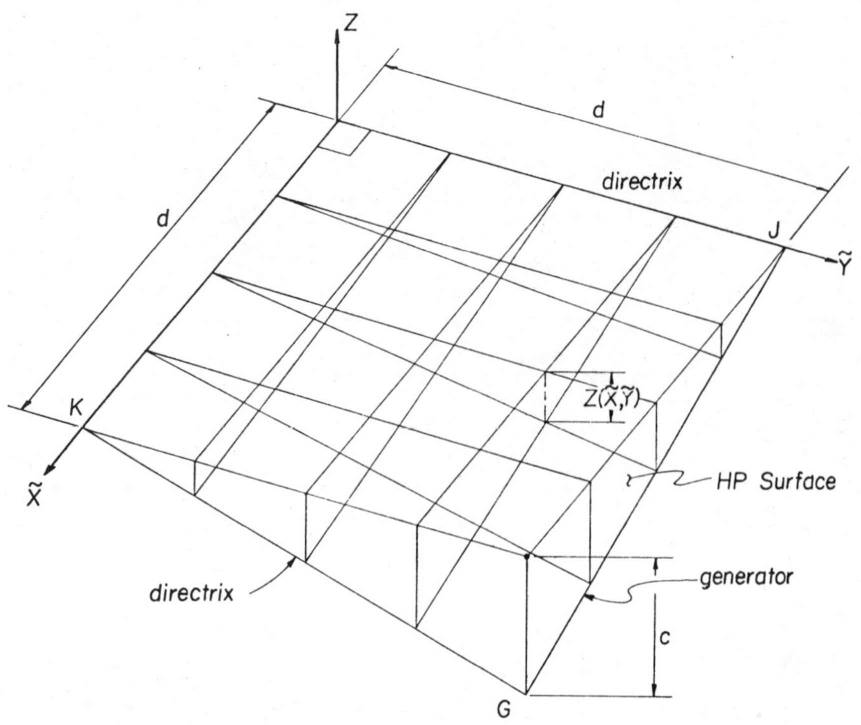

Figure 4–35. Hyperbolic Paraboloid Formed by Straight-Line Generators

4.4.3.2 Stresses in HP Shell with Straight Boundaries. Now we proceed to the stress analysis. We rewrite the governing equation for the Pucher stress function, equation (4.181), to match the $\tilde{X}-\tilde{Y}$ coordinate system:

$$Z_{,\tilde{X}\tilde{X}}\,F_{,\tilde{Y}\tilde{Y}} - 2Z_{,\tilde{X}\tilde{Y}}\,F_{,\tilde{X}\tilde{Y}} + Z_{,\tilde{Y}\tilde{Y}}\,F_{,\tilde{X}\tilde{X}} =$$

$$-q_z + Z_{,\tilde{Y}}q_{\tilde{Y}} + Z_{,\tilde{X}}q_{\tilde{X}} + Z_{,\tilde{X}\tilde{X}}\int q_{\tilde{X}}\,d\tilde{X} + Z_{,\tilde{Y}\tilde{Y}}\int q_{\tilde{Y}}\,d\tilde{Y} \qquad (4.192)$$

with equations (4.180) correspondingly redefined. Then from equation (4.191),

$$Z_{,\tilde{X}} = k\tilde{Y}$$

$$Z_{,\tilde{Y}} = k\tilde{X}$$

$$Z_{,\tilde{X}\tilde{X}} = 0$$

$$Z_{,\tilde{Y}\tilde{Y}} = 0$$

$$Z_{,\tilde{X}\tilde{Y}} = k$$

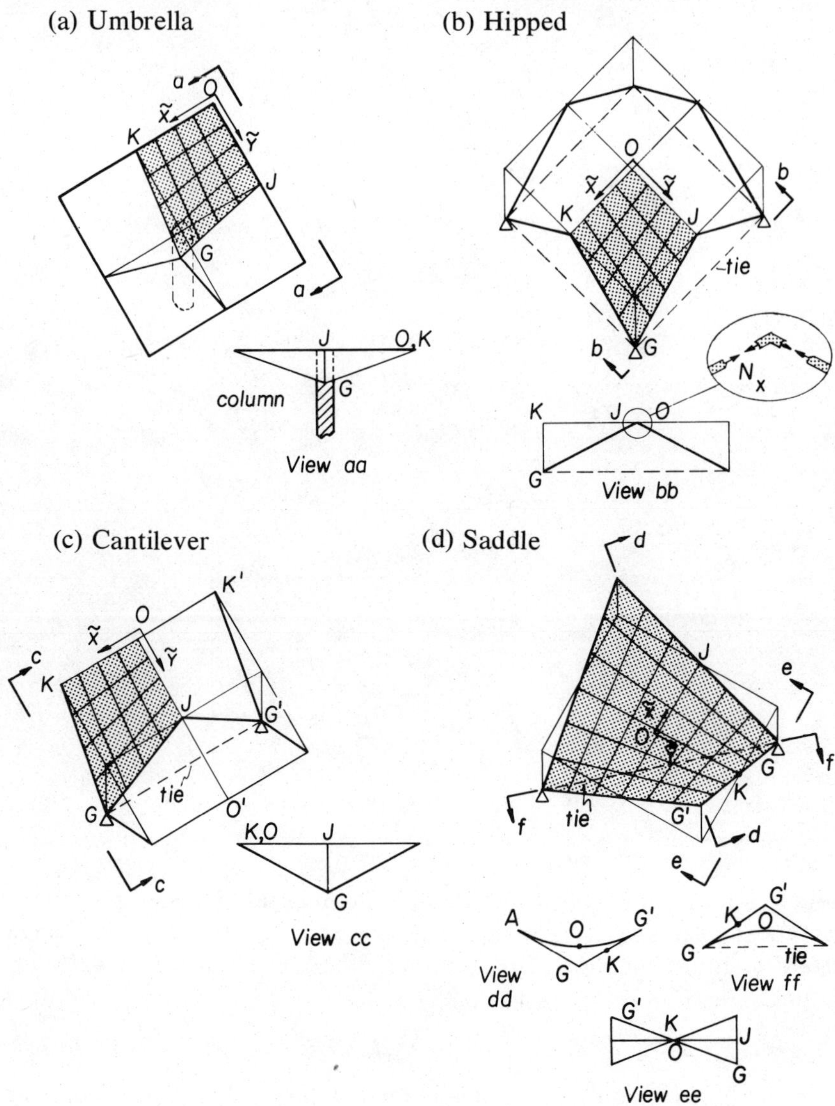

Figure 4–36. Hyperbolic Paraboloids with Straight Boundaries

whereupon equation (4.192) becomes

$$-F_{,XY} = \frac{1}{2k}(-q_Z + k\tilde{X}q_{\tilde{Y}} + k\tilde{Y}q_{\tilde{X}}) \qquad (4.193)$$

Now, referring to equation (4.180c), we see that

$$S = \frac{1}{2k}(-q_z + k\tilde{X}q_{\tilde{Y}} + k\tilde{Y}q_{\tilde{X}}) \tag{4.194}$$

Note that there are no factors in the value of S that can be affected by the boundary conditions. Rather, the boundaries of the shell will have to develop the computed values of S at $\tilde{X} = d$ or $\tilde{Y} = d$.

We now proceed with the solution of equation (4.193) for F. First, we take $\partial/\partial\tilde{X}$ and integrate with respect to $\tilde{Y}$, giving

$$F_{,\tilde{X}\tilde{X}} = -\frac{1}{2k}\int[-q_{z,\tilde{X}} + k(\tilde{X}q_{\tilde{Y}} + \tilde{Y}q_{\tilde{X}})_{,\tilde{X}}]d\tilde{Y} + f_1(\tilde{X})$$

$$= N_{\tilde{Y}} + \int q_{\tilde{Y}}\,d\tilde{Y} \tag{4.195}$$

by equation (4.180a). Similarly,

$$F_{,\tilde{Y}\tilde{Y}} = -\frac{1}{2k}\int[-q_{z,\tilde{Y}} + k(\tilde{X}q_{\tilde{Y}} + \tilde{Y}q_{\tilde{X}})_{,\tilde{Y}}]d\tilde{X} + f_2(\tilde{Y})$$

$$= N_{\tilde{X}} + \int q_{\tilde{X}}\,d\tilde{X} \tag{4.196}$$

We may, from equations (4.194), (4.195), and (4.196), write explicit expressions for the projected stress resultants, and then evaluate the actual stress resultants from equation (4.174) with X and Y replaced by $\tilde{X}$ and $\tilde{Y}$. The functions of integration $f_1(\tilde{X})$ and $f_2(\tilde{Y})$ can then be determined from the specified boundary conditions along $\tilde{X} = \pm d$ and $\tilde{Y} = \pm d$.

Frequently, when the hyperbolic paraboloid is used as a roof and subjected to principally vertically distributed loading, further assumptions are justified. A first step is to consider only the Z component of loading, which greatly simplifies the integrals in equations (4.195) and (4.196). A further step is to assume that the remaining load q_z is constant and uniformly distributed over the projected surface. If this is the case, equations (4.194), (4.195), and (4.196) reduce to

$$S = \frac{-qZ}{2k} \tag{4.197}$$

and

$$N_{\tilde{Y}} = f_1(\tilde{X}) \tag{4.198a}$$

$$N_{\tilde{X}} = f_2(\tilde{Y}) \tag{4.198b}$$

Referring to figure 4–35, if the boundary $\tilde{X} = 0$ or $\tilde{X} = d$ is assumed to be free in the $\tilde{X}$ direction $[N_{\tilde{X}}(0) = 0$ or $N_{\tilde{X}}(d) = 0]$, then as $\tilde{Y}$ varies along that boundary, $N_{\tilde{X}} = f_2(\tilde{Y}) = 0$. Similarly, a free boundary at $\tilde{Y} = 0$ or $\tilde{Y} = d$

gives $f_1(\tilde{X}) = 0$, so that the expressions for the stresses on the hyperbolic paraboloid simplify to

$$S = \frac{-q_z}{2k} \qquad (4.199)$$

and

$$N_{\tilde{x}} = N_{\tilde{y}} = 0 \qquad (4.200)$$

where $\tilde{x}$ and $\tilde{y}$ are curvilinear coordinates on the shell surface corresponding to $\tilde{X}$ and $\tilde{Y}$ on the projected surface.

A loading of the form q_z = constant is frequently termed *live load*, and is often taken to represent snow loading on shell roofs. For shells that are relatively shallow, a uniform distribution for q_z may also be used to closely approximate the dead load. If a uniform load approximation is used for the dead load, an equivalent value q_z should be computed by dividing the total shell weight by the plan area. This will ensure that overall statics will not be violated.

We now reflect on the remarkably simple membrane stress distribution found for a uniform load on a hyperbolic paraboloid. Equation (4.199) indicates a state of pure shear in the shell. To compute S, a designer need only evaluate $k = -c/d^2$ and divide q_z by $2k$. In designers' jargon, this is termed a "back of the envelope" calculation. But one should not be deceived by the apparent simplicity of the mathematics. We must inquire further into the physical boundary conditions which have to be provided to sustain the shell in this simple-to-calculate state of stress.

As an example, consider a uniform load $q_z = -p_u$, constant per unit area of projected surface, acting on the umbrella shell shown in figure 4–36(a). The shell is shown in detail in figure 4–37, where the four segments are numbered. From equation (4.199) and figure 4–37, $k = -c/d^2$ and

$$S = \frac{-(-p_u)}{-2(c/d^2)} = \frac{-p_u d^2}{2c} \qquad (4.201)$$

The sense of S corresponding to the negative value computed in equation (4.201) is obtained by referring to figure 4–32, and is shown in figure 4–37(a). An enlarged view of quadrant *1* is shown in figure 4–37(b), with the shell proper separated from the edges OJ, JG, GK, and KO. Clearly, the uniform shear S must be developed by axially loaded edge members along each of these lines. Edge OJ is subject to a tensile force $T(\tilde{Y})$, which increases from 0 at point O to a maximum value of $T(d) = S \cdot d$ at J, where it is balanced by an equal tension coming from the edge member of adjacent quadrant *2*. An identical tensile force is present along edge OK. The situation along the interior edge members JG and KG is somewhat differ-

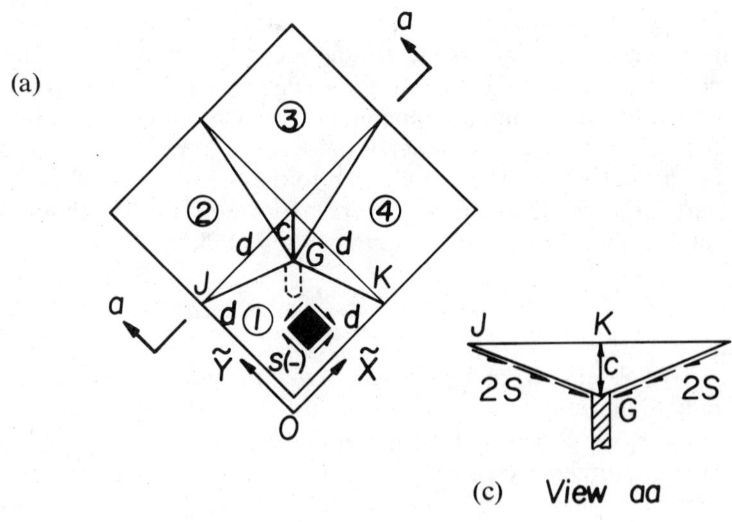

Figure 4–37. Umbrella Shell

ent. Considering JG, the compression $C(X)$ contributed from quadrant *1* builds from 0 at J to a maximum of $S\sqrt{(d^2 + c^2)}$ at G. Adjacent quadrant *2* contributes an identical force, so the maximum compression $C(d)$ =

$2S\sqrt{(d^2 + c^2)}$. An identical situation occurs along KG and the other two interior edge members. While the horizontal component of the maximum compression at G for each interior edge member, $C(d) \cdot d/\sqrt{(d^2 + c^2)}$, is balanced by the similar component from the opposite interior edge member, the vertical components of these four forces, $4 \cdot C(d) \cdot c/\sqrt{(d^2 + c^2)}$, is transmitted through the column to the foundation. Of course, the total vertical force $4[2S\sqrt{(d^2 + c^2)}] \cdot [c/\sqrt{(d^2 + c^2)}] = 8Sc$ should balance the total applied load $4p_u d^2$. From equation (4.201),

$$8Sc = \frac{8p_u d^2}{2c} = 4p_u d^2 \tag{4.202}$$

We observe from figure 4–37(c) that the compressive forces flow naturally to the support. Also note that the internal self-balancing of the maximum tensile forces and the horizontal components of the maximum compressive forces occurs only if the dimensions and loading are identical for the four segments. Any unbalanced loading will produce a resultant shear and overturning moment which would have to be resisted by the single column. For this reason, the single free-standing umbrella is usually restricted to rather modest dimensions. The umbrella has also been used in multiple repetitions to cover large plan areas where occasional columns are acceptable.[53] In such applications, greater lateral stability can be provided through frame action of the interconnected shells.

The stress pattern for any of the other cases shown in figure 4–36 is easily obtained by locating the corresponding quadrant $OJGK$ on each figure. Then the forces in the edge members are computed by starting from a point of assumed zero axial force, usually the junction of the edge member with an external boundary.

Next, consider a dead load p_d, constant per unit area of *middle* surface. Referring to figure 4–32,

$$-p_d \, ds_x ds_y = q_z \, dXdY$$

From equation (4.173), $dX/ds_x = \cos \gamma_x$ and $dY/ds_y = \cos \gamma_y$, so that

$$q_z = \frac{-p_d}{\cos \gamma_X \cos \gamma_Y} \tag{4.203}$$

In terms of the $\tilde{X}$–$\tilde{Y}$ coordinates

$$\tan \gamma_x = Z,_{\tilde{X}} = k\tilde{Y}$$

$$\tan \gamma_y = Z,_{\tilde{Y}} = k\tilde{X}$$

and

$$\left.\begin{array}{l}
\cos \gamma_x = \dfrac{1}{\sqrt{(1 + \tan^2 \gamma_x)}} = \dfrac{1}{\sqrt{(1 + k^2 \tilde{Y}^2)}} \\[3mm]
\cos \gamma_y = \dfrac{1}{\sqrt{(1 + \tan^2 \gamma_y)}} = \dfrac{1}{\sqrt{(1 + k^2 \tilde{X}^2)}}
\end{array}\right\} \qquad (4.204)$$

so that

$$q_z = -p_d[(1 + k^2 \tilde{Y}^2)(1 + k^2 \tilde{X}^2)]^{1/2}$$

$$= -p_d[1 + k^2(\tilde{X}^2 + \tilde{Y}^2) + k^4 \tilde{X}^2 \tilde{Y}^2]^{1/2} \qquad (4.205a)$$

For most practical cases, $k^4 << k^2$, and the last term in equation (2.204) can be neglected. We may also approximate the remaining square root term with the binomial expansion

$$q_z \simeq -p_d\left[1 + \frac{1}{2} k^2(\tilde{X}^2 + \tilde{Y}^2) + \ldots\right] \qquad (4.205b)$$

Substituting equation (4.205b) into equation (4.199) gives

$$S = \frac{-q_z}{2k} = \frac{p_d}{2k}\left[1 + \frac{1}{2} k^2(\tilde{X}^2 + \tilde{Y}^2)\right] \qquad (4.206)$$

Noting that $F,_{\tilde{X}\tilde{Y}} = -S$, we integrate equation (4.206) to get

$$F = \frac{-p_d}{2k}\left[\tilde{X}\tilde{Y} + \frac{k^2}{6}(\tilde{X}^3\tilde{Y} + \tilde{Y}^3\tilde{X}) + f_1(\tilde{Y}) + f_2(\tilde{X})\right] \qquad (4.207)$$

Then we compute the direct stress resultants from equations (4.180a) and (4.180b)

$$N_{\tilde{X}} = F,_{\tilde{Y}\tilde{Y}}$$

$$= \frac{-p_d}{2k}[k^2 \tilde{X}\tilde{Y} + f_1(\tilde{Y}),_{\tilde{Y}\tilde{Y}}] \qquad (4.208)$$

and

$$N_{\tilde{Y}} = F,_{\tilde{Y}\tilde{Y}}$$

$$= \frac{-p_d}{2k}[k^2 \tilde{X}\tilde{Y} + f_2(\tilde{X}),_{\tilde{X}\tilde{X}}] \qquad (4.209)$$

While S is independent of any boundary conditions, as in the case of the uniform loading, the direct stress resultants are dependent on the boundary conditions. Also, since we may specify $N_{\tilde{X}}$ and $N_{\tilde{Y}}$ on only one of the two parallel boundaries, the opposite boundaries must develop the computed stress resultants. If, for example, we consider the umbrella shell shown in figure 4–37, it is reasonable to presume that the direct stress resultants vanish on the exterior boundaries, so that

$$N_{\tilde{X}}(0,\tilde{Y}) = 0 \tag{4.210a}$$

$$N_{\tilde{Y}}(\tilde{X},0) = 0 \tag{4.210b}$$

giving

$$f_1(\tilde{Y}) = f_2(\tilde{X}) = 0 \tag{4.211}$$

The expressions for the dead-load stress resultants in this shell are evaluated from equations (4.211), (4.206), (4.208), (4.209), (4.174), and (4.204).

$$S = \frac{-p_d d^2}{2c}\left[1 + \frac{c^2}{2d^4}(\tilde{X}^2 + \tilde{Y}^2)\right] \tag{4.212a}$$

$$N_x = \frac{p_d c}{2d^2}\,\tilde{X}\tilde{Y}\left[\frac{1 + (c^2/d^4)\tilde{Y}^2}{1 + (c^2/d^4)\tilde{X}^2}\right]^{1/2} \tag{4.212b}$$

$$N_y = \frac{p_d c}{2d^2}\,\tilde{X}\tilde{Y}\left[\frac{1 + (c^2/d^4)\tilde{X}^2}{1 + (c^2/d^4)\tilde{Y}^2}\right]^{1/2} \tag{4.212c}$$

The forces in the edge members are computed in a similar manner as for the uniform load case. However, since $S = S(\tilde{X},\tilde{Y})$, we must proceed more formally. Referring to figure 4–37,

$$T(\tilde{Y}) = \int_0^{\tilde{Y}} S(0,\tilde{Y})\,d\tilde{Y} \tag{4.213a}$$

$$C(\tilde{X}) = 2\int_0^{\tilde{X}} S(\tilde{X},d)\,\frac{d\tilde{X}}{\cos\gamma_x} \tag{4.213b}$$

Note that the compressive force $C(\tilde{X})$ in equation (4.213b) is evaluated by integrating along the inclined edge member. The integrands for equations (4.213a) and (4.213b) are computed by substituting $\tilde{X} = 0$ and $\tilde{Y} = d$, respectively, in equation (4.212a); then, the maximum values of T and C are found by inserting $\tilde{Y} = d$ and $\tilde{X} = d$ as the upper limits of integration.

Earlier, we alluded to the possiblity of replacing the dead loading, constant over the shell surface, with a uniform loading, constant over the projected area. A comparison of the solutions for the loadings p_u and p_d, as given by equations (4.201) and (4.212), provide an obvious motivation, in terms of algebraic simplicity, if the errors introduced are not significant. To satisfy overall statics when we use an equivalent p_u to replace p_d, we should also take an adjusted load $\bar{q}_z$, which represents the total weight divided by the plan area. We find the total weight of the shell, equation (4.205b),

$$Q = \int_0^d \int_0^d p_d \left[1 + \frac{1}{2} k^2 (\tilde{X}^2 + \tilde{Y}^2) \right] d\tilde{X} d\tilde{Y} \qquad (4.214)$$

and then compute an equivalent uniform load

$$\tilde{p}_u = \frac{Q}{d^2} \qquad (4.215)$$

for use in equation (4.201). The effect of this simplification on the accuracy of the ensuing solution is not great for many practical geometries.

An interesting alternative to the free edge boundary conditions stated in equation (4.210) is implied in a design example based on an actual shell having the basic hipped form shown in figure 4–36(b).[54] If the exterior edges KG and GJ were taken as stress free, the integration functions could be readily evaluated by substituting

$$N_{\tilde{X}}(d, \tilde{Y}) = 0$$

and

$$N_{\tilde{Y}}(\tilde{X}, d) = 0$$

into equations (4.208) and (4.209), respectively. However, the subsequent expressions for $N_{\tilde{x}}$ and $N_{\tilde{y}}$, analogous to equations (4.212b) and (4.212c), would give nonzero tensile values along the ridge lines OJ and OK; that is,

$$N_{\tilde{x}}(0, \tilde{Y}) > 0$$

and

$$N_{\tilde{y}}(\tilde{X}, 0) > 0$$

The horizontal components of these stress resultants would be balanced by identical forces from the adjacent segments, but the vertical (Z) components would add to produce an unbalanced force that would have to be sustained by the edge member. This is illustrated in the inset of view bb in figure 4–36(b). To avoid this situation, the designer apparently decided it was preferable to assume that $N_{\tilde{x}}$ and $N_{\tilde{y}}$ vanish along the ridge lines, and then to develop the calculated values of the stress resultants on the exterior perimeter. For this case, $N_{\tilde{x}}$ and $N_{\tilde{y}}$ become compressive as well. We surmise that occasionally there is some latitude in specifying boundary conditions for membrane shells, and that the subsequent problem of providing physical constraints consistent with the assumed conditions should be kept in mind. This is discussed further in section 4.4.3.4.

4.4.3.3 Arch Action. The membrane state of pure shear found for uniformly loaded hyperbolic paraboloid shells can be interpreted from another standpoint. We study the same quadrant $OJGK$ shown in figure

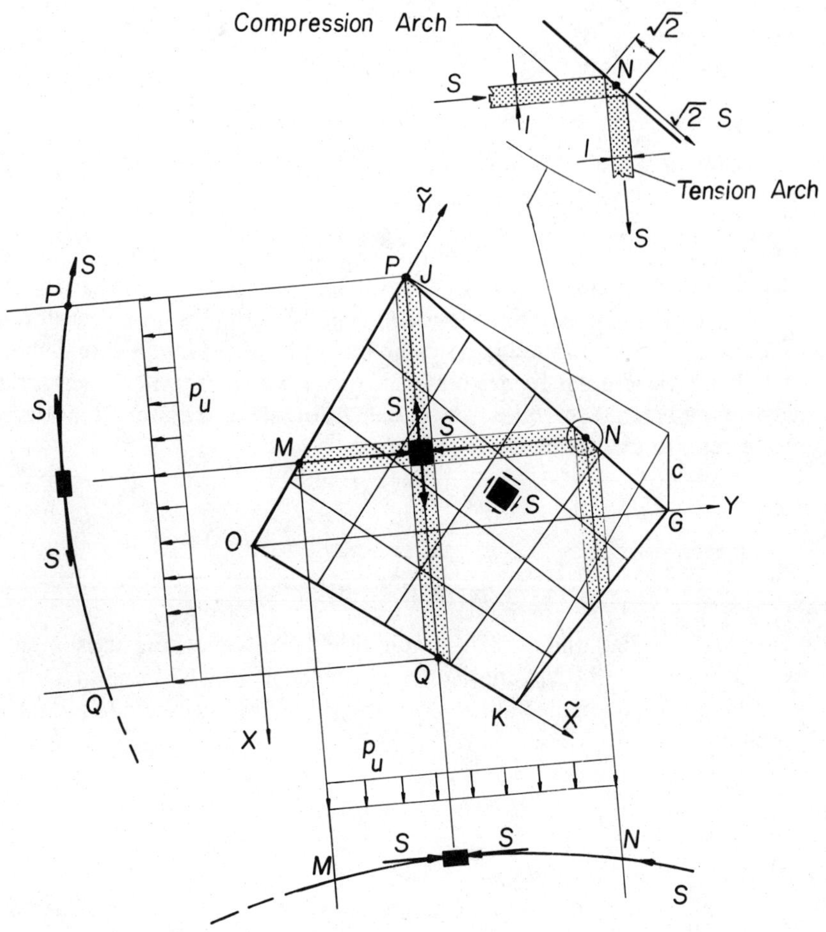

Figure 4–38. Arch Action for a Uniformly Loaded Hyperbolic Paraboloid

4–37, but for variety we consider the saddle shell as illustrated in figure 4–36(d). The membrane stress analysis is the same, and the uniform negative shear S is shown in figure 4–38. From elementary strength of materials, it is easily shown that a state of pure shear with respect to the $\tilde{X}$–$\tilde{Y}$ axes corresponds to principal tensile and compressive stresses of the same magnitude acting in the X and Y directions as shown. If we then take sections through the element parallel to the X and Y axes, we see that we have segments of the parabolic generators of the basic hyperbolic paraboloid shown in figure 4–33. Further, the convex paraboloid parallel to the Y axis acts as a *compression arch* subject to a uniformly distributed load having a constant axial compression of S per unit width. This is the elementary problem of the second-degree parabola subject to a uniform load per unit projected length. Now, examining the parabola parallel to the X axis, we find a similar situation except that we have a concave parabola forming a *tension arch*. Each parabola imparts a constant reaction S, per unit width of shell, on the intersecting edge members, at point N and P on JG. Of course, the other ends of the parabolas, indicated by dashed extensions, will intersect edge members in adjacent quadrants.

To complete the arch representation, we investigate the edge forces. Considering a typical point on the boundary (e.g., N), observe that a unit width compression arch imparts a force of S in the $+Y$ direction, whereas the tension arch acting on the same segment of the edge beam produces a force of S directed along the $+X$ axis. The intersecting arches are coplanar at N, and we may resolve the intersecting forces into components along and normal to the edge member. The normal components cancel, while the axial components produce a resultant force of $\sqrt{2}\,S$. This force acts over a length of edge beam $= \sqrt{2}$, so that we find a constant force S per unit length of edge beam as before. Thus the arch representation of the hyperbolic paraboloid is certainly equivalent to the previous shell solution, and often provides valuable insight—especially if irregular shapes or boundaries are encountered.

4.4.3.4 Edge Members. We have seen that the membrane stress analysis for hyperbolic paraboloids with straight-line boundaries is fairly straightforward. We have also established the necessity of developing the in-plane shearing forces on the boundary. We now look at the requirements for *ideal* edge members:

1. In order to fully develop the shear, the edge members should be inextensible.

2. Since the entire weight of the shell is ultimately carried to the foundation through only four to eight compression members for the cases illustrated in figure 4–36, the edge members should be capable of sustaining sizeable axial forces. These forces will produce corresponding axial strains that will conflict with the first requirement.

3. The edge members should be self-supporting and should not be carried by the shell. Since these members may be rather sizeable, they would impart a sizeable additional load on the shell that might be unacceptable.

From this list, it is obvious that in practice it may be difficult to attain any of these ideal conditions. As a specific illustration, consider the cantilever shell shown in figure 4–36(c). For this shell, we have four different edge members, as seen in table 4–2. Typical sections for such members in rein-

Table 4–2
Maximum Forces in Edge Members

Member	Location	Maximum Force ($+ = $ *tension*)
OK	Exterior	$-Sd$ @ 0
KG	Exterior	$-S\sqrt{(c^2 + d^2)}$ @ G
OJ	Ridge	$+2Sd$ @ 0
JG	Valley	$-2S\sqrt{(c^2 + d^2)}$ @ G

forced concrete shells are shown in figure 4–39. The dashed lines indicate possible smooth transitions between the shell and the edge member that may be desirable for architectural and/or constructional purposes. The edge members may be proportioned as struts under axial tension or compression.[55]

Next, consider the self-weight of the edge members. Referring to figure 4–36(c), the ridge member OJ could be considered as a beam spanning between O and J. The valley member GJ forms a triangular frame GJG' with the corresponding member on the other half of the shell, and resists the beam reactions at O from the two ridge members, OJ and $O'J$, in addition to the self-weight and the membrane forces imparted by the shell. Exterior member KK' is regarded as a beam spanning between K and K' and carrying the beam reaction at O from OJ, the self-weight and the shell membrane forces. This brings us to the exterior members KG and $K'G'$, which are required to take the beam reactions from KK' along with the self-weight and the membrane forces, seemingly as cantilevers.

Other alternatives are possible. In some cases, it is feasible to utilize exterior walls or window wall framing to support the perimeter edge members. The shell shown in figure 2–8(b) admits this possibility, whereas the shell shown in figure 2–8(d) obviously does not. Also, some designers might convert the self-weight of some of the edge members—particularly those which cannot sustain themselves—to equivalent distributed loads, and include these loads in the uniform load q_z. This essentially requires the shell to carry the edge beams.

Perhaps the most challenging cases from the design standpoint are cantilevered compression edge members, such as KG, or the even longer members, such as $G'G$, that occur in saddle shells as shown in figure 4–36(d). A spectacular example is shown in figure 2–8(a).

Referring again to figures 4–36(c) and 4–37, if the neutral axis of the edge beam for member KG does not coincide with the middle surface of

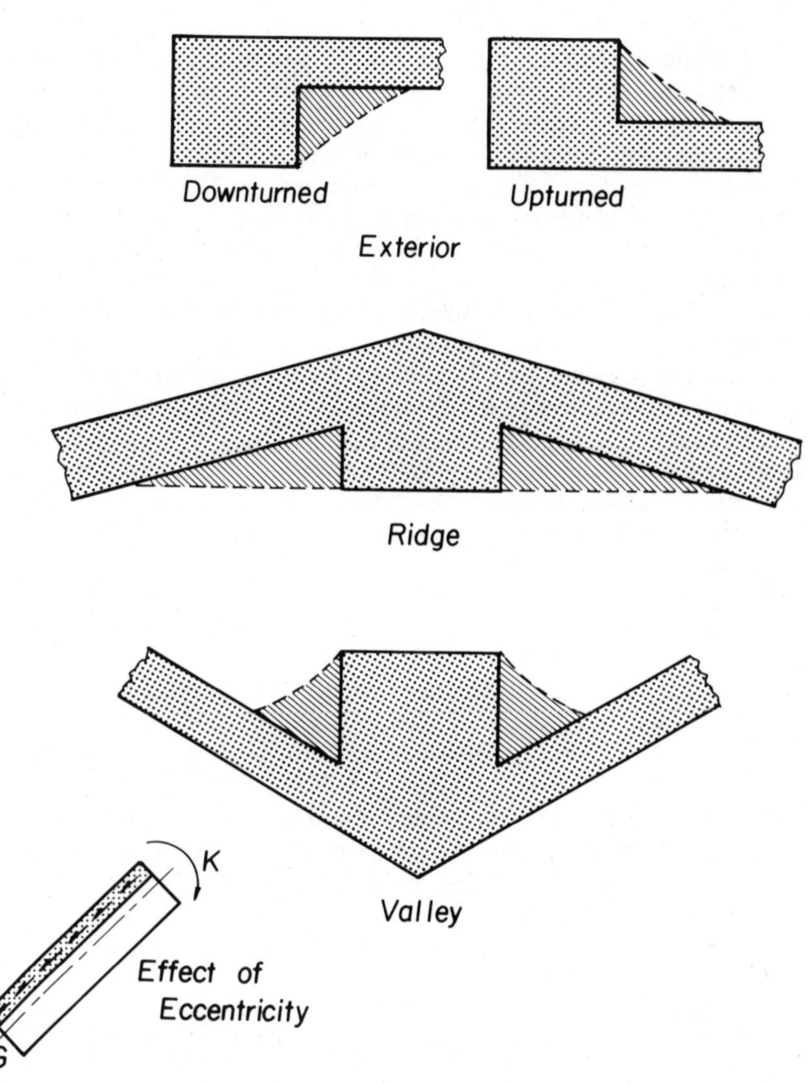

Figure 4–39. Edge Members for Hyperbolic Paraboloids

the shell, the shear force acts through an eccentricity that produces a uniformly distributed moment. For the downturned beam (figure 4–39), the moment is opposite to that produced by the cantilevered self-weight, and thus counters the self-weight somewhat, whereas for the upturned beam, the moment would add to that arising from the self-weight. Depending on whether tension or compression is introduced into the edge member by the shell and the particular exterior support condition, a countering pre-stressing moment can generally be produced by appropriate upturning or downturning of the member to give the required eccentricity.

Another possibility—particularly effective for tension edge members—is to introduce mechanical prestressing by using rods or cables. Also, as shown on the examples of figure 4–36, it is usually desirable to introduce ties between the supported ends of the inclined compression members to counter the horizontal component of the thrust. One of the most detrimental effects, insofar as the reduction of the dominance of membrane action and the subsequent introduction of bending into the shell and the edge member is concerned, has been found to be relative horizontal displacement of the abutments.[56] A comprehensive discussion of the forces in the edge members of a hipped hyperbolic paraboloid and some pertinent design recommendations for such shells are presented in Schnobrich and Shaaban and Ketchum.[57]

Finally, with respect to hyperbolic paraboloids with straight boundaries, note that although the presentation has been pointed somewhat at reinforced concrete shells, there are many interesting applications of this very efficient strucural form in other materials: for example, in steel[58] and wood.[59] A cantilevered, cable-supported hyperbolic paraboloid forms the roof for the jumbo jet hangar shown in figure 2–8(v), and metal hyperbolic paraboloid units were used between long-span arched trusses to cover a large arena in Mexico City constructed for the 1968 Olympic games.

4.4.3.5. Hyperbolic Paraboloid with Curved Boundaries.

We now resume our investigation of the basic hyperbolic paraboloid described by equation (4.182) and shown in figure 4–33. We restrict ourselves to a q_Z loading, for which equation (4.181) becomes

$$Z_{,XX} F_{,YY} - 2Z_{,XY} F_{,XY} + Z_{,YY} F_{,XX} = -q_Z \tag{4.216}$$

From equation (4.182),

$$Z_{,XX} = \frac{2c_X}{a^2}$$

$$Z_{,YY} = -\frac{2c_Y}{b^2}$$

$$Z_{,XY} = 0$$

so that equation (4.216) reduces to

$$\frac{2c_X}{a^2} F_{,YY} - \frac{2c_Y}{b^2} F_{,XX} = -q_Z(X, Y) \tag{4.217}$$

As a specific case, we take the uniform load $q_z = -p_u$. There are obviously multiple particular solutions, since the derivatives on F are uncoupled. Two possibilities are

$$F_1(X) = -\frac{1}{4}p_u\frac{b^2}{c_Y}X^2 \tag{4.218}$$

and

$$F_2(Y) = \frac{1}{4}p_u\frac{a^2}{c_X}Y^2 \tag{4.219}$$

Also, we may have any linear combinations of F_1 and F_2

$$F(X,Y) = \lambda_1 F_1(X) + \lambda_2 F_2(Y)$$

where

$$\lambda_1 + \lambda_2 = 1 \tag{4.220}$$

This presents us with a multiplicity of particular solutions for the projected stress resultants, which can be written using equation (4.180) as

$$N_X(X,Y) = \frac{\lambda_2}{2}p_u\frac{a^2}{c_X} \tag{4.221a}$$

$$N_Y(X,Y) = -\frac{\lambda_1}{2}p_u\frac{b^2}{c_Y} \tag{4.221b}$$

$$S(X,Y) = 0 \tag{4.221c}$$

The actual stress resultants, N_x and N_y, may then be found from equation (4.174).

For a given shell, the solution which will prevail is a function of the relative stiffnesses of the boundaries. Referring to figure 4–33, if the boundaries at $X = \pm a$ are fully developed in the x direction, while the shell is unrestrained at $Y = \pm b$, we would choose $\lambda_1 = 0$ and $\lambda_2 = 1$. This of course, corresponds to a tension arch in the X direction, such as we encountered in a previous section. Similarly, $\lambda_1 = 1$ and $\lambda_2 = 0$ indicate full support in the y direction at $Y = \pm b$, and no restraint at $X = \pm a$, or a compression arch in the Y direction. Obviously, with finite in-plane resistance along all boundaries, we would have nonzero values of both λ_1 and λ_2, but the necessary requirement is that at least one pair of the boundaries must develop the ensuing membrane forces.

The possibility of "directing" the membrane stresses by adjusting the support conditions makes this version of the hyperbolic paraboloid quite versatile. While a designer may prefer the state of uniform compression,

$\lambda_1 = 1$, for a reinforced concrete shell, he might prefer a tensile state, $\lambda_2 = 1$, for a steel structure. Numerous steel structures in this shape have been built using high strength cables as the principal members.

Another variation of the hyperbolic paraboloid with curved boundaries is to leave the edges free of direct stresses and to resist the surface loading by shearing forces along only two parallel boundaries. In turn, the resulting reactions can be provided by a pair of axially loaded boundary arches. As an illustration, consider the hyperbolic paraboloid shown in figure 4–33 with boundary conditions

$$N_X(\pm a, Y) = 0 \tag{4.222a}$$

$$N_Y(X, \pm b) = 0 \tag{4.222b}$$

$$S(X, \pm b) = 0 \tag{4.222c}$$

Moreover, the boundaries AD and BC can develop the in-plane shearing stresses,

$$S(\pm a, Y) \neq 0 \tag{4.222d}$$

A plan view of this shell is shown as figure 4–40(a), with b assumed to be equal to $2a$. Assume $c_Y = 4c_X$, so that $c_X/a^2 = c_Y/b^2 = c/a^2$. Equation (4.182) then becomes

$$Z = \frac{c}{a^2}(X^2 - Y^2) \tag{4.223}$$

These geometrical simplifications facilitate the following development. Also, the applied loading is taken as $q_Z = -p_u$, and the sense of $S(\pm a, Y)$ is shown on the figure so as to provide an upward $(+Z)$ resultant.

This problem is handily treated by first selecting a particular solution, and then enforcing the prescribed boundary conditions through a homogeneous solution.

Initially, the loading is assumed to be resisted by direct stresses along one pair of boundaries (e.g., AB and CD), as shown in figure 4–40(b). This corresponds to $\lambda_1 = 1$ and $\lambda_2 = 0$ in equations 4.220 and 4.221, for which

$$\left. \begin{aligned} N_Y(X,Y) &= -\frac{1}{2}p_u\frac{b^2}{c_Y} = -\frac{p_u}{2} \cdot \frac{a^2}{c} = -N^* \\[2mm] N_X(X,Y) &= S(X,Y) = 0 \end{aligned} \right\} \tag{4.224}$$

as shown on the figure. The actual reactions, which are equivalent to $N_Y(\bar{X}, \pm 2a)$, act in the middle surface and may be found from equations (4.223), (4.171) and (4.174b); however, it is convenient to proceed in terms of the projected resultants. The solution given in equation (4.224)

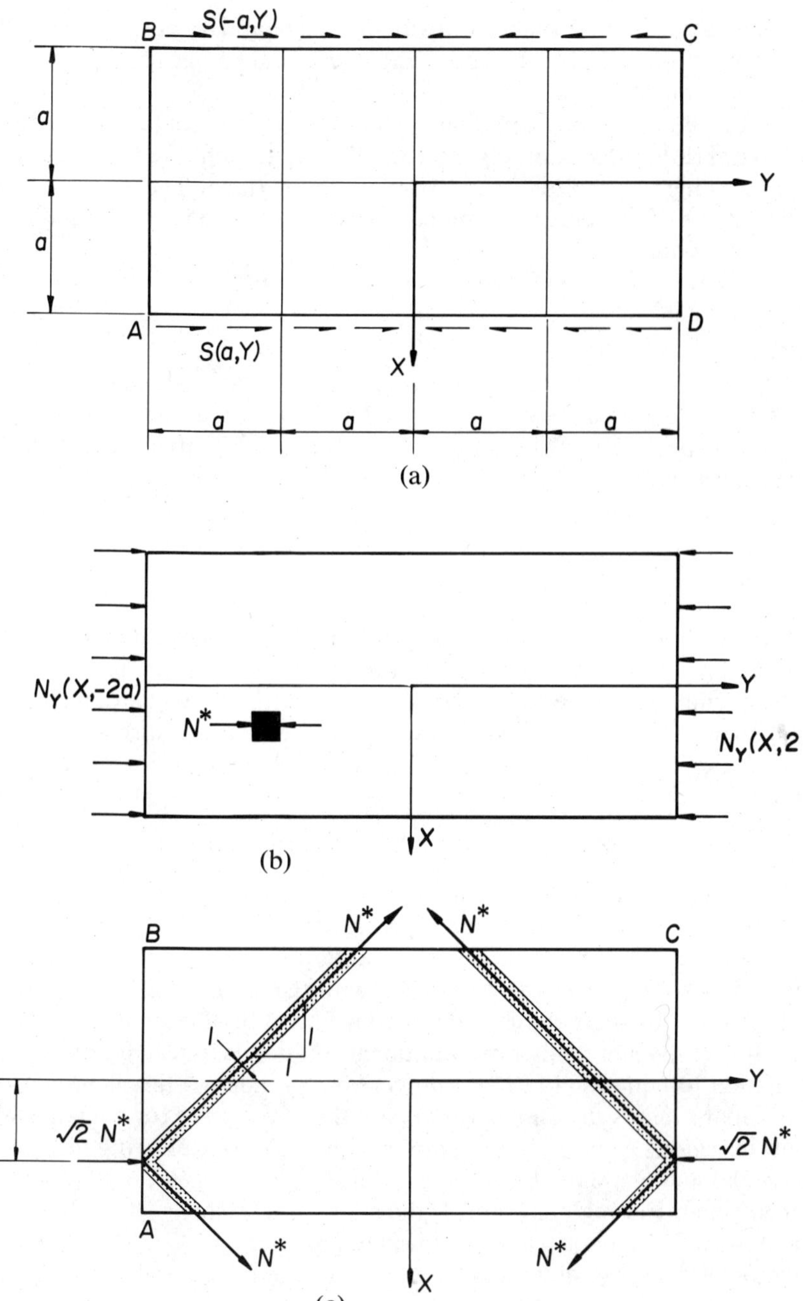

Figure 4–40. Hyperbolic Paraboloid with Curved Boundaries

balances the applied loading but violates the boundary conditions stipulated in equations (4.222). Hence, it represents only a particular solution for this system.

The homogeneous solution may be efficiently obtained through a physical rather than a purely mathematical approach. We first identify the straight lines or real characteristics on the surface. The slopes of these lines in the $X-Y$ plane are found from equation (4.185) specialized for the current geometry:

$$\frac{\bar{a}_1}{\bar{b}_1} = \pm 1 \tag{4.225}$$

Some typical $X-Y$ traces are shown in figure 4–40(c). We concentrate on those which pierce the boundaries AB and CD at the reference points $(\bar{X}, \mp 2a)$, respectively. Taking a segment of length $\sqrt{2}$ along each boundary (see inset, figure 4–38), the edge forces $\sqrt{2}N^*$, are resolved and translated along the characteristics to the $X = \pm a$ boundaries. The translated forces are equal to N^* and act over boundary segments of length $\sqrt{2}$ along BC and CD.

Next, as shown on figure 4–40(d), additional characteristics that pass through the points on BC and AD pierced by the translated forces are added. To produce resultant boundary forces that are parallel to BC and AD, self-equilibrated edge forces N^* acting along the new characteristics are supplied. These forces give a resultant force in the Y direction equal to $2N^* \cdot (\sqrt{2}/2)$. Dividing by the length of the boundary segment, $\sqrt{2}$, we obtain

$$\left| S(\pm a, Y) \right| = N^* = \left| N_Y(\bar{X}, \pm 2a) \right| \tag{4.226}$$

as the projected boundary reactions.

It is obvious that the foregoing argument can be repeated for each point along AB and CD, so that the additional solution represented by figures 4–40(c) and 4–40(d), combined with the solution given by equation (4.224), fulfills the boundary conditions stated in equations (4.222a–c).

It is easily shown that the additional solution is indeed a homogeneous solution to equation (4.217). Allowing for symmetry, the shell is divided into four regions in figure 4–40(e), and the stress state for each is indicated. In regions *1* and *4*, $N_X = N_Y = \pm N^*$ and $S = 0$. Referring to equations (4.180a) and (4.180b), $F_{,YY} = F_{,XX}$. Since $c_X/a^2 = c_Y/b^2$, equation (4.217) is satisfied. For regions *2* and *3*, the state of pure shear—i.e., $N_X = N_Y = 0$; $S = \pm N^*$—obviously is a solution to equation (4.217). Thus the combination of figures 4–40(b) and 4–40(e) gives the final membrane stresses in each region, figure 4–40(f).

Note that the discontinuities of the stresses between the various regions make the membrane state only an idealized possibility for this case.

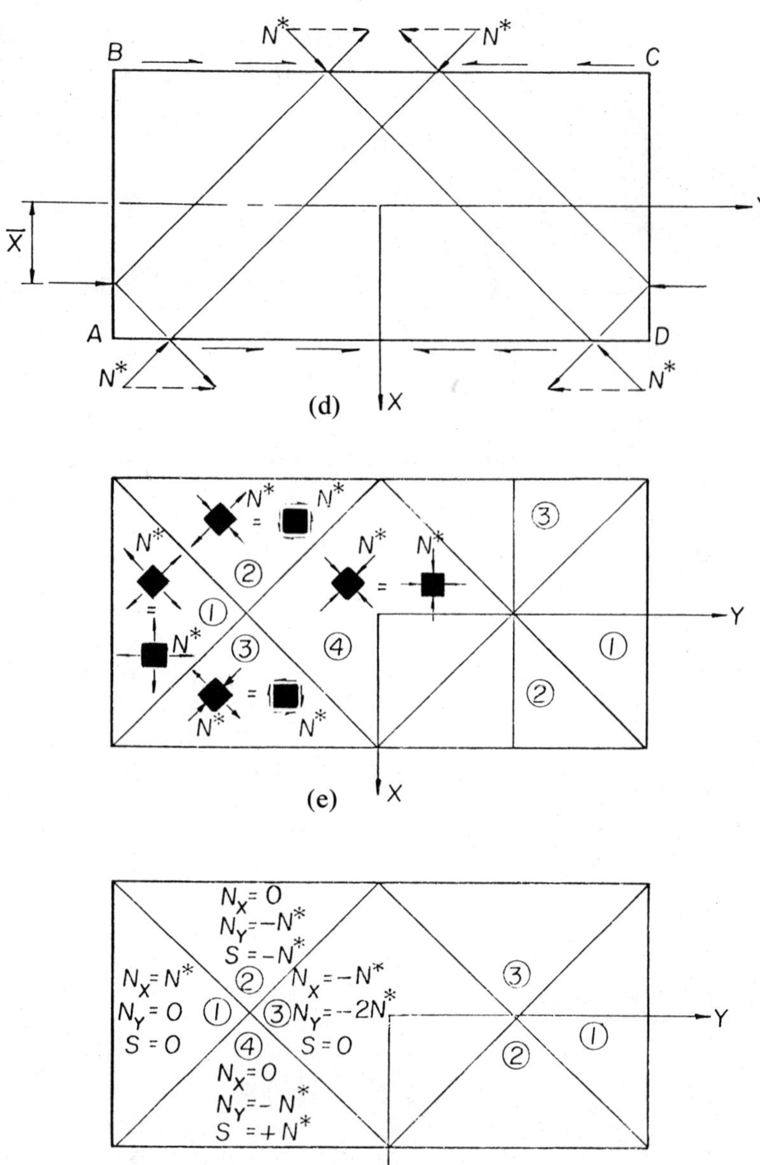

Figure 4–40 (cont.)

Careful detailing and construction are required to approach this stress distribution in an actual structure.

An interesting extension of the basic hyperbolic paraboloidal geome-

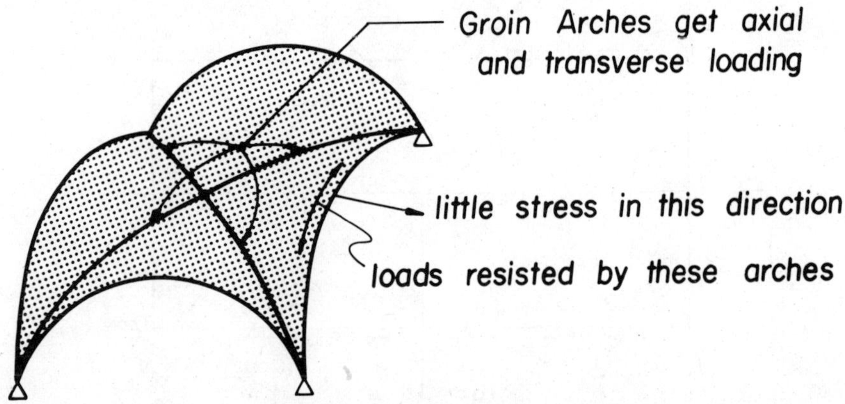

Figure 4–41. Groined Vault

try can be achieved by intersecting two shells, as shown in figure 4–41. Each quadrant is known as a *groined vault*. The basic resistance is essentially in the direction parallel to the exterior boundaries, where the load is transferred by the shell, acting as an assemblage of compression arches, to the groin arches, which pick up reactions from two adjacent vaults and transmit the forces to the foundation. Again, it is desirable to provide horizontal restraint at the base of the groin arches through a system of ties. Many variations on the groined vault concept are found, with a striking example shown in figure 2–8(f).

4.4.4 Elliptic Paraboloid

The elliptic paraboloid is generated by the translation of one convex paraboloid over another, as shown in figure 4–42(a). The equation of the shell is given by

$$Z = -\left(\frac{c_X}{a^2}\right)X^2 - \left(\frac{c_Y}{b^2}\right)Y^2 \tag{4.227}$$

It is easily observed and may be demonstrated numerically using equation (2.38) that this shell has positive Gaussian curvature, and, as such, we do not find straight lines on the surface.

The name *elliptic paraboloid* comes from the fact that if a horizontal plane $Z = \bar{Z} = $ constant $(\bar{Z} < 0)$ is passed through the shell, the equation of the intersection

$$\left(\frac{c_X}{a^2}\right)X^2 + \left(\frac{c_Y}{b^2}\right)Y^2 = -\bar{Z} \tag{4.228}$$

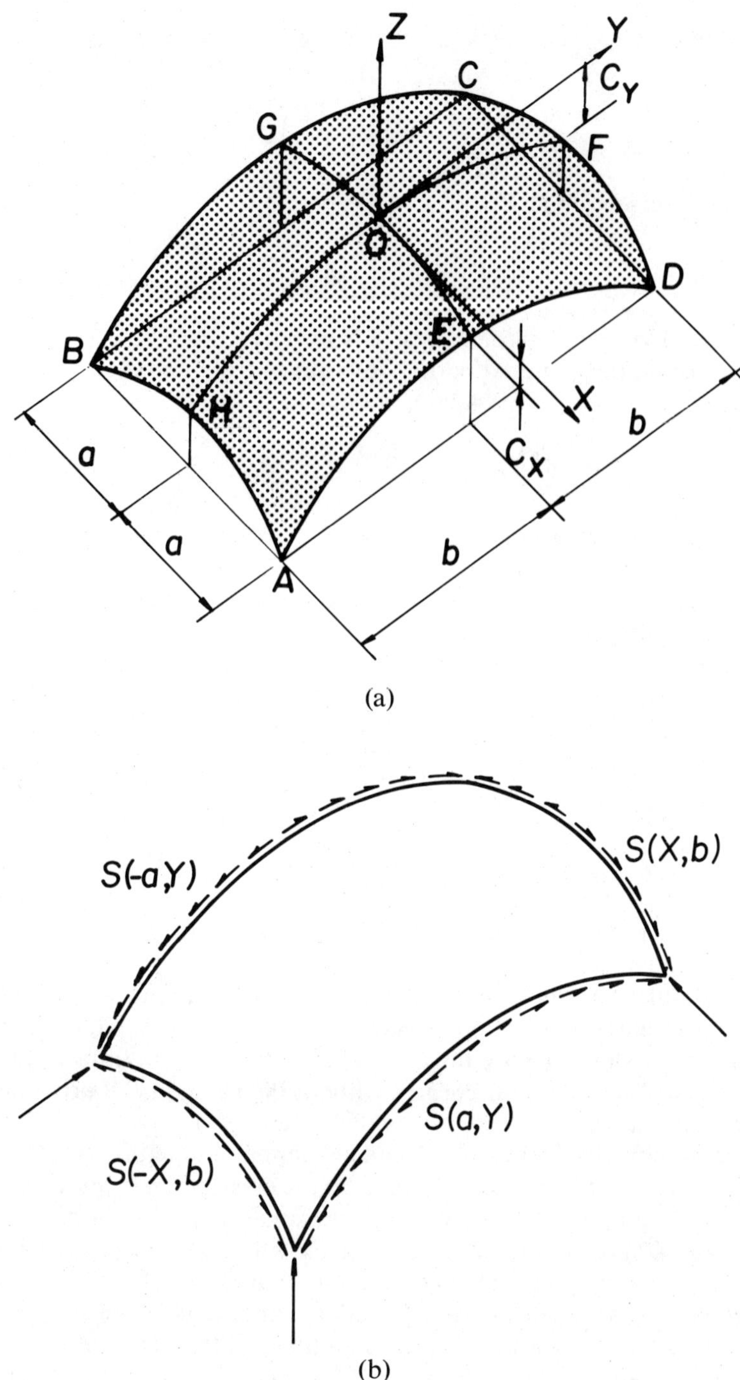

Figure 4–42. Elliptic Paraboloid Shell

is an ellipse. Also if $c_X/a^2 = c_Y/b^2 = c/d^2$, we have a paraboloid of revolution

$$X^2 + Y^2 = -\left(\frac{d^2}{c}\right)Z \qquad (4.229)$$

with the intersection at $Z = \bar{Z}(\bar{Z} < 0)$ given by

$$X^2 + Y^2 = -\left(\frac{d^2}{c}\right)\bar{Z} \qquad (4.230)$$

which is, of course, a circle, since Z is negative.

Evaluating the required partial derivatives from equation (4.227) and substituting into equation (4.216), we obtain

$$\left(\frac{2c_X}{a^2}\right)F_{,YY} + \left(\frac{2c_Y}{b^2}\right)F_{,XX} = q_Z(X, Y) \qquad (4.231)$$

which is very similar to equation (4.217), and suggests that this shell would resist a uniform load $q_Z = -p_u$ as an intersecting system of compression arches parallel to the X and Y axis. It is easily shown by comparison with equations (4.217), (4.218), and (4.221) that particular solutions of the form

$$F_3 = -\frac{1}{4}p_u\frac{b^2}{c_Y}X^2 \qquad (4.232)$$

and

$$F_4 = -\frac{1}{4}p_u\frac{a^2}{c_X}Y^2 \qquad (4.233)$$

will produce projected stresses $N_X = 0$ and $N_Y = 0$, respectively. However, these simple unidirectional patterns are often not compatible with the physical situation, because it may be somewhat difficult to provide the reactions required to develop N_Y along $Y = \pm b$ and/or N_X along $X = \pm a$. Rather, we wish to explore the possibility of this system transmitting the loading directly to the four corners without the necessity of providing in-plane restraint along the exterior boundaries.

The specific requirement is to find a solution to equation (4.231), such that $N_X(\pm a, Y) = N_Y(X, \pm b) = 0$ except at the corners, $(\pm a, \pm b)$. Obviously, it is only possible to satisfy one or the other condition using equation (4.232) or (4.233), so that, at best, F_3 or F_4 will serve as a particular solution. We will add a homogeneous solution in the form of an infinite series that *identically* satisfies the same boundary conditions as the selected particular solution, F_3 or F_4. Then the amplitudes of the terms of the infinite series are adjusted so that the total solution, homogeneous plus particular, will satisfy the remaining boundary conditions.

As an illustration, we select

$$F = F_4 + F_5 \tag{4.234}$$

where $F_5 = \displaystyle\sum_{j=1,3,5\ldots}^{\infty} A_j \cosh mX \cos nY$

$$m = \frac{j\pi}{2a}\sqrt{\frac{c_X}{c_Y}}$$

$$n = \frac{j\pi}{2b}$$

F_5 is easily shown to be a homogeneous solution of equation (4.231), and we have seen that F_4 produces $N_Y = 0$ throughout. From equation (4.180b), we see that due to F_5, $N_Y : \cos (j\pi/2b)\, Y$, which vanishes at $Y = \pm b$. Thus the boundary conditions on $N_Y(X, \pm a)$ are identically satisfied. N_X is then evaluated with the A_j coefficients chosen to satisfy the boundary condition $N_X(\pm a, Y) = 0$. The details are rather involved algebraically and are not given here. Complete solutions, including parametric tabulations, are available.[60]

It is apparent that the membrane solution so obtained results in the boundary arches transmitting the load to the corners by in-plane shears, as shown in figure 4–42(b). This occurs since the enforced boundary conditions prohibit direct stresses along the exterior boundaries. The shear S builds up to a theoretically infinite value at the corners, where a concentrated reaction acts. It is apparent that the actual stress state cannot be fully predicted by membrane theory, and, therefore, bending effects should be considered as well. Also, the details of construction might well include some stiffening and rounding of the corner to spread the high shear forces dictated by membrane equilibrium.

4.4.5 Conoid

A surface produced by the translation of a straight line generator between two directrices—one of which is itself a straight line, and the other a plane curve—is called a *conoid* (figure 4–43). Both directrices are assumed to lie in parallel planes. If the generator is perpendicular to the directrices, we have a square conoid; for a generator not perpendicular to the directirces, the conoid is said to be skewed.[61]

Shells in the form of conoids are used for roof applications when it is

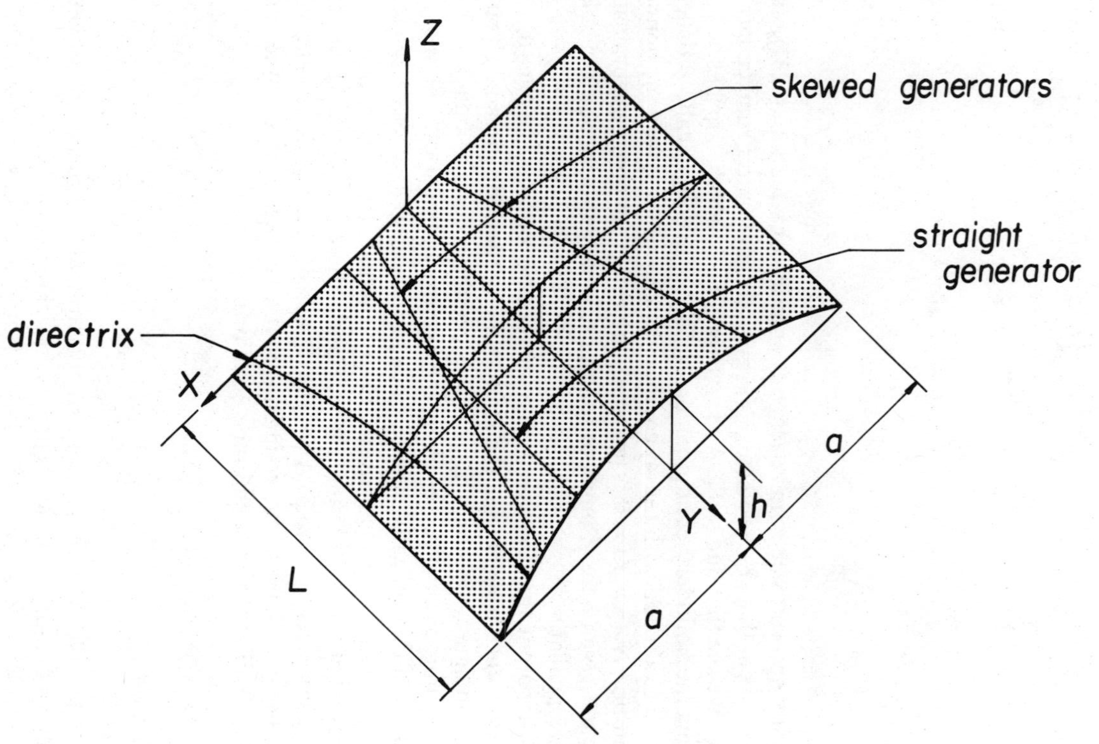

Figure 4–43. Conoidal Shell

desired to maintain a relatively flat profile and still provide an opening to admit natural sunlight.

The coordinate system shown in figure 4–43 has the Y–Z plane normal to the symmetry line of the curve directrix, whereas the X axis is taken along the straight directrix. With respect to this coordinate system, the equation of the curved directrix is of the form $Z = f(X)$. The square concoid is described by

$$Z(X,Y) = \frac{Y}{L} f(X) \qquad (4.235)$$

For a skewed conoid, the surface equation is more complicated. In either case, the conoid is not a second-order surface, and the discriminant test for Gaussian curvature, equation (2.38), is inapplicable. However, it is readily observed from figure 4–43 that the skewed conoid has negative Gaussian curvature, because there are two sets of straight line generators, whereas the square conoid has zero Gaussian curvature, since the two sets of generators become coincident.

Consider a parabolic square conoid

$$\left.\begin{aligned} f(X) &= h\left(1 - \frac{X^2}{a^2}\right) \\[2em] Z &= \frac{h}{L} Y\left(1 - \frac{X^2}{a^2}\right) \end{aligned}\right\} \qquad (4.236)$$

and

We now write equation (4.181) for a uniform load $q_z = -p_u$:

$$-\left(\frac{2h}{a^2 L}\right) Y F_{,YY} + \left(\frac{4h}{a^2 L} X\right) F_{,XY} = p_u \qquad (4.237)$$

Although, as we have observed with respect to the hyperbolic and elliptic paraboloids, multiple solutions for stress functions can frequently be generated, a solution of practical utility must be consistent with the physical boundary conditions. Because of symmetry, we would have S vanishing along $X = 0$, so that $-F_{,XY}(0,Y) = 0$. Also, if the curved directrix cannot sustain forces normal to the plane of the curve, $N_X(X,L) = 0$, implying that $F_{,YY}(X,L) = 0$.

Again, comprehensive results are readily available in the literature,[62] so that we will not dwell on the details of the various solutions for conoidal shells. However, it is of general interest to consider a simplified model of the conoid that is suggested by our earlier study of hyperbolic paraboloids. From figure 4–43, it is apparent that the conoid could possibly transmit the uniform load as a series of arches in the X–Z plane, with the attendant thrusts developed along the boundaries $X = \pm a$. The provision of the appropriate boundary conditions is plausible, since the horizontal components of the thrust would balance along the interior boundaries, and the vertical components could be resisted by walls or beams in the Y direction . Further, the shell would seemingly be entirely in compression.

With respect to equation (4.237), we may easily select a solution of the form

$$F,_{YY} = N_X = -\frac{a^2 L}{2hY} p_u \tag{4.238}$$

which would give the one-way arch action with N_Y and $S = 0$. Then, if we return to equation (4.174a) for the actual stress resultant N_x and compute from equation (4.236)

$$\tan \gamma_x = Z,_X = -\frac{2h}{a^2 L} XY$$

$$\tan \gamma_y = Z,_Y = \frac{h}{a^2 L} X^2$$

we find, from equations (4.204) and (4.38),

$$N_x = -\frac{a^2 L}{2hY} \left\{ \frac{1 + [(-2h/a^2 L)XY]^2}{1 + [(h/a^2 L)X^2]^2} \right\}^{1/2} p_u \tag{4.239}$$

For small values of Y, N_X grows very large, with a singularity at $Y = 0$. Essentially, this reflects the shell becoming very shallow and the arches correspondingly flattening until they degenerate into a straight line along the direction $Y = 0$. Obviously this region of the conoid cannot be sustained in equilibrium by membrane forces alone, and bending must be considered.

Notes

1. W. Flügge, *Stresses in Shells*, 2nd ed. (Berlin: Springer-Verlag, 1973), pp. 100–102.

2. V. V. Novozhilov, *Thin Shell Theory* [translated from 2nd Russian ed. by P. G. Lowe (Groningen, The Netherlands: Noordhoff, 1964), pp. 105–107].

3. P. L. Gould, A. Cataloglu, G. Dhatt, A Chattopadhyay, and R. E. Clark, "Stress Analysis of the Human Heart Valve," *Journal of Computers and Structures*, 3, 1973, pp. 377–384.

4. S. Timoshenko and S. Woinowsky-Krieger, *Theory of Plates and Shells*, 2nd ed. (New York: McGraw-Hill, 1959), pp. 449–450.

5. Novozhilov, *Thin Shell Theory*, pp. 117–119.

6. D. P. Billington, *Thin Shell Concrete Structures* (New York: McGraw-Hill, 1965), pp. 111–121.

7. P. L. Gould and S. L. Lee, "Hyperbolic Cooling Towers under Seismic Design Load," *Journal of the Structural Division, ASCE* 93, no. ST3, (June 1967): 87–109.

8. Closure to ibid., *Journal of the Structural Division, ASCE* 94, no. ST10 (October 1968): 2487–2493; S. L. Lee and P. L. Gould, "Hyperbolic Cooling Towers under Wind Load," *Journal of the Structural Division, ASCE* 93, no. ST5 (October 1967): 487–514.

9. Flügge, *Stresses in Shells*, pp. 171–179.

10. Novozhilov, *Thin Shell Theory*, pp. 130–138.

11. Ibid.

12. A. Fino and R. W. Schneider, "Wrinkling of a Large Thin Code Head under Internal Pressure," *Welding Research Council Bulletin* no. 69 (New York: Welding Research Council, June 1961), pp. 11–13.

13. Novozhilov, *Thin Shell Theory*, pp. 117–119.

14. Ibid, pp. 147–151.

15. H. Kraus, *Thin Elastic Shells* (New York: Wiley, 1967), pp. 307–314.

16. Novozhilov, *Thin Shell Theory*, pp. 151–163.

17. V. Z. Vlasov, *General Theory of Shells and its Application in Engineering*, NASA Technical Translation TTF-99 (Washington, D.C.: National Aeronautics and Space Administration, 1964), pp. 200–201.

18. Flügge, *Stresses in Shells*, pp. 44–46.

19. Timoshenko and Woinowsky-Krieger, *Theory of Plates and Shells*, pp. 449–451.

20. E. P. Popov, "Earthquake Stresses in Spherical Domes and in Cones," *Journal of the Structural Division, ASCE* 82, no. ST3 (May 1956): 974-1–974-14.

21. R. W. Clough, "Earthquake Response of Structures," in R. L. Wiegel, ed., *Earthquake Engineering* (Englewood Cliffs, N.J.: Prentice-Hall, 1970), pp. 307–334.

22. P. L. Gould, S. K. Sen, and H. Suryoutomo, "Dynamic Analysis of Column-Supported Hyperboloidal Shells," *Earthquake Engineering and Structural Dynamics* 2 (1974): 269–280.

23. Ibid.

24. P. L. Gould, "Hyperbolic Cooling Towers under Seismic Design Loading," Proc. of the Fourth Symposium on Earthquake Engineering I., University of Roorkee, Roorkee, India, November 1970.

25. P. L. Gould, "Quasistatic Seismic Loading Distributions for Hyperbolic Cooling Towers," *Bulletin of the Indian Society of Earthquake Technology* 8, no. 4 (December 1971): 163–168.

26. Gould and Lee, Closure to "Hyperbolic Cooling Towers under Seismic Design Load."

27. Building Code Requirements for Reinforced Concrete Structures (ACI 318–71) (Detroit: American Concrete Institute, 1971), pp. 26–27.

28. Timoshenko and Woinowski-Krieger, *Theory of Plates and Shells*, pp. 453–456.

29. P. L. Gould and S. L. Lee, "Column-Supported Hyperboloids under Wind Load," Publications of the International Association for Bridge and Structural Engineering, Zurich, Switzerland 31–11, 1971, pp. 47–64; S. K. Sen and P. L. Gould, "Hyperboloidal Shells on Discrete Supports," technical note, *Journal of the Structural Division, ASCE* 99, no. ST3 (March 1973): 595–603.

30. Flügge, *Stresses in Shells*, pp. 71–79.

31. R. F. Rish and T. F. Steel, "Design and Selection of Hyperbolic Cooling Towers," *Journal of the Power Division, ASCE* 85, no. PO5, (October 1959): 89–117.

32. Ibid.

33. C. O. Oakley, *The Calculus* (New York: Barnes & Noble, 1952), p. 133.

34. Lee and Gould, "Hyperbolic Cooling Towers under Wind Load."

35. P. L. Gould, "Unsymmetrically Loaded Hyperboloids of Revolution," *Journal of the Engineering Mechanics Division, ASCE* 94, no. EM5, (October 1968): 1029–1043.

36. Novozhilov, *Thin Shell Theory*, pp. 315–319.

37. Lee and Gould, "Hyperbolic Cooling Towers under Wind Load."

38. G. W. Hill, *Collected Mathematical Works*, vol. 1 (Washington, D.C.: Carnegie Institute of Washington, 1905–1907), pp. 243–270; E. T. Whittaker and G. N. Watson, *A Course of Modern Analysis*, 4th ed. (Cambridge: Cambridge University Press, 1935), pp. 412–417.

39. Whittaker and Watson, *Course of Modern Analysis*.

40. Lee and Gould, "Hyperbolic Cooling Towers under Wind Load."

41. E. Ingerslev, "Design of Cooling Tower Shells," Proceedings of the June 1966 Bratislava, Czechoslovakia, Symposium on Tower Shaped Steel and Reinforced Concrete structures, IASS, Madrid, Spain, 1968.

42. L. J. Brombolich and P. L. Gould, "A High-Precision Curved Shell Finite Element," *Synoptic, AIAA Journal* 10, no. 6 (June 1972): 727–728.

43. E. H. Baker, L. Kovalevsky, and F. H. Rish, *Structural Analysis of Shells* (New York: McGraw-Hill, 1972).

44. "Design of Cylindrical Concrete Roofs," *ASCE Manual of Engineering Practice* no. 31 (New York: American Society of Civil Engineers, 1952); Billington, *Thin Shell Concrete Structures*, chaps. 5 and 6.

45. Billington, ibid, p. 163.

46. Billington, ibid, pp 182–197.

47. Ibid, pp. 240–245.

48. A. L. Parme, "Shells of Double Curvature," *Trans. ASCE*, vol. 123, 1958, pp. 990–1025.

49. Billington, *Thin Shell Concrete Structures*, p. 244.

50. Ibid., pp. 245–249.

51. Ibid.

52. C. Faber, *Candela: The Shell Builder* (New York: Reinhold, 1963).

53. "The New Newark Airport," *Civil Engineering* 44, no. 9 (September 1974): 74–76.

54. Billington, *Thin Shell Concrete Structures*, pp. 250–254.

55. Building Code Requirements for Reinforced Concrete Structures (ACI 318–71) (Detroit: American Concrete Institute, 1971).

56. W. C. Schnobrich, "Analysis of Hyperbolic Paraboloid Shells," Symposium on Concrete Thin Shells, ACI Special Publication SP-28, paper SP-28-13 (Detroit: American Concrete Institute, 1971), pp. 275–311.

57. W. C. Schnobrich, "Analysis of Hipped Roof Hyperbolic Paraboloid Structures," *Journal of the Structural Division, ASCE* 98, no. ST7 (July 1972): 1575–1583; discussion by M. S. Ketchum, vol. 99, no. ST4 (April 1973): 796–797; closure, vol. 100, no. ST2 (February 1974): 467–469; A. Shaaban and M. S. Ketchum "Design of Hipped Hypar Shells," *Journal of the Structural Division, ASCE* 102, no. ST11 (November 1976): 2151–2161.

58. S. S. Tezcan, K. M. Agrawal, and G. Kostro, "Finite Element Analysis of Hyperbolic Paraboloid Shells," *Journal of the Structural Division, ASCE* 97, no. ST1 (January 1971): 407–424; P. V. Banavalkar and P. Gergely, "Thin-Steel Hyperbolic Paraboloid Shells," *Journal of the Structural Division, ASCE* 98, no. ST11 (November 1973): 2605–2621.

59. R. C. Liu and N. C. Teter, "Hyperbolic Paraboloid Shells Built with Wood Products," *Bulletin of IASS* 30 (June 1967): 3–8.

60. Billington, *Thin Shell Concrete Structures*, pp. 254–260.

61. A. M. Haas, *Thin Concrete Shells*, vol. 2 (New York: Wiley, 1967), p. 103.

62. Ibid., p. 103–133.

Nondimensional Geometry

a = throat radius

$$s = \frac{S}{a}$$

$$\xi = \frac{\Xi}{a}$$

$$r_\theta = \frac{R_\theta}{a} = \frac{\sqrt{(k^2 - 1)}}{[k^2 \sin^2 \phi - 1]^{1/2}}$$

$$r_\phi = \frac{R_\phi}{a} = \frac{-\sqrt{(k^2 - 1)}}{[k^2 \sin^2 \phi - 1]^{3/2}}$$

$$r_0 = r_\theta \sin \phi$$

Surface Loading

$$\bar{q}(\phi) = \frac{q(\phi)}{q_{DL}}$$

$$\bar{q}_\phi^1 = \bar{q}(\phi) \cos \phi$$

$$\bar{q}_\theta^1 = \bar{q}(\phi)$$

$$\bar{q}_n^1 = -\bar{q}(\phi) \sin \phi$$

Stress Resultants

$$n_\phi^1 = \frac{N_\phi^1}{q_{DL}a} = \frac{1}{r_0^2 \sin \phi}\, \Phi_1(\phi)$$

$$n_\theta^1 = \frac{N_\theta^1}{q_{DL}a} = r_\theta\left(\bar{q}_n^1 - \frac{n_\phi^1}{r_\phi}\right)$$

$$s^1 = \frac{S^1}{q_{DL}a} = n_\phi^1 \cos \phi - \frac{\Phi_2(\phi)}{r_0}$$

Functions

M Distribution

$$q(\phi) = C q_{DL}$$

$$\Phi_1(\phi) = C\{2\zeta_1(\phi_t)[\zeta_3(\phi) - \zeta_3(\phi_t)] - [\zeta_4(\phi) - \zeta_4(\phi_t)]\}$$

$$\Phi_2(\phi) = 2C[\zeta_1(\phi) - \zeta_1(\phi_t)]$$

MH Distribution

$$q(\phi) = C\bar{\zeta}_1\xi$$

$$\Phi_1 = \frac{2C\bar{\zeta}_1}{k^2 - 1}\left\{\frac{1}{12k^2}\,(3\zeta_1 - \zeta_3 r_\theta^3) - [\zeta_2(\phi_t) + s\zeta_1(\phi)]\right.$$

$$\left. \times\, r_\theta \cos \phi + s\,\frac{k^2 - 1}{3k^2}\,r_\phi\right\}$$

$$\Phi_2 = 2C\bar{\zeta}_1[\zeta_2(\phi) - \zeta_2(\phi_t)]$$

MH² Distribution

$$q(\phi) = C\bar{\zeta}_2\xi^2$$

$$\Phi_1 = 2C\bar{\zeta}_2[\zeta_1\zeta_8 + \zeta_9 + \xi\zeta_5(\phi_t)]_{\phi_t}^{\phi}$$

$$\Phi_2 = 2C\bar{\zeta}_2[\zeta_5(\phi) - \zeta_5(\phi_t)]$$

General

$$\zeta_1(\phi) = \frac{1}{4(k^2 - 1)} \left[\frac{\sqrt{(k^2 - 1)}}{k} \, ln \, K - 2r_\theta^2 \cos \phi \right]$$

$$K = \frac{\sqrt{(k^2 - 1)} - k \cos \phi}{\sqrt{(k^2 - 1)} + k \cos \phi}$$

$$\zeta_2(\phi) = \frac{r_\phi}{3k^2} - \zeta_1 s$$

$$\zeta_3(\phi) = \frac{\cos \phi \, r_\theta}{k^2 - 1}$$

$$\zeta_4(\phi) = \frac{1}{2k^2} \, \zeta_3 \left[\frac{1}{\sqrt{(k^2 - 1)}} \, ln \, K + \frac{2}{k \cos \phi} + \frac{r_\phi}{3k^2} \right]$$

$$\zeta_5(\phi) = \zeta_1 \zeta_6 + \xi \zeta_2 + \zeta_7$$

$$\zeta_6(\phi) = \frac{1}{4k^2(k^2 - 1)} - \zeta_3 s$$

$$\zeta_7(\phi) = \frac{r_\phi}{12k^2} \, [\zeta_3 + 4s]$$

$$\zeta_8(\phi) = \frac{\zeta + s}{4k^2(k^2 - 1)} + \zeta_3 \left[\frac{1}{2k^2(k^2 - 1)} - s^2 \right]$$

$$\zeta_9(\phi) = \frac{r_\phi}{3k^2} \left[\frac{\zeta_3(\xi + s)}{4} + s^2 - \frac{1}{2k^2(k^2 - 1)} + \frac{r_\theta^2}{10k^2(k^2 - 1)} \right]$$

$$\bar{\zeta}_1 = \frac{\zeta_1(\phi_b) - \zeta_1(\phi_t)}{\zeta_2(\phi_b) - \zeta_2(\phi_t)}$$

$$\bar{\zeta}_2 = \frac{\zeta_1(\phi_b) - \zeta_1(\phi_t)}{\zeta_5(\phi_b) - \zeta_5(\phi_t)}$$

5 Deformations

5.1 General

In the earlier chapters, we developed a geometric description of the middle surface of a shell that proved to be adequate to derive the equations of equilibrium. In turn, a simplified subset of the equilibrium equations formed the basis for the membrane theory of shells, for which many important practical applications were illustrated. Although membrane action in a shell is desirable from the dual standpoints of mathematical simplification and material efficiency, the requisite conditions for membrane behavior cannot always be simulated in an actual structure. Consequently, to expand our base of understanding of shell behavior, we must develop relationships between the forces and the deformations of the shell. The first step is a description of the displacements, where we follow the vector approach suggested by Novozhilov.[1]

5.2 Displacement

5.2.1 Displacement Vector

In figure 5–1(a), a differential element of the middle surface is shown before and after deformation. The reference point o located on the middle surface moves to the position o', and this displacement is denoted by $\mathbf{\Delta}$. Considering a second point $o(\zeta)$, located a distance from ζ from o along the original unit normal $\mathbf{t}_n$, the corresponding position after deformation is $o_i'(\zeta)$ ($i = 1$ or 2), and the displacement between $o(\zeta)$ and $o_i'(\zeta)$ is $\mathbf{\Delta}_i(\zeta)$.

We have shown two possible positions of $o_i'(\zeta)$, $o_1'(\zeta)$ and $o_2'(\zeta)$. Point $o_1'(\zeta)$ lies on the unit normal to the deformed middle surface, $\mathbf{t}_n'$, at the same distance ζ from the middle surface. The point $o_1'(\zeta)$ is located on $\mathbf{t}_n'$ because of assumption [3], table 1–1, and remains ζ from the middle surface because of assumption [4]. On the other hand, point $o_2'(\zeta)$ is also located ζ from the deformed middle surface, but no longer necessarily lies on the normal $\mathbf{t}_n'$; that is, the location of $o_2'(\zeta)$ only requires the enforcement of assumption [4]. Since assumption [3] refers to the suppression of transverse shearing strains, we may surmise that the difference in $o_1'(\zeta)$ and $o_2'(\zeta)$ is the effect of the transverse shearing strain. This is clearly

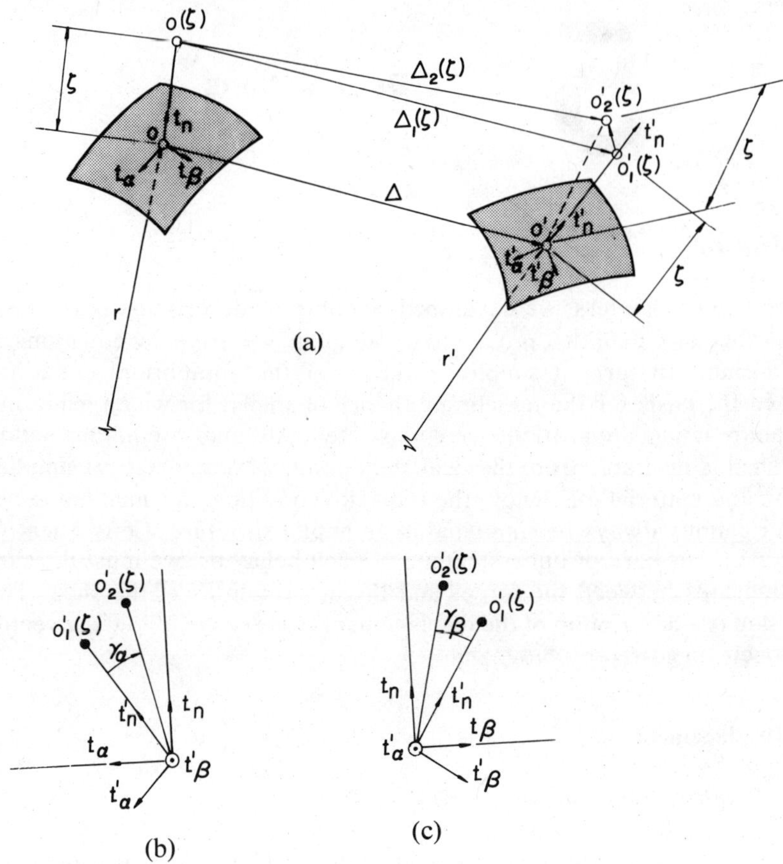

Figure 5–1. Deformation of a Middle Surface Element

shown in figures 5–1(b) and 5–1(c) where views normal to t'_β and t'_α are shown with the transverse shearing strains defined as γ_a and γ_β. The positive sense shown is consistent with the positive sense of the transverse shear stress resultants Q_α and Q_β, figure 3–2(a).

It was stated in the introduction that most classical work in shell and plate theories has been based on the suppression of transverse shearing deformations to achieve mathematical simplification. Currently, however, the analysis of complex plates and shells is frequently carried out by powerful numerical techniques that are not necessarily dependent on this simplification. Also, some authors have expressed an opinion that the inclusion of transverse shearing strains extends the bounds of the theory to include somewhat thicker plates and shells. This is somewhat difficult to quantify, since a true thick shell theory should account for transverse normal stresses as well, but it appears to be correct in a heuristic sense. In the interest of generality, therefore, we will retain the transverse shearing strains and treat $o_2'(\zeta)$ as the deformed point away from the middle surface, and $\Delta_2(\zeta)$ as the corresponding displacement. Subsequently, we will show that the theory which neglects transverse shearing strains is easily obtained from the more general theory.

From figure 5–1,

$$\zeta t_n + \Delta_2(\zeta) = \Delta + \zeta t_n' - \zeta\gamma_\alpha t_\alpha' - \zeta\gamma_\beta t_\beta' \tag{5.1}$$

or

$$\Delta_2(\zeta) = \Delta + \zeta(t_n' - t_n) - \zeta(\gamma_\alpha t_\alpha' + \gamma_\beta t_\beta') \tag{5.2}$$

Equation (5.2) expresses the deformation of a point off the middle surface in terms of the deformation of the corresponding middle surface point and the unchanging distance between the middle surface and the point in question. This establishes the pattern for subsequent developments: to relate the behavior of a point away from the middle surface to the behavior of the corresponding point on the middle surface.

5.2.2 Displacements in Terms of Middle Surface Parameters

Now we express the displacement vectors in terms of the unit tangent vectors of the undeformed middle surface:

$$\Delta = D_\alpha t_\alpha + D_\beta t_\beta + D_n t_n \tag{5.3a}$$

$$\Delta_2(\zeta) = D_\alpha(\zeta)t_\alpha + D_\beta(\zeta)t_\beta + D_n(\zeta)t_n \tag{5.3b}$$

Here, D_α, D_β, and D_n are the components of the middle surface displacement in the respective directions, and $D_\alpha(\zeta)$, $D_\beta(\zeta)$, and $D_n(\zeta)$ are the corresponding components away from the middle surface. Also the radius vector to the deformed middle surface is expressed in terms of the displacement vector by

$$r' = r + \Delta$$

$$= r + D_\alpha t_\alpha + D_\beta t_\beta + D_n t_n \tag{5.4}$$

In equation (5.2), the unit vectors of the deformed middle surface, t_α', t_β', and t_n', are present. In line with equation (5.4), for the radius vector r', we seek to express t_α', t_β', and t_n' in terms of the unit vectors of the undeformed middle surface. Since

$$t_n' = t_\alpha' \times t_\beta' \tag{5.5}$$

it is sufficient to develop suitable expressions for t_α' and t_β'.

196

By analogy with the definitions in equations (2.16) and (2.8), we have

$$t'_\alpha = \frac{\mathbf{r'},_\alpha}{A'} \tag{5.6a}$$

$$t'_\beta = \frac{\mathbf{r'},_\beta}{B'} \tag{5.6b}$$

where

$$A' = \pm(\mathbf{r'},_\alpha \cdot \mathbf{r'},_\alpha)^{1/2} \tag{5.7a}$$

and

$$B' = \pm(\mathbf{r'},_\beta \cdot \mathbf{r'},_\beta)^{1/2} \tag{5.7b}$$

Thus we have defined A' and B' as the Lamé parameters for the deformed middle surface. Now, considering equation (5.4)

$$\mathbf{r'},_\alpha = \mathbf{r},_\alpha + (D_\alpha t_\alpha),_\alpha + (D_\beta t_\beta),_\alpha + (D_n t_n),_\alpha \tag{5.8}$$

The first term in equation (5.8) is replaced using equation (2.16a), while the remaining terms are differentiated by the product rule with the derivatives of the unit tangent vectors obtained from equation (2.17). After carrying out the indicated operations, we have

$$\mathbf{r'},_\alpha = A[(1 + \epsilon_\alpha)t_\alpha + \epsilon_{\alpha\beta}t_\beta - \psi_\alpha t_n] \tag{5.9}$$

where the coefficients of the unit vectors are given by

$$\epsilon_\alpha = \frac{1}{A} D_{\alpha,\alpha} + \frac{1}{AB} A,_\beta D_\beta + \frac{1}{R_\alpha} D_n \tag{5.10a}$$

$$\epsilon_{\alpha\beta} = \frac{1}{A} D_{\beta,\alpha} - \frac{1}{AB} A,_\beta D_\alpha \tag{5.10b}$$

$$\psi_\alpha = -\left(\frac{1}{A} D_{n,\alpha} - \frac{1}{R_\alpha} D_\alpha\right) \tag{5.10c}$$

Similarly,

$$\mathbf{r'},_\beta = B[\epsilon_{\beta\alpha}t_\alpha + (1 + \epsilon_\beta)t_\beta - \psi_\beta t_n] \tag{5.11}$$

in which

$$\epsilon_\beta = \frac{1}{B} D_{\beta,\beta} + \frac{1}{AB} B,_\alpha D_\alpha + \frac{1}{R_\beta} D_n \tag{5.12a}$$

$$\epsilon_{\beta\alpha} = \frac{1}{B} D_{\alpha,\beta} - \frac{1}{AB} B,_\alpha D_\beta \tag{5.12b}$$

$$\psi_\beta = -\left(\frac{1}{B} D_{n,\beta} - \frac{1}{R_\beta} D_\beta\right) \tag{5.12c}$$

The coefficients defined in equations (5.10) and (5.12) are extremely important in what follows and are interpreted from a physical standpoint at a later stage of this chapter.

We are ready to substitute equations (5.9)–(5.12) into equations (5.6) and (5.7). As we carry out this operation, recall assumption [2] of table 1–1, which restricted the development to the domain of comparatively small deformations. This limitation enables products of the coefficients defined in equations (5.10) and (5.12) to be neglected, and ensures that the subsequent relationships will be linear. We wish to emphasize at this point that although assumption [2] greatly simplifies the problem from the mathematical standpoint, the justification is based on verification of the magnitudes of the neglected terms for a particular problem or class of problems. In most practical applications, the linear theory has been verified to be adequate, but, if not, a variety of nonlinear theories are available.[2]

Performing the indicated substitutions, we have from equations (5.7a) and (5.9)

$$A' = A[(1 + \epsilon_\alpha)^2 + \epsilon_{\alpha\beta}^2 + \psi_\alpha^2]^{1/2}$$

$$\simeq A(1 + \epsilon_\alpha) \tag{5.13}$$

and from equations (5.7b) and (5.11),

$$B' = B[\epsilon_{\beta\alpha}^2 + (1 + \epsilon_\beta)^2 + \psi_\beta^2]^{1/2}$$

$$\simeq B(1 + \epsilon_\beta) \tag{5.14}$$

after the product terms are dropped. Then, from equation (5.6a),

$$\mathbf{t}_\alpha' = \frac{A[(1 + \epsilon_\alpha)\mathbf{t}_\alpha + \epsilon_{\alpha\beta}\mathbf{t}_\beta - \psi_\alpha\mathbf{t}_n]}{A(1 + \epsilon_\alpha)}$$

$$\simeq \mathbf{t}_\alpha + \epsilon_{\alpha\beta}\mathbf{t}_\beta - \psi_\alpha\mathbf{t}_n \tag{5.15}$$

and from equation (5.6b)

$$\mathbf{t}_\beta' = \frac{B[\epsilon_{\beta\alpha}\mathbf{t}_\alpha + (1 + \epsilon_\beta)\mathbf{t}_\beta - \psi_\beta\mathbf{t}_n]}{B(1 + \epsilon_\beta)}$$

$$\simeq \epsilon_{\beta\alpha}\mathbf{t}_\alpha + \mathbf{t}_\beta - \psi_\beta\mathbf{t}_n \tag{5.16}$$

where

$$\frac{\epsilon_{ij}}{1 + \epsilon_i} \simeq \epsilon_{ij} \quad \text{and} \quad \frac{\psi_i}{1 + \epsilon_i} \simeq \psi_i \left(\begin{array}{l} i = \alpha, \beta \\ j = \beta, \alpha \end{array} \right)$$

Finally, we compute the vector product $\mathbf{t}_\alpha' \times \mathbf{t}_\beta'$, which, after simplification, is

$$\mathbf{t}_n' = \mathbf{t}_n + \psi_\alpha\mathbf{t}_\alpha + \psi_\beta\mathbf{t}_\beta \tag{5.17}$$

198

Equation (5.17) is the desired expression of the unit normal to the *deformed* middle surface, in terms of the tangent vectors to the *undeformed* middle surface. Substituting equations (5.15), (5.16), and (5.17) into equation (5.2) gives

$$\Delta_2(\zeta) = \Delta + \zeta[(\psi_\alpha - \gamma_\alpha - \gamma_\alpha\epsilon_{\beta\alpha})\mathbf{t}_\alpha + (\psi_\beta - \gamma_\beta - \gamma_\beta\epsilon_{\alpha\beta})\mathbf{t}_\beta$$

$$+ (\gamma_\alpha\psi_\alpha + \gamma_\beta\psi_\beta)\mathbf{t}_n] \tag{5.18a}$$

Again, products of the deformation parameters, such as $\gamma_\alpha\epsilon_{\beta\alpha}$, etc., are neglected, so that equation (5.18a) simplifies to

$$\Delta_2(\zeta) = \Delta + \zeta[(\psi_\alpha - \gamma_\alpha)\mathbf{t}_\alpha + (\psi_\beta - \gamma_\beta)\mathbf{t}_\beta] \tag{5.18b}$$

In view of equation (5.3a), equation (5.18b) may be rewritten as

$$\Delta_2(\zeta) = [D_\alpha + \zeta(\psi_\alpha - \gamma_\alpha)]\mathbf{t}_\alpha + [D_\beta + \zeta(\psi_\beta - \gamma_\beta)]\mathbf{t}_\beta + D_n\mathbf{t}_n \tag{5.19}$$

so that the components of $\Delta(\zeta)$ are

$$D_\alpha(\zeta) = D_\alpha + \zeta(\psi_\alpha - \gamma_\alpha) \tag{5.20a}$$

$$D_\beta(\zeta) = D_\beta + \zeta(\psi_\beta - \gamma_\beta) \tag{5.20b}$$

$$D_n(\zeta) = D_n \tag{5.20c}$$

Here we have expressed the components of the *displacement of an arbitrary point* on the shell in terms of the *displacement of the corresponding point on the middle surface* and the *unchanging separation of the points along the unit normal vector*. Equation (5.20c) is, of course, a direct restatement of assumption [4]. At this point, we observe that the popular shell theories which incorporate assumption [3], as well, may be obtained by simply setting $\gamma_\alpha = \gamma_\beta = 0$ in the previous equations.

5.2.3 Rotations

Recalling that the coefficients D_α and D_β represent the respective displacements of the middle surface along the α and β coordinate lines, the terms ψ_α and ψ_β may be viewed as rotations of the normal $\mathbf{t}'_n$ about the $\mathbf{t}'_\beta$ and $\mathbf{t}'_\alpha$ axes. To show this, we construct figures 5–2(a) and 5–2(b), which are identical to the views shown in figures 5–1(b) and 5–1(c), respectively. On figure 5–2, the projections of the unit vectors to the undeformed surface $\mathbf{t}_\alpha$ and $\mathbf{t}_\beta$ may be considered to remain unit vectors within the scope of this theory; i.e., $\mathbf{t}_\alpha \cdot \mathbf{t}'_\alpha \simeq 1$ from equation (5.15). If we scalar multiply equation (5.17) by $\mathbf{t}_\alpha$, and then $\mathbf{t}_\beta$, we get

$$\psi_\alpha = \mathbf{t}_\alpha \cdot \mathbf{t}'_n \tag{5.21a}$$

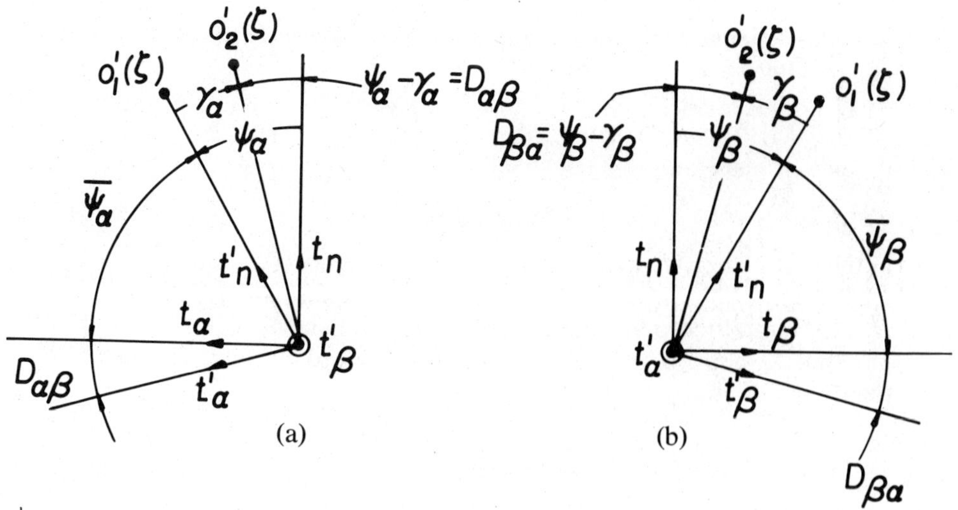

Figure 5–2. Transverse Shearing Strains

and

$$\psi_\beta = \mathbf{t}_\beta \cdot \mathbf{t}_n' \tag{5.21b}$$

Equation (5.21a) is interpreted in figure 5–2(a). Continuing,

$$\mathbf{t}_\alpha \cdot \mathbf{t}_n' = \cos \bar{\psi}_\alpha$$

$$= \sin\!\left(\frac{\pi}{2} - \bar{\psi}_\alpha\right) \tag{5.21c}$$

For small rotations

$$\sin\!\left(\frac{\pi}{2} - \bar{\psi}_\alpha\right) \simeq \frac{\pi}{2} - \bar{\psi}_\alpha = \psi_\alpha \tag{5.21d}$$

or, ψ_α is the rotation of the normal to the deformed middle surface about the $\mathbf{t}_\beta'$ axis. Similarly, from figure 5–2(b), ψ_β is the rotation of the normal about the $\mathbf{t}_\alpha'$ axis. Within the scope of the small deformation theory, ψ_α and ψ_β may be visualized as rotations about $\mathbf{t}_\beta$ and $\mathbf{t}_\alpha$ as well.

We now consider the transverse shearing strains, initially defined in figures 5–1(b) and 5–1(c), superimposed on the rotations ψ_α and ψ_β in figures 5–2(a) and 5–2(b). We see that the terms $\psi_\alpha - \gamma_\alpha$ and $\psi_\beta - \gamma_\beta$, which initially appeared in equation (5.19), may be interpreted as rotations. If we consider a line connecting the point on the *deformed* middle surface and the corresponding middle surface point and a second line along the normal to the *undeformed* middle surface, the angles formed by the intersection of these lines are

$$D_{\alpha\beta} = \psi_\alpha - \gamma_\alpha \tag{5.22a}$$

in the α–n plane and

$$D_{\beta\alpha} = \psi_\beta - \gamma_\beta \tag{5.22b}$$

in the β–n plane. We also may observe from figure 5–2 that $D_{\alpha\beta}$ and $D_{\beta\alpha}$ are, respectively, the angles between $\mathbf{t}_\alpha$ and $\mathbf{t}'_\alpha$, and $\mathbf{t}_\beta$ and $\mathbf{t}'_\beta$. Thus $D_{\alpha\beta}$ and $D_{\beta\alpha}$ are logically termed middle surface rotations and, together with D_α, D_β, and D_n, constitute the set of *generalized displacements* for the shell theory under study.

5.3 Strain

5.3.1 Strain on Middle Surface

Referring to figure 5–3, consider an arc *op* on the s_α coordinate line that has an initial length of

$$ds_\alpha = A\, d\alpha \tag{5.23}$$

After deformation, the points *o* and *p* move to *o'* and *p'*, and the arc length becomes

$$ds'_\alpha = A'\, d\alpha$$

$$= A(1 + \epsilon_\alpha)\, d\alpha \tag{5.24}$$

from equation (5.13). The *linear strain* is defined as the relative increase in length of a curvilinear line element, or the final length minus the initial length, divided by the initial length. So, in the α direction, we have

$$\frac{ds'_\alpha - ds_\alpha}{ds_\alpha} = \frac{A(1 + \epsilon_\alpha)\, d\alpha - A\, d\alpha}{A\, d\alpha}$$

$$= \epsilon_\alpha \tag{5.25}$$

Thus ϵ_α, which appeared in the coefficient of $\mathbf{t}_\alpha$ in equation (5.9) and is defined by equation (5.10a), is the linear strain in the α direction. Correspondingly, in the β direction, the strain is given by ϵ_β, which is defined in equation (5.12a). We have thus found that the linear strains occur quite naturally in the expressions for the derivatives of the radius vectors to the deformed middle surface.

The *in-plane* or *membrane shearing strain* is defined as the change in

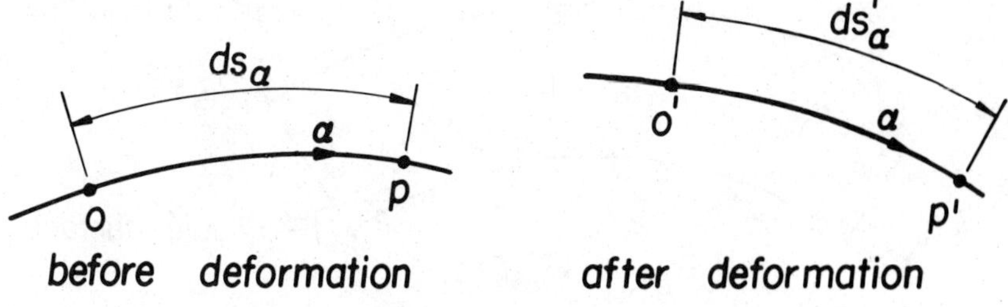

Figure 5–3. Deformation of a Coordinate Line

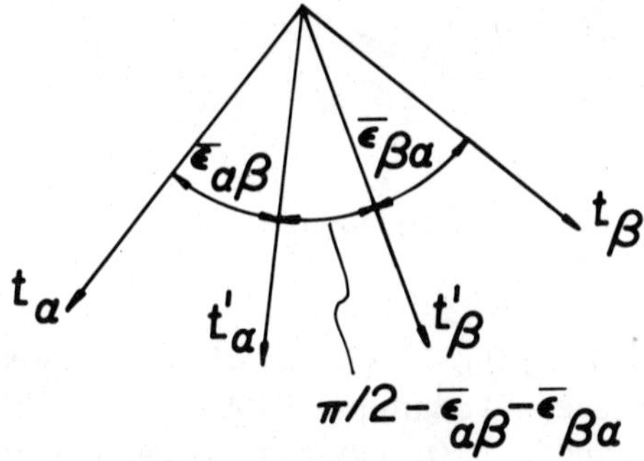

Figure 5–4. In-Plane Shearing Strain

angle between $\mathbf{t}_\alpha$ and $\mathbf{t}_\beta$ during deformation. Referring to figure 5–4, the shear strain is defined as

$$\omega = \bar{\epsilon}_{\alpha\beta} + \bar{\epsilon}_{\beta\alpha} = \sin(\bar{\epsilon}_{\alpha\beta} + \bar{\epsilon}_{\beta\alpha})$$

$$= \cos\left(\frac{\pi}{2} - \bar{\epsilon}_{\alpha\beta} - \bar{\epsilon}_{\beta\alpha}\right)$$

$$= \mathbf{t}'_\alpha \cdot \mathbf{t}'_\beta \tag{5.26}$$

We form the indicated scalar product using equations (5.15) and (5.16), which gives

$$\omega = \mathbf{t}'_\alpha \cdot \mathbf{t}'_\beta = \epsilon_{\beta\alpha} + \epsilon_{\alpha\beta} + \psi_\alpha\psi_\beta \tag{5.27}$$

Since the last term is negligible, we have

$$\omega = \epsilon_{\alpha\beta} + \epsilon_{\beta\alpha} \tag{5.28}$$

as the shearing strain. Again, $\epsilon_{\alpha\beta}$ and $\epsilon_{\beta\alpha}$ are found among the coefficients

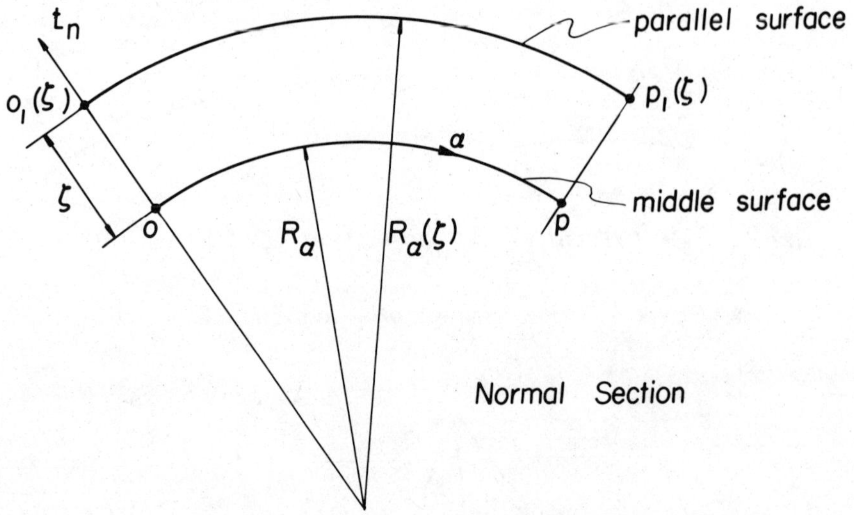

Figure 5–5. Displacement of a Parallel Surface

of the equations defining the derivatives of the radius vectors of the deformed middle surface as equations (5.10b) and (5.12b). The transverse shearing strains γ_α and γ_β have previously been defined.

5.3.2 Strain on Parallel Surface

The differential arc length along coordinate line s_α of the parallel surface is

$$ds_\alpha(\zeta) = A(\zeta)\,d\alpha \tag{5.29}$$

Referring to the normal section, figure 5–5,

$$\frac{ds_\alpha(\zeta)}{ds_\alpha} = \frac{R_\alpha(\zeta)}{R_\alpha} \tag{5.30a}$$

Since

$$R_\alpha(\zeta) = R_\alpha + \zeta \tag{5.30b}$$

and

$$ds_\alpha = A\,d\alpha \tag{5.30c}$$

then

$$ds_\alpha(\zeta) = \frac{R_\alpha + \zeta}{R_\alpha} A \, d\alpha$$

$$= A\left(1 + \frac{\zeta}{R_\alpha}\right) d\alpha \tag{5.31}$$

Comparing equations (5.29) and (5.31), the Lamé parameter for the parallel surface is

$$A(\zeta) = A\left(1 + \frac{\zeta}{R_\alpha}\right) \tag{5.32}$$

For a parallel surface on a normal section along the β coordinate line

$$R_\beta(\zeta) = R_\beta + \zeta \tag{5.33}$$

$$ds_\beta(\zeta) = B\left(1 + \frac{\zeta}{R_\beta}\right) d\beta \tag{5.34}$$

and

$$B(\zeta) = B\left(1 + \frac{\zeta}{R_\beta}\right) \tag{5.35}$$

We recognize that the parallel surface is described by a system of curvilinear coordinates that is identical to those which describe the middle surface. Therefore, the relationships between the strains and displacements defined for the middle surface in equations (5.10a) and (5.10b) and (5.12a) and (5.12b) may be written for the parallel surface by replacing R_α, R_β, A, and B by $R_\alpha(\zeta)$, $R_\beta(\zeta)$, $A(\zeta)$, and $B(\zeta)$, respectively, and by replacing D_α, D_β, and D_n by $D_\alpha(\zeta)$, $D_\beta(\zeta)$, and $D_n(\zeta)$, as defined in equation (5.20).

For the extensional strains, we get

$$\epsilon_\alpha(\zeta) = \epsilon_\alpha + \zeta\kappa_\alpha \tag{5.36}$$

$$\epsilon_\beta(\zeta) = \epsilon_\beta + \zeta\kappa_\beta \tag{5.37}$$

after dropping terms of order (h/R) compared to 1; the latter order of magnitude comparison is widely used in shell theory and is concisely stated as $O(h/R){:}1$. The coefficients of ζ are

$$\kappa_\alpha = \frac{1}{A} D_{\alpha\beta,\alpha} + \frac{1}{AB} A_{,\beta} D_{\beta\alpha} \tag{5.38}$$

and

$$\kappa_\beta = \frac{1}{B} D_{\beta\alpha,\beta} + \frac{1}{AB} B_{,\alpha} D_{\alpha\beta} \tag{5.39}$$

and $D_{\alpha\beta}$ and $D_{\beta\alpha}$ are defined in equation (5.22). The terms κ_α and κ_β correspond to the familiar curvature terms in linear beam theory. For plates, they are indeed curvatures but for shells, they are more properly known as *changes in curvature*, since shells by definition are initially curved.

For the in-plane shear strain $\omega(\zeta)$, we first combine $\epsilon_{\alpha\beta}$ and $\epsilon_{\beta\alpha}$ as indicated in equation (5.28). Adding equations (5.10b) and (5.12b), we have

$$\omega = \frac{1}{A} D_{\beta,\alpha} - \frac{1}{AB} A_{,\beta} D_\alpha + \frac{1}{B} D_{\alpha,\beta} - \frac{1}{AB} B_{,\alpha} D_\beta \qquad (5.40)$$

Now, referring to the parallel surface, we perform the indicated substitutions for the radii of curvature, Lamé parameters, and displacements and drop terms of $O(h/R):1$. The resulting expression for the shear strain on the parallel surface is

$$\omega(\zeta) = \epsilon_{\alpha\beta} + \zeta\tau_\alpha + \epsilon_{\beta\alpha} + \zeta\tau_\beta \qquad (5.41)$$

where

$$\tau_\alpha = \frac{1}{A} D_{\beta\alpha,\alpha} - \frac{1}{AB} A_{,\beta}D_{\alpha\beta} \qquad (5.42)$$

$$\tau_\beta = \frac{1}{B} D_{\alpha\beta,\beta} - \frac{1}{AB} B_{,\alpha} D_{\beta\alpha} \qquad (5.43)$$

It is convenient to introduce

$$\tau = \frac{1}{2}(\tau_\alpha + \tau_\beta) \qquad (5.44)$$

Then, equation (5.41) is written as

$$\omega(\zeta) = \omega + 2\zeta\tau \qquad (5.45)$$

where τ is called the *twist* or the *torsion* of the middle surface.

We have thus completely defined each strain on the parallel surface, except for the transverse shear strain, as a *linear combination of two terms*: a) the *corresponding strain* on the *middle surface*, which is expressed in terms of the *middle surface displacements* by equations (5.10) and (5.12); and b) a change in *curvature* or a *twist*, which are in turn given in terms of the *middle surface displacements* by equations (5.38), (5.39), and (5.42)–(5.44), respectively. The transverse shearing strains are also defined in terms of middle surface displacements by equations (5.22), (5.10c), and (5.12c), but remain constant over the shell depth. These relationships between strains, changes in curvature, and twist on the one hand, and the displacements of the middle surface on the other, are of primary importance in the theory of shells and are called generalized *strain-*

displacement equations or *compatibility* equations. We have therefore established the second basic component of the elasticity problem as outlined in section 1.1, the conditions of *compatibility,* suitably specialized for the shell theory under consideration.

$$\epsilon_\alpha = \frac{1}{A} D_{\alpha,\alpha} + \frac{1}{AB} A_{,\beta} D_\beta + \frac{1}{R_\alpha} D_n \tag{5.46a}$$

$$\epsilon_\beta = \frac{1}{B} D_{\beta,\beta} + \frac{1}{AB} B_{,\alpha} D_\alpha + \frac{1}{R_\beta} D_n \tag{5.46b}$$

$$\omega = \frac{1}{A} D_{\beta,\alpha} - \frac{1}{AB} A_{,\beta} D_\alpha + \frac{1}{B} D_{\alpha,\beta} - \frac{1}{AB} B_{,\alpha} D_\beta \tag{5.46c}$$

$$\kappa_\alpha = \frac{1}{A} D_{\alpha\beta,\alpha} + \frac{1}{AB} A_{,\beta} D_{\beta\alpha} \tag{5.46d}$$

$$\kappa_\beta = \frac{1}{B} D_{\beta\alpha,\beta} + \frac{1}{AB} B_{,\alpha} D_{\alpha\beta} \tag{5.46e}$$

$$\tau = \frac{1}{2}\left[\frac{1}{A} D_{\beta\alpha,\alpha} + \frac{1}{B} D_{\alpha\beta,\beta} - \frac{1}{AB} (A_{,\beta} D_{\alpha\beta} + B_{,\alpha} D_{\beta\alpha})\right] \tag{5.46f}$$

$$\gamma_\alpha = -\left(\frac{1}{A} D_{n,\alpha} - \frac{1}{R_\alpha} D_\alpha\right) - D_{\alpha\beta} \tag{5.46g}$$

$$\gamma_\beta = -\left(\frac{1}{B} D_{n,\beta} - \frac{1}{R_\beta} D_\beta\right) - D_{\beta\alpha} \tag{5.46h}$$

If we choose to neglect transverse shearing strains, we set $\gamma_\alpha = \gamma_\beta = 0$ in equation (5.46), so that $D_{\alpha\beta} = \psi_\alpha$ and $D_{\beta\alpha} = \psi_\beta$. Then, replacing $D_{\alpha\beta}$ and $D_{\beta\alpha}$ in equations (5.46d)–(5.46f) by ψ_α and ψ_β, as given by equations (5.10c) and (5.12c), we get the curvatures directly in terms of the middle surface displacements:

$$\kappa_\alpha = -\frac{1}{A}\left(\frac{1}{A} D_{n,\alpha} - \frac{1}{R_\alpha} D_\alpha\right)_{,\alpha} - \frac{1}{AB} A_{,\beta}\left(\frac{1}{B} D_{n,\beta} - \frac{1}{R_\beta} D_\beta\right) \tag{5.47a}$$

$$\kappa_\beta = -\frac{1}{B}\left(\frac{1}{B} D_{n,\beta} - \frac{1}{R_\beta} D_\beta\right)_{,\beta} - \frac{1}{AB} B_{,\alpha}\left(\frac{1}{A} D_{n,\alpha} - \frac{1}{R_\alpha} D_\alpha\right) \tag{5.47b}$$

$$\tau = \frac{1}{2}\left\{-\frac{1}{A}\left(\frac{1}{B} D_{n,\beta} - \frac{1}{R_\beta} D_\beta\right)_{,\alpha} - \frac{1}{B}\left(\frac{1}{A} D_{n,\alpha} - \frac{1}{R_\alpha} D_\alpha\right)_{,\beta}\right.$$

$$\left. + \frac{1}{AB}\left[A_{,\beta}\left(\frac{1}{A} D_{n,\alpha} - \frac{1}{R_\alpha} D_\alpha\right) + B_{,\alpha}\left(\frac{1}{B} D_{n,\beta} - \frac{1}{R_\beta} D_\beta\right)\right]\right\} \tag{5.47c}$$

Note with respect to equations (5.46) and (5.47) that no force-deformation relationships are involved, and hence linear material behavior is not a prerequisite for the application of these equations. Also, as previously demonstrated for the equilibrium equations, a considerable simplification of these equations often occurs for specific geometries through the elimination of some of the coupling terms.

5.4 Strain-Displacement Relations for Shells of Revolution

5.4.1 Specialization of Equations

The specialization of the curvilinear coordinate system to the shell of revolution geometry, shown in figure 2–11, is carried out in section 3.3.1. The corresponding strain-displacement equations for $\alpha = \phi$ and $\beta = \theta$ are

$$\epsilon_\phi = \frac{1}{R_\phi} (D_{\phi,\phi} + D_n) \tag{5.48a}$$

$$\epsilon_\theta = \frac{1}{R_0} (D_{\theta,\theta} + \cos \phi \, D_\phi + \sin \phi \, D_n) \tag{5.48b}$$

$$\omega = \frac{1}{R_\phi} D_{\theta,\phi} + \frac{1}{R_0} (D_{\phi,\theta} - \cos \phi \, D_\theta) \tag{5.48c}$$

$$\kappa_\phi = \frac{1}{R_\phi} D_{\phi\theta,\phi} \tag{5.48d}$$

$$\kappa_\theta = \frac{1}{R_0} D_{\theta\phi,\theta} + \frac{\cos \phi}{R_0} D_{\phi\theta} \tag{5.48e}$$

$$\tau = \frac{1}{2} \left[\frac{1}{R_\phi} D_{\theta\phi,\phi} + \frac{1}{R_0} \left(D_{\phi\theta,\theta} - \cos \phi \, D_{\theta\phi} \right) \right] \tag{5.48f}$$

$$\gamma_\phi = -\frac{1}{R_\phi} (D_{n,\phi} - D_\phi) - D_{\phi\theta} \tag{5.48g}$$

$$\gamma_\theta = -\frac{1}{R_0} (D_{n,\theta} - \sin \phi \, D_\theta) - D_{\theta\phi} \tag{5.48h}$$

If the transverse shearing strains are neglected, we have the alternate form

$$\kappa_\phi = -\frac{1}{R_\phi}\left[\frac{1}{R_\phi}(D_{n,\phi} - D_\phi)\right] \tag{5.49a}$$

$$\kappa_\theta = -\frac{1}{R_0^2}(D_{n,\theta} - \sin\phi\, D_\theta)_{,\theta} - \frac{\cos\phi}{R_\phi R_0}(D_{n,\phi} - D_\phi) \tag{5.49b}$$

$$\tau = \frac{1}{2}\left\{-\frac{1}{R_\phi}\left[\frac{1}{R_0}(D_{n,\theta} - \sin\phi\, D_\theta)\right]_{,\phi}\right.$$

$$\left. -\frac{1}{R_\phi R_0}(D_{n,\phi} - D_\phi)_{,\theta} + \frac{\cos\phi}{R_0^2}(D_{n,\theta} - \sin\phi\, D_\theta)\right\} \tag{5.49c}$$

We may also write the strains and displacements for the shell of revolution in the Fourier series form that proved expedient in the treatment of nonsymmetric loading on shells of revolution. Following the procedure of section 4.3.5.2, we have

$$\begin{Bmatrix} \epsilon_\phi \\ \epsilon_\theta \\ \omega \\ \kappa_\phi \\ \kappa_\theta \\ \tau \\ \gamma_\phi \\ \gamma_\theta \end{Bmatrix} = \sum_{j=0}^{\infty} \begin{Bmatrix} \epsilon_\phi^j \cos j\theta \\ \epsilon_\theta^j \cos j\theta \\ \omega^j \sin j\theta \\ \kappa_\phi^j \cos j\theta \\ \kappa_\theta^j \cos j\theta \\ \tau^j \sin j\theta \\ \gamma_\phi^j \cos j\theta \\ \gamma_\theta^j \sin j\theta \end{Bmatrix} \tag{5.50}$$

and

$$\begin{Bmatrix} D_\phi \\ D_\theta \\ D_n \\ D_{\phi\theta} \\ D_{\theta\phi} \end{Bmatrix} = \sum_{j=0}^{\infty} \begin{Bmatrix} D_\phi^j \cos j\theta \\ D_\theta^j \sin j\theta \\ D_n^j \cos j\theta \\ D_{\phi\theta}^j \cos j\theta \\ D_{\theta\phi}^j \sin j\theta \end{Bmatrix} \tag{5.51}$$

Equations (5.50) and (5.51) are substituted into equations (5.48) and (5.49) to derive the compatibility equations for general harmonic j:

$$\epsilon^j_\phi = \frac{1}{R_\phi} (D^j_{\phi,\phi} + D^j_n) \tag{5.52a}$$

$$\epsilon^j_\theta = \frac{1}{R_0} (jD^j_\theta + \cos \phi \, D^j_\phi + \sin \phi \, D^j_n) \tag{5.52b}$$

$$\omega^j = \frac{1}{R_\phi} D^j_{\theta,\phi} - \frac{1}{R_0} (jD^j_\phi + \cos \phi \, D^j_\theta) \tag{5.52c}$$

$$\kappa^j_\phi = \frac{1}{R_\phi} D^j_{\phi\theta,\phi} \tag{5.52d}$$

$$\kappa^j_\theta = \frac{j}{R_0} D^j_{\theta\phi} + \frac{\cos \phi}{R_0} D^j_{\phi\theta} \tag{5.52e}$$

$$\tau^j = \frac{1}{2}\left[\frac{1}{R_\phi} D^j_{\theta\phi,\phi} - \frac{1}{R_0} (jD^j_{\phi\theta} + \cos \phi \, D^j_{\theta\phi})\right] \tag{5.52f}$$

$$\gamma^j_\phi = -\frac{1}{R_\phi} (D^j_{n,\phi} - D^j_\phi) - D^j_{\phi\theta} \tag{5.52g}$$

$$\gamma^j_\theta = \frac{1}{R_0} (jD^j_n + \sin \phi \, D^j_\theta) - D^j_{\theta\phi} \tag{5.52h}$$

or with transverse shear strains neglected

$$\kappa^j_\phi = -\frac{1}{R_\phi}\left[\frac{1}{R_\phi} (D^j_{n,\phi} - D^j_\phi)\right]_{,\phi} \tag{5.53a}$$

$$\kappa^j_\theta = \frac{j}{R_0^2} (jD^j_n + \sin \phi \, D^j_\theta) - \frac{\cos \phi}{R_\phi R_0} (D^j_{n,\phi} - D^j_\phi) \tag{5.53b}$$

$$\tau^j = \frac{1}{2}\left[\frac{1}{R_\phi}\left[\frac{1}{R_0} (jD^j_n + \sin \phi \, D^j_\theta)\right]_{,\phi} + \frac{j}{R_\phi R_0} (D^j_{n,\phi} - D^j_\phi)\right.$$

$$\left. - \frac{\cos \phi}{R_0^2} (jD^j_n - \sin \phi \, D^j_\theta)\right] \tag{5.53c}$$

5.4.2 *Physical Interpretation*

The first two of equations (5.48a–h), the extensional strains, may be physically interpreted by considering figure 5–6, which shows a segment of the meridian of a shell of revolution. The first term of equation (5.48a) follows from the basic definition of strain as the change in length of the differential segment op, $D_{\phi,\phi} \, d\phi$, divided by the initial length $R_\phi d\phi$. The second term

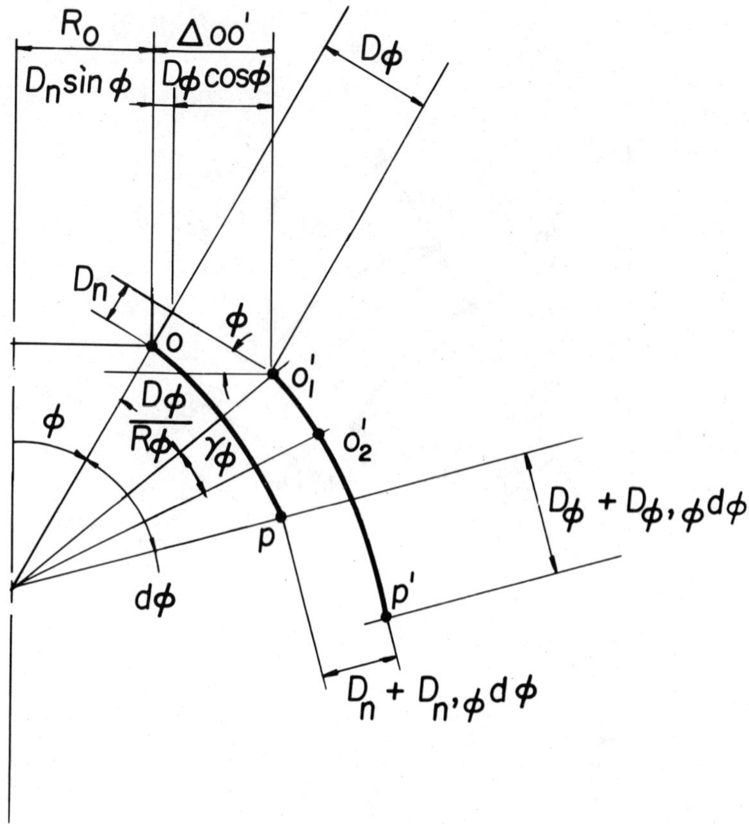

Figure 5–6. Displacement of the Meridian of a Shell of Revolution

of the equation represents the strain due to the normal displacement D_n; this is easily visualized as a change in the arc length op due to a change in R_ϕ, $(R_\phi + D_n)\,d\phi - R_\phi\,d\phi$, again divided by the initial length, $R_\phi\,d\phi$. In equation (5.48b), the first term is again straightforward. The remaining two terms occur because of the change in the radius of the parallel circle due to D_ϕ and D_n, which is shown as $\Delta_{oo'}$ on figure 5–6. The horizontal projection of D_ϕ is $D_\phi \cos \phi$, and that of D_n is $D_n \sin \phi$. Together, we have the change in length, $(D_\phi \cos \phi + D_n \sin \phi)\,d\theta$ divided by the original length $R_0\,d\theta$.

Next, consider equations (5.48d) and (5.48e), the changes in curvature. These are defined as the change in the surface rotations over the segments $d\phi$ and $d\theta$, respectively, divided by the length of the segment.

Equation (5.48d) is the change in the rotation in the ϕ direction, $D_{\phi\theta,\phi}\,d\phi$ divided by the arc length, $R_\phi\,d\phi$, or $(1/R_\phi)D_{\phi\theta,\phi}$. We may also

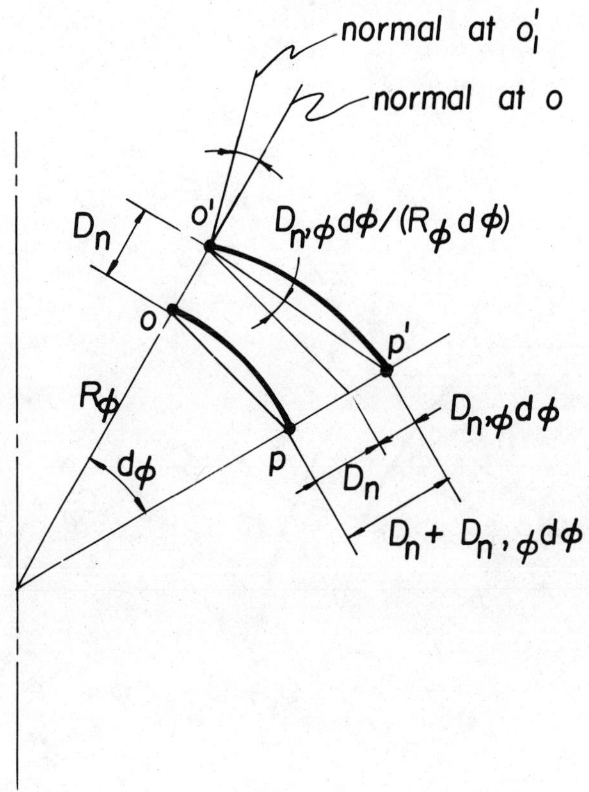

Figure 5–7. Normal Displacement of the Meridian

interpret this result directly in terms of the displacements by using figure 5–6. The rotation at the initial point in the segment, point o, as it moves to o_1' is $(1/R_\phi)D_\phi$. From o_1' to o_2', we have the transverse shear strain contribution of γ_ϕ. There is also a rotation of o_1' due to D_n. This is illustrated in figure 5–7, where the same differential segment is shown with only an exaggerated D_n displacement. The rotation at o_1' due to D_n is seen to be $(1/R_\phi)D_{n,\phi}$—in a sense opposite that of the rotation due to D_ϕ. Hence the total change in rotation over the segment

$$\left(\frac{1}{R}\,D_\phi - \gamma_\phi - \frac{1}{R_\phi}\,D_{n,\phi}\right)_{,\phi} d\phi$$

is divided by the segment length $R_\phi\,d\phi$ to derive the meridional change in curvature. This expression is verified by substituting equation (5.48g) into (5.48d).

Equation (5.48e) has two components. The first term is the change in the rotation $D_{\theta\phi}$, $D_{\theta\phi,\theta}\,d\theta$, divided by the arc length, $R_0\,d\theta$, or $(1/R_0)D_{\theta\phi,\theta}$. Again, if a segment of the shell along a parallel circle is examined, this term may be interpreted directly in terms of the displacements. Note that there is a second term present in the expression for κ_θ. This term is $(\cos\phi/R_0)D_{\phi\theta}$, and indicates that there is a contribution from $D_{\phi\theta}$ which acts about $\mathbf{t}_\phi$. The resolution of the rotation at o may be followed by referring to figure 3–6 and replacing the term $N_\theta R_\phi d\phi$ by $D_{\phi\theta}$ on the force element and on view $HV1$. Then, moving to view $MV1$, we see the resolution of this vector into the meridional direction and the normal direction. The component corresponding to rotation about the normal is not admissible in the present theory. The meridional component, acting in the negative ϕ direction, is the rotation $D_{\phi\theta}\cos\phi$ and represents a change in slope along the parallel circle over the segment. The sign of this component is positive, since the positive sense of the rotation ψ_θ is in the negative ϕ direction, as shown in figure 5–2(b). After division by the arc length, $R_0\,d\theta$, we have the second term of equation (5.48e).

The preceding use of an argument from the equilibrium equations to explain a compatibility relationship is not an isolated coincidence. This is a representative part of a complete static-geometric analogy that exists between the equations of equilibrium and the equations of compatibility. This analogy is sometimes useful in constructing dual solutions. A complete development of the static-geometric analogy may be found in Gol'denveizer.[3]

We now turn our attention to the shearing strain ω, which is given by equation (5.48c) and is shown on figure 5–8 as the sum of ω_ϕ and ω_θ. The term ω_ϕ is given by $D_{\phi,\theta}\,d\theta/(R_0\,d\theta) = (1/R_0)\,D_{\phi,\theta}$. To evaluate ω_θ, note that point p' has moved in the θ direction a distance $D_\theta + D_{\theta,\phi}\,d\phi$. When the θ displacement of point o, D_θ, is projected onto the corresponding displacement of p, it becomes

$$D_\theta\left(\frac{R_0 + R_{0,\phi}\,d\phi}{R_0}\right) = D_\theta\left(1 + \frac{R_\phi\cos\phi}{R_0}\,d\phi\right)$$

due to the change in the horizontal radius R_0. The difference,

$$\frac{[D_{\theta,\phi} - D_\theta\,(R_\phi/R_0)\cos\phi]\,d\phi}{R_\phi\,d\phi} = \omega_\theta$$

and the sum of ω_ϕ and ω_θ is ω, as given by equation (5.48c).

Last, we examine the twist τ, which is given by equation (5.48f). Since τ is obtained from τ_ϕ and τ_θ as defined in equation (5.44), we concentrate on the basic terms τ_ϕ and τ_θ. Referring to figures 5–8 and 5–9(a), the twist in the ϕ direction is given by the change in the rotation in the θ direction at

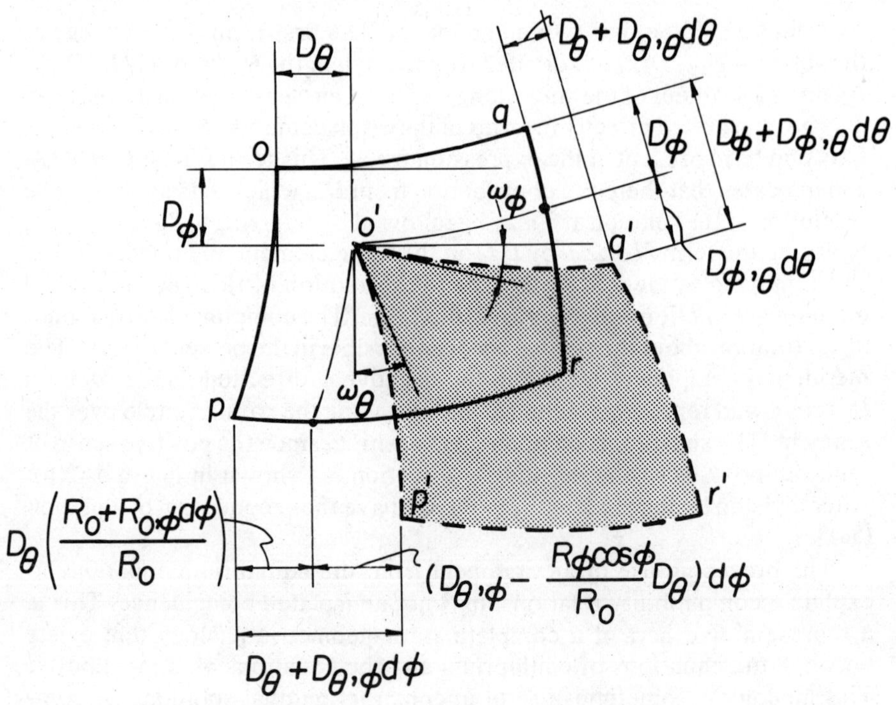

Figure 5–8. Shearing Displacements

o' as ϕ varies, $D_{\theta\phi,\phi}\,d\phi$, divided by the arc length, $R_\phi\,d\phi$, which is the first term in equation (5.48f). The twist in the θ direction consists partially of the corresponding change in rotation in the ϕ direction at o' as θ varies, $D_{\phi\theta,\theta}\,d\theta$, divided by the arc length $R_0\,d\theta$, shown in figure 5–9(b). Another contribution to the twist in the θ direction is due to the change in the horizontal radius R_0 as ϕ changes. From equation (5.48h), $D_{\theta\phi}$ is approximately inversely proportional to R_0 so that $D_{\theta\phi}$ is reduced to

$$\frac{R_0}{R_0 + R_{0,\phi}\,d\phi}\,D_{\theta\phi}$$

over the segment, or we have a change of

$$-\frac{R_{0,\phi}\,d\phi}{R_0}\,D_{\theta\phi} = -\frac{R_\phi}{R_0}\cos\phi\,D_{\theta\phi}\,d\phi$$

After division by the arc length over which the change in horizontal radius occurs, $R_\phi\,d\phi$, we have the third term of equation (5.48f).

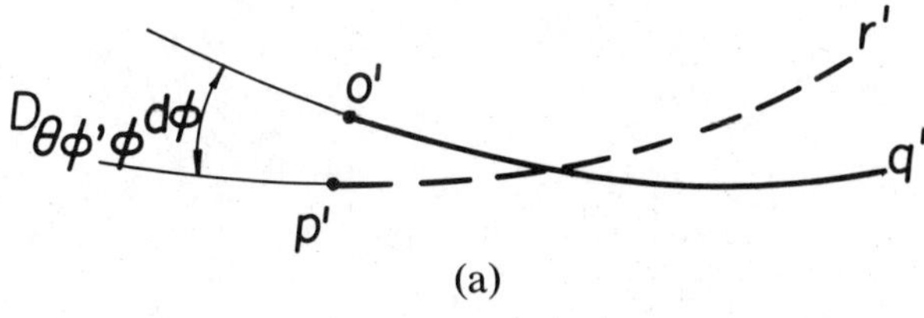

(a)

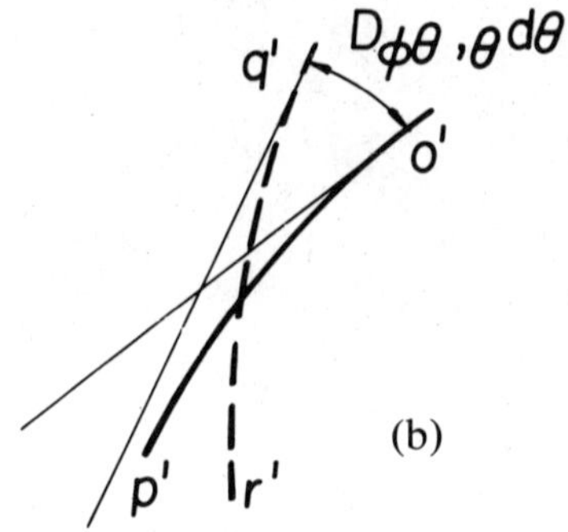

(b)

Figure 5–9. Twisting Displacements

We have attempted to provide an explanation of the individual terms of the compatibility equations by using physical arguments for the specialized shell of revolution geometry. In a general shell, the physical illustrations would become untenable; yet, the initial derivation by vector calculus was straightforward. In the past, considerable effort has been expended in deriving compatibility relationships for specialized geometries using strictly physical reasoning. Here, in both the equilibrium and compatibility treatments, we have sought to vary the approach somewhat by initially deemphasizing the physical in favor of the mathematical approach, but then closely supporting the results with specific physical illustrations.

5.5 Strain-Displacement Relations for Plates

As initially set forth in section 3.4, the shell equations are transformed into a sufficiently general form for medium-thin plates by letting the radii of curvature approach infinity. From equation (5.46),

$$\epsilon_\alpha = \frac{1}{A} D_{\alpha,\alpha} + \frac{1}{AB} A,_\beta D_\beta \tag{5.54a}$$

$$\epsilon_\beta = \frac{1}{B} D_{\beta,\beta} + \frac{1}{AB} B,_\alpha D_\alpha \tag{5.54b}$$

$$\omega = \frac{1}{A} D_{\beta,\alpha} - \frac{1}{AB} A,_\beta D_\alpha + \frac{1}{B} D_{\alpha,\beta} - \frac{1}{AB} B,_\alpha D_\beta \tag{5.54c}$$

$$\kappa_\alpha = \frac{1}{A} D_{\alpha\beta,\alpha} + \frac{1}{AB} A,_\beta D_{\beta\alpha} \tag{5.54d}$$

$$\kappa_\beta = \frac{1}{B} D_{\beta\alpha,\beta} + \frac{1}{AB} B,_\alpha D_{\alpha\beta} \tag{5.54e}$$

$$\tau = \frac{1}{2}\left[\frac{1}{A} D_{\beta\alpha,\alpha} + \frac{1}{B} D_{\alpha\beta,\beta} - \frac{1}{AB}(A,_\beta D_{\alpha\beta} + B,_\alpha D_{\beta\alpha})\right] \tag{5.54f}$$

$$\gamma_\alpha = -\left(\frac{1}{A} D_{n,\alpha} + D_{\alpha\beta}\right) \tag{5.54g}$$

$$\gamma_\beta = -\left(\frac{1}{B} D_{n,\beta} + D_{\beta\alpha}\right) \tag{5.54h}$$

When the transverse shearing strains are neglected, we have

$$\kappa_\alpha = -\left[\frac{1}{A}\left(\frac{1}{A} D_{n,\alpha}\right)_{,\alpha} + \frac{A,_\beta}{AB^2} D_{n,\beta}\right] \tag{5.55a}$$

$$\kappa_\beta = -\left[\frac{1}{B}\left(\frac{1}{B} D_{n,\beta}\right)_{,\beta} + \frac{B,_\alpha}{A^2B} D_{n,\alpha}\right] \tag{5.55b}$$

$$\tau = -\frac{1}{2}\left\{\frac{1}{A}\left(\frac{1}{B} D_{n,\beta}\right)_{,\alpha} + \frac{1}{B}\left(\frac{1}{A} D_{n,\alpha}\right)_{,\beta}\right.$$

$$\left. - \frac{1}{AB}\left[\frac{A,_\beta}{A} D_{n,\alpha} + \frac{B,_\alpha}{B} D_{n,\beta}\right]\right\} \tag{5.55c}$$

It has been noted several times that the adoption of assumption [3] and the subsequent suppression of transverse shearing strains generally leads to significant mathematical simplifications. An example of this may be seen for the plate case by comparing equations (5.54d)–(5.54f) with equations (5.55a)–(5.55c). In the latter equations, the curvatures and hence the entire flexural behavior of the shell are dependent on only the normal displacement D_n and are essentially uncoupled from the extensional behavior. The form of the curvature-displacement relations is identical to that encountered in elementary beam theory. On the other hand the retention of transverse shearing strains leads to a considerably more complicated mathematical formulation.[4]

Notes

1. V. V. Novozhilov, *Thin Shell Theory* [translated from 2nd Russian ed. by P. G. Lowe (Groningen, The Netherlands: Noordhoff, 1964), pp. 14–27].

2. H. Kraus, *Thin Elastic Shells* (New York: Wiley, 1967), chap. 1.

3. A. L. Gol'denveizer, *Theory of Elastic Thin Shells* [translated from Russian ed. (New York Pergamon Press, 1961), pp. 92–96].

4. S. Timoshenko and S. Woinowsky-Krieger, *Theory of Plates and Shells*, 2nd ed. (New York: McGraw-Hill, 1959), pp. 165–173.

6   Constitutive Laws, Boundary Conditions and Displacements

6.1 Constitutive Laws

The constitutive law or stress-strain relationship is the third basic component of the elasticity problem. This subject is quite broad, but the detailed examination of various alternatives that are dependent on the characteristics of particular engineering materials is not properly within the scope of this book. Rather, it remains within the purview of the theory of elasticity, since we may accommodate a variety of material laws within our formulation of the shell or plate problem. Initially, we use the basic Hooke's law for isotropic materials, and then we illustrate how some extended material laws can be accommodated.

6.1.1 Isotropic Material Law

6.1.1.1 Strain-Stress Relationship Including Temperature Effects. The strains, as defined in section 5.1, are related to the stresses introduced in section 3.1 by

$$\epsilon_\alpha(\alpha,\beta,\zeta) = \frac{1}{E}\,(\sigma_{\alpha\alpha} - \mu\sigma_{\beta\beta}) + \bar{\alpha}T(\alpha,\beta,\zeta) \tag{6.1a}$$

$$\epsilon_\beta(\alpha,\beta,\zeta) = \frac{1}{E}\,(\sigma_{\beta\beta} - \mu\sigma_{\alpha\alpha}) + \bar{\alpha}T(\alpha,\beta,\zeta) \tag{6.1b}$$

$$\omega(\alpha,\beta,\zeta) = \frac{1}{G}\sigma_{\alpha\beta} = \frac{1}{G}\sigma_{\beta\alpha} \tag{6.1c}$$

$$\gamma_\alpha(\alpha,\beta) = \frac{1}{G_n}\sigma_{\alpha n} \tag{6.1d}$$

$$\gamma_\beta(\alpha,\beta) = \frac{1}{G_n}\sigma_{\beta n} \tag{6.1e}$$

In equation (6.1), E = Young's modulus, G and G_n = in-plane and normal shearing moduli, and μ = Poisson's ratio. Also, $\bar{\alpha}$ represents the coefficient of thermal expansion, which will be taken as piecewise constant, and $T(\alpha,\beta,\zeta)$ = the temperature, measured from a specified reference value. Thermal effects are included in the basic derivation, since they are a very important design condition in some applications.[1]

We find it convenient to invert equation (6.1) and substitute the expressions $\epsilon_\alpha(\alpha,\beta,\zeta)$, $\epsilon_\beta(\alpha,\beta,\zeta)$, and $\omega(\alpha,\beta,\zeta)$, as given by equations (5.36), (5.37), and (5.45). After performing the indicated substitutions we have, assuming for the present that $G_n = G = E/[2(1 + \mu)]$, matrix 6–1 can be developed. Matrix 6–1 may be written as

$$\{\boldsymbol{\sigma}(\alpha,\beta,\zeta)\} = [\mathbf{C}]\{\boldsymbol{\epsilon}(\alpha,\beta,\zeta)\} - \{\boldsymbol{\sigma}_T(\alpha,\beta,\zeta)\} \tag{6.2}$$

We use the matrix representation to illustrate that any other consistent stress-strain relationship may be accommodated by altering the elements of $[\mathbf{C}]$ and $\{\boldsymbol{\sigma}_T\}$, which are essentially independent from the remainder of the development.

6.1.1.2 Stress Resultant-Strain and Stress Couple-Curvature Relations. Returning to the surface structure problem, recognize that the constitutive equations are required to relate the *stress resultants* and *stress couples*, instead of merely stresses, to the corresponding strains and curvatures. These relationships are obtained in a straightforward fashion. We substitute the stress-strain law, matrix 6–1, into equations (3.6)–(3.9). For example, consider the expression for N_α in equation (3.6):

$$N_\alpha = \int_{-h/2}^{h/2} \sigma_{\alpha\alpha}\left(1 + \frac{\zeta}{R_\beta}\right) d\zeta$$

$$= E\int_{-h/2}^{h/2} \left(\frac{\epsilon_\alpha}{1 - \mu^2} + \frac{\zeta\kappa_\alpha}{1 - \mu^2} + \frac{\mu\epsilon_\beta}{1 - \mu^2} + \frac{\mu\zeta\kappa_\beta}{1 - \mu^2} - \frac{\bar{\alpha}T}{1 - \mu}\right)$$

$$\cdot \left(1 + \frac{\zeta}{R_\beta}\right) d\zeta \tag{6.3}$$

Since $\zeta/R_\beta \ll 1$, it may be neglected in this integration, and we may split the integral into

$$N_\alpha = \frac{E}{1 - \mu^2} \int_{-h/2}^{h/2} (\epsilon_\alpha + \zeta\kappa_\alpha + \mu\epsilon_\beta + \mu\zeta\kappa_\beta)\, d\zeta - \frac{E\bar{\alpha}}{1 - \mu} \int_{-h/2}^{h/2} T(\zeta)\, d\zeta$$

$$\tag{6.4}$$

In this development the cross section is assumed to be symmetric about the middle plane or surface, so that

$$\int_{-h/2}^{h/2} \zeta\, d\zeta = 0$$

and the second and fourth terms of the first integral vanish, leaving

Matrix 6–1

$$
\begin{Bmatrix}
\sigma_{\alpha\alpha}(\alpha,\beta,\zeta) \\
\sigma_{\beta\beta}(\alpha,\beta,\zeta) \\
\sigma_{\alpha\beta}(\alpha,\beta,\zeta) \\
\sigma_{\alpha n}(\alpha,\beta) \\
\sigma_{\beta n}(\alpha,\beta)
\end{Bmatrix}
= E
\begin{bmatrix}
\dfrac{1}{1-\mu^2} & \dfrac{\mu}{1-\mu^2} & 0 & 0 & 0 \\[2ex]
\dfrac{\mu}{1-\mu^2} & \dfrac{1}{1-\mu^2} & 0 & 0 & 0 \\[2ex]
0 & 0 & \dfrac{1}{2(1+\mu)} & 0 & 0 \\[2ex]
0 & 0 & 0 & \dfrac{1}{2(1+\mu)} & 0 \\[2ex]
0 & 0 & 0 & 0 & \dfrac{1}{2(1+\mu)}
\end{bmatrix}
\begin{Bmatrix}
\epsilon_\alpha + \zeta\kappa_\alpha \\
\epsilon_\beta + \zeta\kappa_\beta \\
\omega + 2\zeta\tau \\
\gamma_\alpha \\
\gamma_\beta
\end{Bmatrix}
- E\bar{\alpha}T(\alpha,\beta,\zeta)
\begin{Bmatrix}
\dfrac{1}{1-\mu} \\[2ex]
\dfrac{1}{1-\mu} \\[2ex]
0 \\
0 \\
0
\end{Bmatrix}
$$

220

$$N_\alpha = \frac{Eh}{1 - \mu^2}(\epsilon_\alpha + \mu\epsilon_\beta) - N_{\alpha T} \tag{6.5}$$

where

$$N_{\alpha T} = \frac{E\bar{\alpha}}{1 - \mu}\int_{-h/2}^{h/2} T(\zeta)\,d\zeta$$

which may be called the *thermal load*, is left in general form to accommodate the possibility that $T = T(\zeta)$. The remainder of the stress resultant–strain equations are obtained in the same fashion.

Now, consider the expression for M_α in equation (3.8)

$$M_\alpha = \int_{-h/2}^{h/2} \sigma_{\alpha\alpha}\zeta\left(1 + \frac{\zeta}{R_\beta}\right)d\zeta \tag{6.6}$$

which, referring to equations (6.3) and (6.4), becomes

$$M_\alpha = \frac{E}{1 - \mu^2}\int_{-h/2}^{h/2}(\zeta\epsilon_\alpha + \zeta^2\kappa_\alpha + \mu\zeta\epsilon_\beta + \mu\zeta^2\kappa_\beta)\,d\zeta - \frac{E\bar{\alpha}}{1 - \mu}\int_{-h/2}^{h/2} T(\zeta)\zeta\,d\zeta$$

$$\tag{6.7}$$

For a symmetric cross section, the first and third terms of the first integral vanish, leaving

$$M_\alpha = \frac{Eh^3}{12(1 - \mu^2)}(\kappa_\alpha + \mu\kappa_\beta) - M_{\alpha T} \tag{6.8}$$

where

$$M_{\alpha T} = \frac{E\alpha}{1 - \mu}\int_{-h/2}^{h/2} T(\zeta)\zeta\,d\zeta$$

is known as the *thermal moment*. If T is constant through the thickness, there are no thermal moments. The remainder of the stress couple-curvature equations are obtained in an identical fashion. Collecting the equations, we can develop matrix 6–2. In matrix 6–2,

$$N_{\alpha T} = N_{\beta T} = N_T = \frac{E\bar{\alpha}}{1 - \mu}\int_{-h/2}^{h/2} T(\zeta)\,d\zeta \tag{6.9a}$$

$$M_{\alpha T} = M_{\beta T} = M_T = \frac{E\bar{\alpha}}{1 - \mu}\int_{-h/2}^{h/2} T(\zeta)\zeta\,d\zeta \tag{6.9b}$$

and $\lambda = $ a shear shape factor, such that $G_n = \lambda G$. The factor λ is relatively complex to define explicitly, but is commonly taken as 5/6 for isotropic shells.

Matrix 6–2 is in the form

Matrix 6–2

$$
\begin{Bmatrix} N_\alpha \\ N_\beta \\ N_{\alpha\beta} \\ M_\alpha \\ M_\beta \\ M_{\alpha\beta} \\ Q_\alpha \\ Q_\beta \end{Bmatrix}
= E
\begin{bmatrix}
\dfrac{h}{1-\mu^2} & \dfrac{\mu h}{1-\mu^2} & 0 & 0 & 0 & 0 & 0 & 0 \\[2ex]
\dfrac{\mu h}{1-\mu^2} & \dfrac{h}{1-\mu^2} & 0 & 0 & 0 & 0 & 0 & 0 \\[2ex]
0 & 0 & \dfrac{h}{2(1+\mu)} & 0 & 0 & 0 & 0 & 0 \\[2ex]
0 & 0 & 0 & \dfrac{h^3}{12(1-\mu^2)} & \dfrac{\mu h^3}{12(1-\mu^2)} & 0 & 0 & 0 \\[2ex]
0 & 0 & 0 & \dfrac{\mu h^3}{12(1-\mu^2)} & \dfrac{h^3}{12(1-\mu^2)} & 0 & 0 & 0 \\[2ex]
0 & 0 & 0 & 0 & 0 & \dfrac{h^3}{12(1+\mu)} & 0 & 0 \\[2ex]
0 & 0 & 0 & 0 & 0 & 0 & \dfrac{\lambda h}{2(1+\mu)} & 0 \\[2ex]
0 & 0 & 0 & 0 & 0 & 0 & 0 & \dfrac{\lambda h}{2(1+\mu)}
\end{bmatrix}
\begin{Bmatrix} \epsilon_\alpha \\ \epsilon_\beta \\ \omega \\ \kappa_\alpha \\ \kappa_\beta \\ \tau \\ \gamma_\alpha \\ \gamma_\beta \end{Bmatrix}
-
\begin{Bmatrix} N_{\alpha T} \\ N_{\beta T} \\ 0 \\ M_{\alpha T} \\ M_{\beta T} \\ 0 \\ 0 \\ 0 \end{Bmatrix}
$$

$$\{\mathbf{N}\} = [\mathbf{D}]\{\boldsymbol{\epsilon}\} - \{\mathbf{N}_T\} \tag{6.10}$$

which serves to illustrate again that more complex material properties, expressed through a modification of the stress-strain law [C], can be represented through a corresponding generalization of [D] without affecting the remaining development. Similarly, the $\{\mathbf{N}_T\}$ matrix can be suitably altered if $\{\boldsymbol{\sigma}_T\}$ is changed. We consider some of these possibilities in the next section. Furthermore, we must emphasize that two terms of [D], D_{33} and D_{66}, often appear in slightly different forms, modified by a factor of 2. This arises because of the somewhat ambiguous definitions of ω and τ as given in equations (5.28) and (5.44); there is little consensus on whether ω should be defined as in equation (5.28), or as one-half of that quantity. Similarly, the factor of ½ introduced into the definition of τ could be omitted. The only requirement is to maintain consistency in the constitutive matrix and [D], as given in equation (6.10), is correct in this regard.

Also, note that for plates with no in-plane loading, the first three rows and columns of matrix 6–2 may be deleted. If transverse shearing strains are not included in the formulation—either for shells or plates—the last two rows and columns are not relevant; rather, the transverse shear resultants are obtained from the fourth and fifth equilibrium equations, (3.22a) and (3.22b), once the stress couples are computed.

6.1.2 Specifically Orthotropic Materials

A generalization of the isotropic condition which is of interest to us is the specifically orthotropic material. The strict definition of this configuration is that elastic properties are specified along two mutually orthogonal *material* axes, while the shell still retains the piecewise constant thickness and cross-sectional symmetry presumed so far in this text. Further, it is assumed in our development that the orthogonal material axes coincide with the coordinate lines of middle surface of the *shell*. The development of composite material technology has made it feasible to produce very efficient shells and plates having such orthotropic properties.

With this as a preface, we introduce a modified matrix [C] into equation (6.2)

$$[\mathbf{C}_{or}] = \begin{bmatrix} c_{11} & c_{12} & 0 & 0 & 0 \\ c_{12} & c_{22} & 0 & 0 & 0 \\ 0 & 0 & c_{44} & 0 & 0 \\ 0 & 0 & 0 & c_{55} & 0 \\ 0 & 0 & 0 & 0 & c_{66} \end{bmatrix} \tag{6.11}$$

The elements of matrix $[\mathbf{C}_{or}]$ are generally determined experimentally.

The elements of $[\mathbf{C}_{or}]$ may be used to derive a corresponding stress resultant-strain matrix $[\mathbf{D}_{or}]$ and a corresponding thermal load vector $\{\mathbf{N}_{Tor}\}$ by following the same procedure described in section 6.1.1.2. We also modify the thermal coefficients and transverse shear-shape factors to permit different values in each direction. The resulting generalization of equation (6.10) is

$$\{\mathbf{N}_{or}\} = [\mathbf{D}_{or}]\{\boldsymbol{\epsilon}\} - \{\mathbf{N}_{Tor}\} \tag{6.12}$$

where $[\mathbf{D}_{or}]$ is given in general form as shown in matrix 6–3, and the elements of $\{\mathbf{N}_{tor}\}$ are

$$N_{\alpha Tor} = \int_{h} (c_{11}\bar{\alpha}_{\alpha} + c_{12}\bar{\alpha}_{\beta})T(\zeta)\, d\zeta \tag{6.13a}$$

$$N_{\beta Tor} = \int_{h} (c_{12}\bar{\alpha}_{\alpha} + c_{22}\bar{\alpha}_{\beta})T(\zeta)\, d\zeta \tag{6.13b}$$

$$M_{\alpha Tor} = \int_{h} (c_{11}\bar{\alpha}_{\alpha} + c_{12}\bar{\alpha}_{\beta})T(\zeta)\zeta\, d\zeta \tag{6.13c}$$

$$M_{\beta Tor} = \int_{h} (c_{12}\bar{\alpha}_{\alpha} + c_{22}\bar{\alpha}_{\beta})T(\zeta)\zeta\, d\zeta \tag{6.13d}$$

For homogeneous shells, the integrals in matrix 6–3 and equations 6.13a–d) are easily evaluated in explicit form. For multilayered sections, the integrals may be computed by summing over the thickness. Multilayered shells and plates are widely used in applications of high-performance materials. For such layered shells, the shape factors λ_{α} and λ_{β} are determined for specific cases through correlation with selected reference solutions based on the theory of elasticity.[2] A complete bibliography on the calculation of shape factors for layered plates and shells is also given in Dong and Tso.

6.1.3 Stiffened Shells and Plates, and Shell-like Frames

There are two additional structural systems which, although not shells or plates by the strict definition, may be conveniently treated as such. These are:(a) shells or plates to which discrete stiffeners, surface undulations, or folds [figure 2–8(j)] have been added to supplement the strength and/or rigidity of the basic constant thickness section; and (b) frames composed of intersecting grids of closely spaced members that follow the topography of a curved surface. Such frameworks may be termed plate-like or shell-like structures and combine the efficient structural characteristics of the baisc geometric form with the use of prefabricated components requir-

Matrix 6–3

$$[\mathbf{D}_{or}] = \begin{bmatrix} \int_h c_{11}\,d\zeta & \int_h c_{12}\,d\zeta & 0 & 0 & 0 & 0 & 0 & 0 \\ \int_h c_{12}\,d\zeta & \int_h c_{22}\,d\zeta & 0 & 0 & 0 & 0 & 0 & 0 \\ 0 & 0 & \int_h c_{44}\,d\zeta & 0 & 0 & 0 & 0 & 0 \\ 0 & 0 & 0 & \int_h c_{11}\zeta^2\,d\zeta & \int_h c_{12}\zeta^2\,d\zeta & 0 & 0 & 0 \\ 0 & 0 & 0 & \int_h c_{12}\zeta^2\,d\zeta & \int_h c_{22}\zeta^2\,d\zeta & 0 & 0 & 0 \\ 0 & 0 & 0 & 0 & 0 & \int_h c_{44}\zeta^2\,d\zeta & 0 & 0 \\ 0 & 0 & 0 & 0 & 0 & 0 & \lambda_\alpha\int_h c_{55}\,d\zeta & 0 \\ 0 & 0 & 0 & 0 & 0 & 0 & 0 & \lambda_\beta\int_h c_{66}\,d\zeta \end{bmatrix}$$

ing little or no falsework for erection.[3] A well-known example of a shell-like structure is the Astrodome in Houston, Texas [figures 2–8(m) and (n)].

Although it is theoretically possible to model these relatively complex systems as ensembles of the various constituent components, it is often feasible, considerably simpler and more economical to use surface structure equations. Naturally, the question arises as to what to take for the "thickness" of the surface for, by definition, a shell or plate has a piecewise constant or smoothly varying thickness. It appears to be proper to consider this question at the level of the *stress resultant-strain* relationships, matrix [**D**], rather than at the *stress-strain* level, matrix [**C**], since the latter matrix is based on the *stress-at-a-point* focus of the Theory of Elasticity—which obviously has little meaning when we must consider equivalent thicknesses.

A fairly common and quite illustrative example is that of a shell or plate with equispaced stiffening ribs running in one direction, as shown in figure 6–1, or in both directions. A common method of treating this problem is to combine the properties of the basic shell or plate and the stiffener over a repeating interval of the cross section, such as d_β, and then to compute an equivalent thickness in the direction of the stiffener. This process is often referred to by the colloquial title of *verschmieren*, after the German verb "to smear over." For a shell with only in-plane or membrane forces considered, the equivalent thickness is logically computed with respect to the extensional rigidity, whereas for a plate subjected to bending, it would be based on the flexural rigidity.

To illustrate such a computation, consider the situation shown in figure 6–1. For a shell, we compute the area of the repeating cross section $hd_\beta + \bar{h}b$ and divide by the spacing, d_β, to get

$$h_\alpha = h + \frac{b\bar{h}}{d_\beta} \tag{6.14}$$

In the other direction, we use $h_\beta = h$. If the stiffened structure is a plate, we compute the moment of inertia of the repeating T section about its centroidal axis, $I_{h\bar{h}}$. Then, the equivalent thickness h_α is found from setting

$$\frac{1}{12} \cdot d_\beta \cdot h_\alpha^3 = I_{h\bar{h}} \tag{6.15a}$$

or

$$h_\alpha = \left(\frac{12I_{h\bar{h}}}{d_\beta}\right)^{1/3} \tag{6.15b}$$

The thickness in the β direction is again taken as $h_\beta = h$.

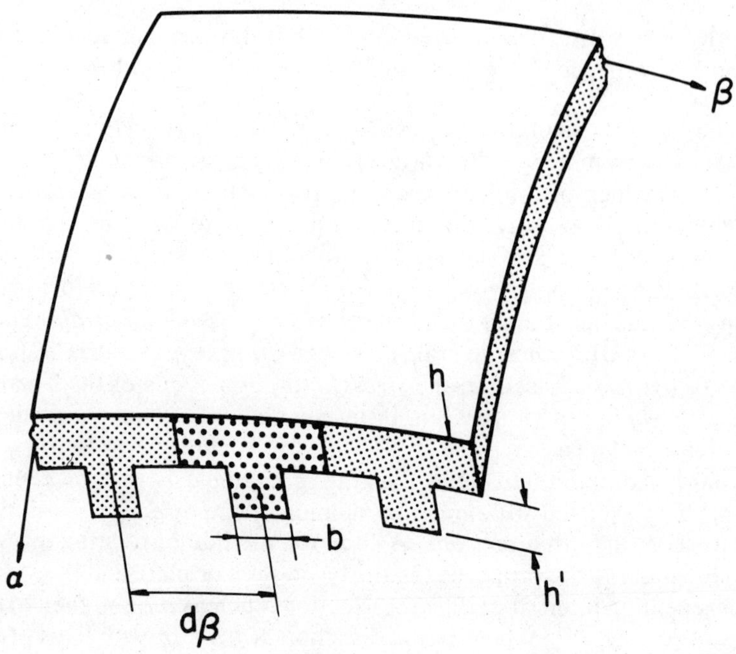

Figure 6–1. Shell with Stiffening Ribs

The equivalent thicknesses are used to compute the corresponding terms of matrix [**D**]. Referring to the elements of matrix 6–2, we would use h_α to compute the terms in the first and fourth rows, and h_β for the terms in the second and fifth rows. It has been suggested[4] that the terms corresponding to the shearing rigidity (rows three, seven, and eight) be evaluated by using the basic thickness h, whereas the twisting rigidity (row six) may be computed using the basic thickness or, in some cases, be neglected altogether. This technique may be easily extended to cases in which the rib stiffeners run in both directions by computing appropriate h_β thicknesses from formulas analogous to equations (6.14) and (6.15). A form of a plate with stiffeners running in two directions known as a *waffle slab* [figure 2–8(w)] is quite widely used in reinforced concrete building construction.

This simple illustration of a *verschmieren* procedure raises several questions which we should consider at this point. If only one equivalent thickness is defined in each direction, should it be the extensional or the flexural thickness for a shell that may have significant bending? We have opted for the extensional, but will this equivalent thickness adequately represent the flexural characteristics of the stiffened shell? Also, what about the shearing and twisting thicknesses? To include only the basic shell or plate, as mentioned in the preceding paragraph, seems to be a rather crude approximation. Finally, the use of different equivalent thicknesses in the α and β directions for the extensional and bending rigidities will cause matrix $[\mathbf{D}]$ to by unsymmetric because of the off-diagonal terms in rows one, two, four, and five. This leads to contradictions and computational difficulties that should be avoided if possible.

From the preceding questions, it is obvious that a procedure which emobidies more of the characteristics of the actual structure is desirable. This is especially true as the stiffeners become more prominent with respect to the basic cross section and, in the case of plate-like and shell-like structures, may constitute practically the entire means of resisting the major forces and moments.

A very promising technique to accomplish this objective is the *split rigidity* method introduced by Buchert[5] in his generalization of the shell stability theory of E. Reissner.[6] This approach admits different equivalent thicknesses for the extensional, shearing, bending, and twisting terms. To avoid difficulties with nonsymmetry in $[\mathbf{D}]$, Poisson's ratio may be neglected. This leads to a diagonal matrix $[\mathbf{D}_{eq}]$, which may be used in place of $[\mathbf{D}]$ in equation (6.10):

$$[\mathbf{D}_{eq}] = E\left[h_{\alpha(m)} \quad h_{\beta(m)} \quad \frac{h_{\alpha\beta(m)}}{2} \quad \frac{h^3_{\alpha(b)}}{12} \quad \frac{h^3_{\beta(b)}}{12} \quad \frac{h^3_{\alpha\beta(b)}}{12} \quad \frac{\lambda h_{\alpha(m)}}{2} \quad \frac{\lambda h_{\beta(m)}}{2} \right] \qquad (6.16)$$

The additional subscripts (m) and (b) indicate membrane or extensional and bending or flexural, respectively.

To illustrate the split rigidity concept, we chose a shell-like structure composed of a gridwork of structural members, as shown in figure 6–2. The members are assumed to have locally constant spacing, d_α and d_β; areas A_α and A_β; moments of inertia I_α and I_β; and St. Venant torsion constants K_α and K_β, in each direction. Any infilled material between the main members is assumed to either be restricted to local loading distribution and therefore negligible in this analysis, or to be included in the section properties of the main members.

With these idealizations, we may immediately compute

$$h_{\alpha(m)} = \frac{A_\alpha}{d_\beta} \qquad (6.17)$$

and

$$h_{\beta(m)} = \frac{A_\beta}{d_\alpha} \qquad (6.18)$$

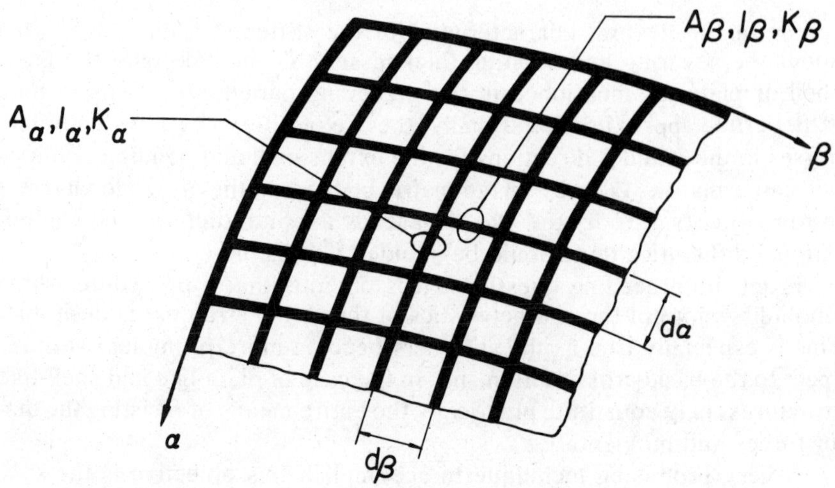

Figure 6–2. Shell-like Gridwork

The equivalent in-plane shearing thickness is a bit harder to evaluate. A first approximation, considering only the primary members of the frame, would be to take the average of the equivalent membrane thickness, $\frac{1}{2} \cdot [h_{\alpha(m)} + h_{\beta(m)}]$, so that

$$h_{\alpha\beta(m)} = \frac{1}{2}\left(\frac{A_\alpha}{d_\beta} + \frac{A_\beta}{d_\alpha}\right) \tag{6.19}$$

Note that there is the possibility of increasing the in-plane shearing resistance considerably through the integration of the infilling material with the primary frame structure.

Now we turn to the equivalent bending thicknesses. Considering first the α direction, we equate the moments of inertia per unit width of the frame member and the equivalent shell

$$\frac{I_\alpha}{d_\beta} = \frac{1}{12} \cdot 1 \cdot h^3_{\alpha(m)}$$

from which

$$h_{\alpha(b)} = \left(12\,\frac{I_\alpha}{d_\beta}\right)^{1/3} \tag{6.20}$$

Similarly, in the β direction

$$h_{\beta(b)} = \left(12\,\frac{I_\beta}{d_\alpha}\right)^{1/3} \tag{6.21}$$

Last, we turn to the equivalent twisting thickness. It is in this mode of resistance that the shell-like structure will often differ most from the true shell. Whereas a true shell or plate has relatively great twisting rigidity as compared with the flexural rigidity, and the shell-like or plate-like structure is likely to have comparatively little twisting rigidity, especially if it is comprised of thin-walled members with open cross section. Huffington, in his study of orthotropic plates, has suggested a procedure based on the St. Venant torsional rigidity of the members.[7] When applied to the problem at hand, we get

$$h_{\alpha\beta(b)} = \left[\frac{3}{8}\left(\frac{K_\alpha}{d_\beta} + \frac{K_\beta}{d_\alpha}\right)\right]^{1/3} \tag{6.22}$$

Substituting equations (6.17) through (6.22) into equation (6.16), we have

$$[\mathbf{D}_{eq}] = E\left[\frac{A_\alpha}{d_\beta} \frac{A_\beta}{d_\alpha} \frac{1}{4}\left(\frac{A_\alpha}{d_\beta} + \frac{A_\beta}{d_\alpha}\right)\right.$$

$$\left.\frac{I_\alpha}{d_\beta} \frac{I_\beta}{d_\alpha} \frac{1}{16}\left(\frac{K_\alpha}{d_\beta} + \frac{K_\beta}{d_\alpha}\right) \frac{\lambda A_\alpha}{2d_\beta} \frac{\lambda A_\beta}{2d_\alpha}\right] \tag{6.23}$$

The use of an equivalent stress resultant-strain matrix, such as the one defined here, enables a large variety of curved frames and gridworks to be efficiently analyzed by using solutions based on shell and plate theories. We must remember, however, that the stress resultants and couples are computed per unit length of middle surface from such solutions and must subsequently be converted to forces and moments in the discrete members. A logical procedure for accomplishing this conversion is to take the average value of the stress resultant or couple over a repeating interval, d_α or d_β, multiplied by the interval.

6.2 Boundary Conditions

6.2.1 Role of Boundary Conditions in the Formulation of Shell and Plate Theories

In the preceding section, we have presented the *constitutive* laws that serve to connect the *equilibrium* equations derived in chapter 3 and the *compatibility* relations developed in chapter 5. The classical formulation of the solid mechanics problem embodies these three components, which may be collected as a set of *partial differential equations* or as an *energy principle*. We explore both of these formats in depth in the succeeding chapters; however, there is one more ingredient necessary to complete the theory in either case: a suitable set of *boundary conditions*. The possibilities may be classified as follows: (a) *static* (a stress resultant or stress couple is specified; (b) *kinematic* (a displacement or rotation is specified); and (c) *elastic* (a linear combination of a static and a kinematic quantity is specified). This may represent a spring-type support.

In the following discussion, we place two restrictions on the boundary conditions to simplify the formulation somewhat. These are: (a) all boundaries coincide with either an s_α or s_β coordinate line; and (b) all boundary conditions are homogeneous. Exceptions to the first restriction include such problems as skew intersections and cutouts, and are best treated using discrete approximation techniques such as the finite element method. Nonhomogeneous boundary conditions are not necessarily more difficult to deal with than the homogeneous variety, but since they occur comparatively infrequently, it is usually expedient to exclude this possibility from the general formulation and to treat such situations that may arise as special cases.

6.2.2 Boundary Conditions

We now look at two representative boundaries of a surface in figure 6–3. The boundary coinciding with the s_α coordinate line is located by $\beta = \bar{\beta}$, and the boundary along the s_β coordinate line is designated by $\alpha = \bar{\alpha}$. On the boundary $\alpha = \bar{\alpha}$, there are five static quantities acting: N_α, $N_{\alpha\beta}$, Q_α, M_α, and $M_{\alpha\beta}$; also, there are five kinematic components present: D_α, D_β, D_n, $D_{\alpha\beta}$, and $D_{\beta\alpha}$. Note that *each static quantity* performs work through *one and only one kinematic quantity* within the small deformation theory. We group the associated quantities as *static-kinematic correspondents* as listed on figure 6–3. For the boundary $\beta = \bar{\beta}$, the correspondents are also given on the figure. This pairing is quite important, for at any point on either boundary, only *one* of *each pair* of static-kinematic correspondents may be specified—except for an elastic boundary represented by a linear combination of the two. In no case, however, may *both* or *neither* of the correspondents by designated on a boundary. Further, when one correspondent is specified, the other must be developed by the boundary; e.g., if a displacement, such as D_α, is set equal to 0, the boundary must develop whatever value of N_α is calculated from the subsequent solution; or if the stress resultant $N_\alpha = 0$, the boundary must be free to displace through the computed D_α.

Before examining some specialized forms of the boundary conditions, we should mention one further requirement, the elimination of *rigid body* displacements. To accomplish this, it is sufficient that D_α, D_β, D_n, $D_{\alpha\beta}$,

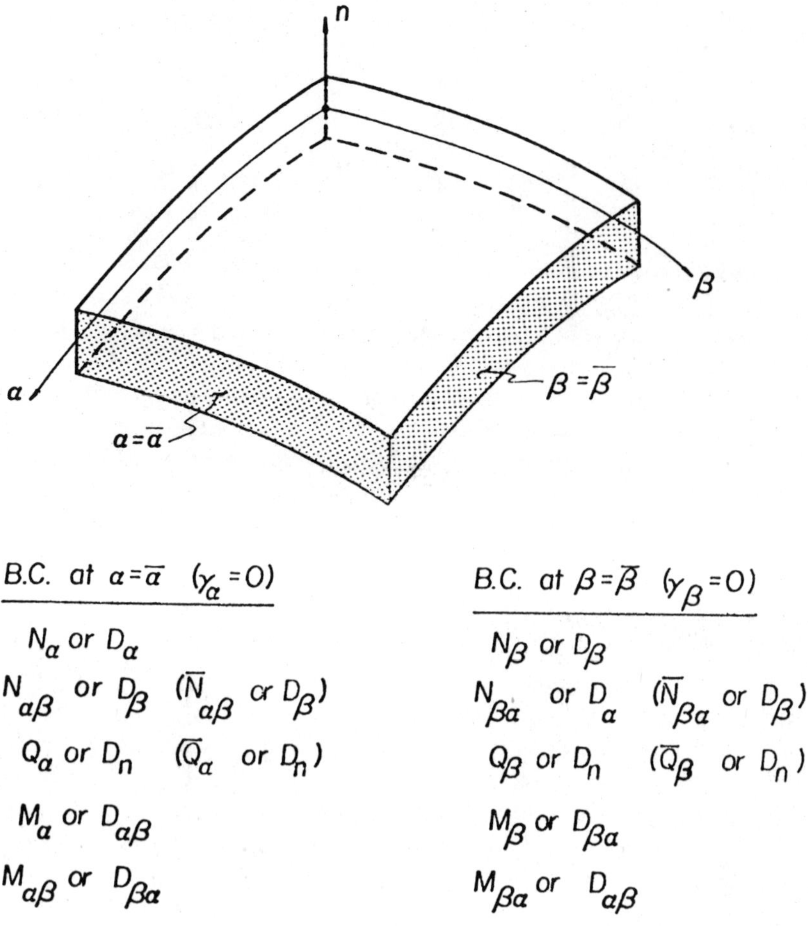

B.C. at $\alpha = \bar{\alpha}$ $(\gamma_\alpha = 0)$

N_α or D_α

$N_{\alpha\beta}$ or D_β $\quad(\bar{N}_{\alpha\beta}$ or $D_\beta)$

Q_α or D_n $\quad(\bar{Q}_\alpha$ or $D_n)$

M_α or $D_{\alpha\beta}$

$M_{\alpha\beta}$ or $D_{\beta\alpha}$

B.C. at $\beta = \bar{\beta}$ $(\gamma_\beta = 0)$

N_β or D_β

$N_{\beta\alpha}$ or D_α $\quad(\bar{N}_{\beta\alpha}$ or $D_\beta)$

Q_β or D_n $\quad(\bar{Q}_\beta$ or $D_n)$

M_β or $D_{\beta\alpha}$

$M_{\beta\alpha}$ or $D_{\alpha\beta}$

Figure 6–3. Boundary Conditions

and $D_{\beta\alpha}$ each be restrained to 0 at one point or more on the surface. Seemingly, we would also need to restrict the rotation about the normal, but this degree of freedom already has been eliminated from the theory. There are also certain cases when the loading is self-equilibrated, such as those corresponding to the $j > 1$ harmonics in section 4.3.7, for which the rigid body displacements do not have to be eliminated explicitly.

As specific examples of easily visualized boundary conditions which are frequently encountered, we consider the boundary $\alpha = \bar{\alpha}$:

Condition Desired	Specified Quantities
Free	$\underline{N_\alpha} = N_{\alpha\beta} = \underline{Q_\alpha} = \underline{M_\alpha} = M_{\alpha\beta} = 0$
Fixed	$\underline{D_\alpha} = D_\beta = \underline{D_n} = \underline{D_{\alpha\beta}} = D_{\beta\alpha} = 0$
Hinged	$\underline{M_\alpha} = 0 \,;\, \underline{D_\alpha} = D_\beta = \underline{D_n} = \underline{D_{\beta\alpha}} = 0$
Roller	$\underline{Q_\alpha} = \underline{M_\alpha} = 0 \,;\, \underline{D_\alpha} = D_\beta = \underline{D_{\beta\alpha}} = 0$
Simply Supported	$\underline{N_\alpha} = \underline{M_\alpha} = 0 \,;\, D_\beta = \underline{D_n} = \underline{D_{\beta\alpha}} = 0$

Recognize that the terms used to describe the boundaries—e.g., hinged, roller, etc.—are essentially two-dimensional in implication and do not fully describe the constraint in the β direction. The terms underscored in the above listing are definitely *implied* by the stated condition, whereas the others in question, pertaining to quantities acting *parallel* to the boundary, are merely *surmised*. A similar set of implied and surmised conditions can be derived from the boundary $\beta = \bar{\beta}$.

6.2.3 Kirchhoff Boundary Conditions

We have mentioned several times that much of the classical theories of shells and plates is based on the suppression of the transverse shearing strains. From the statement of the admissible boundary conditions presented in the preceding section, we can anticipate that the most general system of governing differential equations is of tenth order in each direction, permitting prescription of five conditions on each of the four boundaries. When transverse shearing strains are not included, the governing equations are reduced to eighth order in each direction, and a modified set of boundary conditions is required.

The derivation of the static conditions consistent with the suppression of transverse shearing strains is one of the more interesting theoretical problems in the theories of shells and plates. These modified boundary conditions are commonly known as the Kirchhoff boundary conditions, after the noted mathematician, G. Kirchhoff, who first established the correct relationship for a plate using an energy formulation.[8] Although Kirchhoff's derivation served to clarify the situation somewhat, it was mathematical in basis, and it remained for two distinguished nineteenth-century scientists, Lord Kelvin and P. Tait,[9] to provide a physical interpretation of the Kirchhoff conditions.

To illustrate the Kelvin–Tait argument and to provide a logical exten-

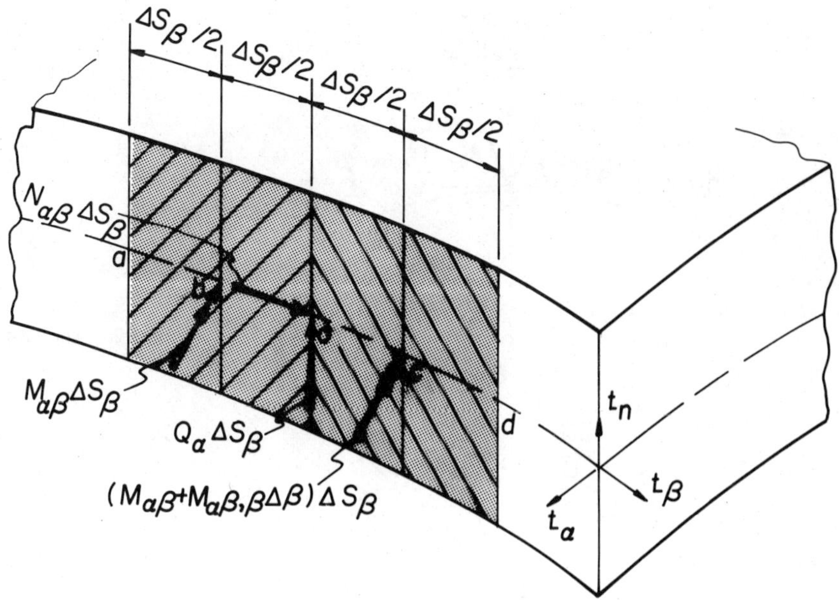

Figure 6–4. Kelvin–Tate Argument

sion into our general shell formulation, refer to the $\alpha = \bar{\alpha}$ boundary as shown on figure 6–4. We show four adjacent differential half-segments ab, bo, oc and cd, each of length $\Delta s_\beta/2$, and indicate the resultant forces acting on segment boc due to $N_{\alpha\beta}$ and Q_α. These resultants act at point o. Also, we show the resultant couples on two adjacent segments, abo and ocd, due to $M_{\alpha\beta}$. These act at points b and c, respectively. The directions shown correspond to the sign convention of figure 3–2. Now we examine a normal section showing the middle surface of the differential segment in figure 6–5, and replace the moments due to the stress couples by equivalent forces at a, o, and d. The replacement of a moment on the boundary by a pair of equal and opposite forces separated by a differential distance is justified by the St. Venant principle,[10] which is just one of the many enduring contributions of this outstanding nineteenth-century scientist. From $M_{\alpha\beta}$, we have the forces $(M_{\alpha\beta}\Delta s_\beta)/\Delta s_\beta = M_{\alpha\beta}$ acting at a and o normal to the chord ao. From $M_{\alpha\beta} + M_{\alpha\beta,\beta}\Delta\beta$, we derive a similar pair of forces acting normal to chord od. Also shown by dashed vectors are the *effective* in-plane shear $\bar{N}_{\alpha\beta}$ and the *effective* transverse shear $\bar{Q}_\alpha$ that we wish to evaluate at point o.

Summing forces in the β direction, we have

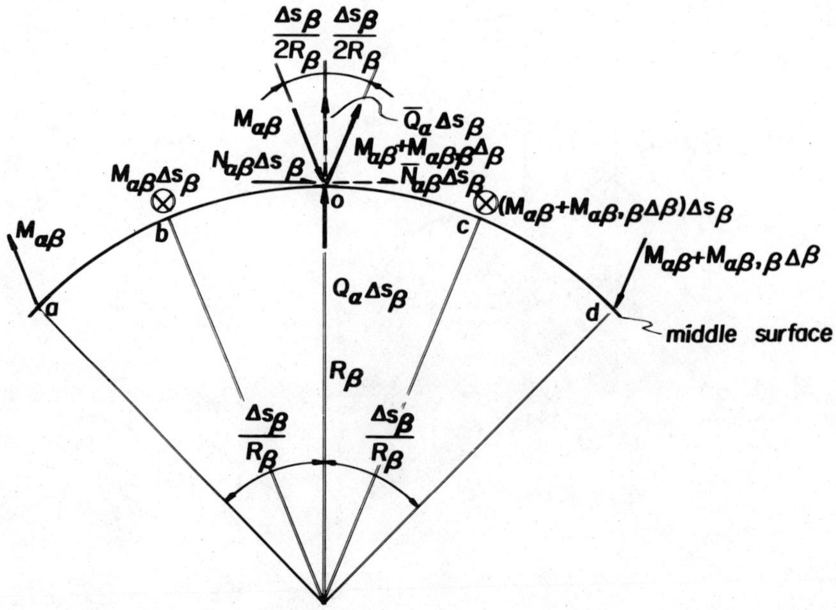

Figure 6–5. Kelvin–Tate Argument

$$\bar{N}_{\alpha\beta}\Delta s_\beta = N_{\alpha\beta}\Delta s_\beta + M_{\alpha\beta}\sin\left(\frac{\Delta s_\beta}{2R_\beta}\right)$$

$$+ (M_{\alpha\beta} + M_{\alpha\beta,\beta}\Delta\beta)\sin\left(\frac{\Delta s_\beta}{2R_\beta}\right)$$

Since $M_{\alpha\beta,\beta}\Delta\beta$ is of higher order than the remaining terms, and $\sin(\Delta s_\beta/2R_\beta) \simeq \Delta s_\beta/2R_\beta$, the effective in-plane shear is

$$\bar{N}_{\alpha\beta} = N_{\alpha\beta} + \frac{M_{\alpha\beta}}{R_\beta} \tag{6.24}$$

Summing forces in the n direction, we have

$$\bar{Q}_\alpha\Delta s_\beta = Q_\alpha\Delta s_\beta - M_{\alpha\beta}\cos\left(\frac{\Delta s_\beta}{2R_\beta}\right) + (M_{\alpha\beta} + M_{\alpha\beta,\beta}\Delta\beta)\cos\left(\frac{\Delta s_\beta}{2R_\beta}\right)$$

Noting that $\cos(\Delta s_\beta/2R_\beta) \simeq 1$, and that $\Delta s_\beta = B\,\Delta\beta$, the effective transverse shear is

$$\bar{Q}_\alpha = Q_\alpha + \frac{M_{\alpha\beta,\beta}}{B} \tag{6.25}$$

Along the boundary $\beta = \bar{\beta}$, we can develop a similar set of effective shears

$$\bar{N}_{\beta\alpha} = N_{\beta\alpha} + \frac{M_{\beta\alpha}}{R_\alpha} \tag{6.26}$$

and

$$\bar{Q}_\beta = Q_\beta + \frac{M_{\beta\alpha,\alpha}}{A} \tag{6.27}$$

Now, referring to figure 6–3, we have the five static boundary conditions contracted to four along each boundary.

For the corresponding kinematic conditions, we first examine equations (5.46g) and (5.46h) with $\gamma_\alpha = \gamma_\beta = 0$. Then $D_{\alpha\beta}$ and $D_{\beta\alpha}$, corresponding to the bending couples, are equal to ψ_α and ψ_β, respectively. For the effective shearing stresses which incorporate the twisting couples, we have D_β and D_n on the $\alpha = \bar{\alpha}$ boundary, and D_α and D_n on the $\beta = \bar{\beta}$ boundary. These are also listed on figure 6–3.

6.2.4 Boundary Conditions for Plates with No Transverse Shearing Strains

Although the general boundary conditions described in section 6.2.2 are applicable for plates as well as shells, the Kirchhoff conditions are somewhat simplified for plates. An examination of equations (6.24) and (6.26) reveals R_β and R_α in the denominators of the twisting stress couple terms, $M_{\alpha\beta}$ and $M_{\beta\alpha}$. Since these radii $\to \infty$, the effective in-plane shear is superfluous, and we are left with only the effective transverse shears, as given by equations (6.25) and (6.27).

6.2.5 Boundary Conditions for Shells of Revolution

The shell of revolution geometry was specialized in section 2.8.2, with the meridians corresponding to the α coordinate lines and the parallel circles to the β coordinate lines. For this shell form, we only have boundary conditions along the parallel circles. If the meridional coordinate is taken as ϕ, and the circumferential coordinate as θ, a parallel circle boundary corresponds to $\phi = \bar{\phi}$. The Kirchhoff conditions follow from equations (6.24) and (6.25) as

$$\bar{N}_{\phi\theta} = N_{\phi\theta} + \frac{M_{\phi\theta}}{R_\theta} \tag{6.28}$$

and

$$\bar{Q}_\phi = Q_\phi + \frac{M_{\phi\theta,\theta}}{R_0} \tag{6.29}$$

6.2.6 Symmetry

In addition to external boundary conditions such as those we have discussed in the preceding paragraphs, there may be equations of condition that arise because of symmetry properties of both the geometry and the loading. The practical manifestation of the recognition of symmetry conditions is that only a part of the entire continuum must be analyzed.

A fairly general modern treatment of the topic of symmetry is presented by Glockner,[11] and we focus here on the applications to shells and plates. Of course, we have already used the concept of symmetry extensively in our treatment of shells of revolution as a distinct class of surface structures. Such shells are distinguished *geometrically* by their rotational symmetry. Further, in the analysis by the Fourier series technique in chapter 4, symmetric and antisymmetric components of the *loading* were identified. Formally, we may state that a loading function $q(x)$ is symmetric about $x = 0$ if $q(-x) = q(x)$, and antisymmetric if $q(-x) = -q(x)$.

As a first illustration, consider a complete rotational shell, such as a sphere (figure 4–3), or an ellipsoid of revolution subject to a uniform normal pressure $q_n = q$. Using the notation of figures 4–2 and 6–3, the following symmetry conditions may be enforced at $\phi = \pi/2$:

$$D_\phi = 0; \quad Q_\phi = 0$$

These conditions require only half of the shell to be represented in the ensuing analysis. This may not be important in a membrane theory analysis which is statically determinate—e.g., equations (4.27) and (4.28)—but may result in considerable savings when a numerical analysis using the general theory is performed. In some numerical solutions, it also may be necessary to enforce such symmetry conditions to eliminate rigid body displacement.

Another case of interest is a shell of translation, such as the elliptic paraboloid shown in figure 4–42(a) subject to a uniform vertical loading $q_z = q$. We may consider only the quadrant $OEDF$ and, using the notation of figure 4–32, take $\alpha = x$ and $\beta = y$ on figure 6–3. Then, we have

$$D_x = 0; \quad N_{xy} = 0; \quad Q_x = 0; \quad D_{xy} = 0; \quad M_{xy} = 0$$

along OF and

$$D_y = 0; \quad N_{yx} = 0; \quad Q_y = 0; \quad D_{yx} = 0; \quad M_{yx} = 0$$

along *OE*. It is possible to effect further savings by defining symmetry along the diagonal *OD*, but since this involves displacements along non-principal lines of curvature, it lies outside our present scope. The conditions for the shell of translation are also applicable to uniformly loaded, symmetrically supported plates.

As a final example, we choose the same elliptic paraboloid illustrated in figure 4–42(a) subject to a hydrostatic loading acting normal to the *X–Y* plane. The distribution in the *Y* direction is given by

$$q_z(X,Y) = q\frac{(b - Y)}{2}$$

while the loading is assumed to be uniform across the *X* direction. This loading is readily decomposed into a constant component ($j = 0$)

$$q_z^0(X,Y) = q\frac{b}{2}$$

and a linear component ($j = 1$)

$$q_z^1(X,Y) = -q\frac{Y}{2}$$

The constant component q_z^0 is symmetrical with respect to both the *X* and *Y* axes, and is treated identically to the previous example. The linear component q_z^1 remains symmetric with respect to the *Y* axis, so the conditions along *OF* are unaltered. However, this component is antisymmetric with respect to the *X* axis, since $q_z^1(X, -Y) = -q_z^1(X, Y)$. The resulting conditions along *OE* are

$$N_x = 0; \quad D_y = 0; \quad D_n = 0; \quad M_x = 0; \quad D_{yx} = 0$$

which are the correspondents of the conditions on a line of loading symmetry in each case. Again the antisymmetry conditions are applicable to plates as well.

6.3 Membrane Theory Displacements

6.3.1 General Aspects

In the preceding sections of this chapter, we have completed the derivation of the individual components required to formulate the general theories of shells and plates. In chapter 4, we investigated a very important class of problems, the membrane theory of shells, which requires only the solution of a reduced set of equilibrium equations to determine the in-plane stress resultants. The membrane theory is generally statically determinate as opposed to the general theory, which is statically indeterminate and therefore requires simultaneous consideration of equilibrium, compatibility, and constitutive relations. Nevertheless, we cannot say that the membrane solution is complete until the displacements are determined. As is the usual procedure for statically determinate problems, we may compute these displacements, assuming that the stress field is known, *a priori*, from the equilibrium solution.

At the outset, it is instructive to consider the relevance of the membrane theory displacements in overall shell analysis and design. First, since many shells are designed on the basis of the membrane theory, the magnitudes of the associated displacements are required to verify the applicability of small displacement theory (assumption [2], table 1–1). Also, knowledge of the expected displacements can be important in designing a supporting system to accommodate the required boundary deformations, as mentioned in the previous section.

Next, even if a solution is obtained using a bending theory, the values of the in-plane stress resultants are often not changed significantly from those computed from a membrane analysis except, perhaps, in localized regions. Since certain of the membrane stress resultants, such as N_ϕ in symmetrically and antisymmetrically loading shells of revolution, are dependent on the *overall* equilibrium of the shell, they cannot be altered very much by bending action. This means that the middle surface displacements, which are primarily functions of the in-plane stress resultants, may be estimated quite closely from the membrane theory solution.

Additionally, we may anticipate that a logical procedure to analyze shells which do not fully meet the requirements of the membrane theory because of inappropriate boundary conditions, will be to use the flexibility method. An example would be a spherical shell, such as that illustrated in figure 4–2, with the lower boundary pinned or fixed, as defined in section 6.2. Briefly, the flexibility method consists of introducing a sufficient number of kinematic *releases* into the system to make it statically determinate. Then, the conditions of deformation compatibility consistent with the original system are enforced on the superposed primary and secondary systems to determine the static correspondents of the kinematic releases, known as the redundants. This procedure is familiar to students of structural mechanics and is described in most standard texts on the analysis of indeterminate structures, so we need not elaborate here. Briefly, we note that the compatibility equations take the form

$$\{\Delta\} = \{\bar{\Delta}\} + [F]\{\chi\} \tag{6.30}$$

where $\{\Delta\}$ = the vector of the displacements of each of the releases on the original structure. The elements of $\{\Delta\}$ are usually 0 unless the support is elastic.

$\{\bar{\Delta}\}$ = the vector of the displacements of each of the releases due to the applied loading. These displacements are computed on the released, statically determinate, structure.

$\{\chi\}$ = the vector of the statical correspondents of the kinematic releases, the redundants, initially unknown. For plates and shells, the elements of $\{\chi\}$ are usually transverse shear forces and moments on the boundary.

$[\mathbf{F}]$ = the matrix of flexibility influence coefficients, each element of which is a displacement of a release due to a unit value of one of the elements of $\{\chi\}$. The influence coefficients are computed on the released, statically determinate, structure.

Once $\{\bar{\Delta}\}$ and $[\mathbf{F}]$ are computed and $\{\Delta\}$ is specified, we solve for

$$\{\chi\} = [\mathbf{F}^{-1}][\{\Delta\} - \{\bar{\Delta}\}] \tag{6.31}$$

For shells, $\{\bar{\Delta}\}$ can usually be evaluated with sufficient accuracy by using the membrane theory, as discussed in the previous paragraph. The elements of $[\mathbf{F}]$, being displacements due to boundary transverse shears and moments, must still be computed by using the bending theory; however, since no surface loads are involved, only the homogeneous bending equations need be considered, which frequently simplifies the solution. A specific illustration of the superposition method is given in chapter 9 for the bending analysis of cylindrical shells.

Thus we have suggested a variety of reasons for studying the middle surface displacements of shells in equilibrium under a known internal stress field computed by the membrane theory.

6.3.2 Governing Equations

To connect the middle surface displacements D_α, D_β, and D_n with the membrane stress resultants N_α, N_β, and $N_{\alpha\beta}$, compare equations (5.46a)–(5.46c) with the first three rows of matrix 6–2. Since matrix 6–2 gives the stress resultants as functions of strain, we must invert the relations

$$N_\alpha = D_1\epsilon_\alpha + \mu D_1\epsilon_\beta - N_{\alpha T} \tag{6.32a}$$

$$N_\beta = \mu D_1\epsilon_\alpha + D_1\epsilon_\beta - N_{\beta T} \tag{6.32b}$$

$$S = D_2\omega \tag{6.32c}$$

recalling that $N_{\alpha\beta}$ is replaced by S in the membrane theory. From matrix 6–2, $D_1 = Eh/(1 - \mu^2)$ and $D_2 = Eh/[2(1 + \mu)]$. Solving for the strains in view of equations (6.9a) and (6.9b), we have

$$\epsilon_\alpha = \frac{1}{Eh}\left[(N_\alpha - \mu N_\beta) + (1 - \mu)N_T\right] \tag{6.33a}$$

$$\epsilon_\beta = \frac{1}{Eh}\left[(N_\beta - \mu N_\alpha) + (1 - \mu)N_T\right] \tag{6.33b}$$

$$\omega = \frac{2(1 + \mu)}{Eh}S \tag{6.33c}$$

By equating the r.h.s. of equations (5.46a–c) and equations (6.33a–c), we arrive at the complete set of equations relating to the displacements and the in-plane stress resultants. With these stress resultants determined by using the methods of chapter 4, we have three partial differential equations in the three unknown middle surface displacements, D_α, D_β, and D_n.

We may also obtain expressions for the rotations $D_{\alpha\beta}$ and $D_{\beta\alpha}$ in terms of the middle surface displacements from equations (5.46g) and (5.46h), if we were to take the transverse shearing strains as 0. This is consistent with the membrane theory where Q_α and Q_β are assumed to be 0.

Many practical applications of shells and plates allow simplifications in the functional dependence of the geometrical parameters. Therefore, it is expedient now to specialize the equations for specific geometrical forms.

6.3.3 Shells of Revolution

6.3.3.1 Transformation of Governing Equations. Since we have already specialized the strain-displacement relations for shells of revolution with $= \phi$ and $\beta = \theta$ in section 5.4, we have from equations (5.48a–c) and (6.33a–c)

$$R_\phi\epsilon_\phi = D_{\phi,\phi} + D_n = \frac{R_\phi}{Eh}\left[(N_\phi - \mu N_\theta) + (1 - \mu)N_T\right] \tag{6.34a}$$

$$R_0\epsilon_\theta = D_{\theta,\theta} + \cos\phi D_\phi + \sin\phi D_n = \frac{R_0}{Eh}\left[(N_\theta - \mu N_\phi) + (1 - \mu)N_T\right]$$

$$\tag{6.34b}$$

$$R_0\omega = \frac{R_0}{R_\phi}D_{\theta,\phi} + D_{\phi,\theta} - \cos\phi D_\theta = \frac{2(1 + \mu)R_0}{Eh}S \tag{6.34c}$$

Since the stress resultants and thermal loads are both presumed to be known at this stage, we drop the thermal terms which, if present, can simply be absorbed into N_ϕ and N_θ.

Now we return to the technique of section 4.3.1, whereby a judiciously chosen set of auxiliary variables was used to reduce the set of equations. Again, we follow Novozhilov,[12] introducing

$$\bar{\psi} = \frac{D_\theta}{R_0} \tag{6.35a}$$

$$\bar{\xi} = \frac{D_\phi}{\sin \phi} \tag{6.35b}$$

Next, we substitute equations (6.35a) and (6.35b) into equations (6.34a–c) and eliminate D_n between the first two equations, giving

$$\bar{\xi},_\phi - \frac{R_\theta}{\sin \phi} \, \bar{\psi},_\theta = \frac{1}{Eh \sin \phi} \left[(R_\phi + \mu R_\theta)N_\phi - (R_\theta + \mu R_\phi)N_\theta \right] \tag{6.36a}$$

$$\frac{R_\theta^2 \sin \phi}{R_\phi} \, \bar{\psi},_\phi + \bar{\xi},_\theta = \frac{2(1 + \mu)}{Eh} R_\theta S \tag{6.36b}$$

These transformed equations may be compared with the transformed equilibrium equations, (4.6b) and (4.6a), respectively. We see that the l.h.s. of both would be identical if we were to interchange ψ and $\bar{\psi}$, and ξ and $\bar{\xi}$; whereas the r.h.s., in both cases, consist entirely of known functions. Therefore, whatever solution strategies were deduced for the membrane equilibrium equations are applicable for the equations governing the membrane displacements. Also, we recognize that once $\bar{\xi}$ and $\bar{\psi}$ are determined, D_ϕ and D_θ follow from equations (6.35a), and (6.35b) and D_n may be calculated from equation (6.34a) or (6.34b).

As a practical matter, when we attempt to use the various solutions already derived in chapter 4 for the determination of the corresponding displacements, we soon realize that the r.h.s. of equations (6.34a–c) are likely to be more complicated, algebraically, than the corresponding functions in the equilibrium equations. This is because the expressions for the membrane stress resultants are already quite involved for some cases that we have solved in chapter 4 and, combined with the radii of curvature expressions, often produce integrals that only can be evaluated numerically. This slight complication notwithstanding, the similarity of the system of governing equations for the membrane theory stress resultants in terms of the known applied loading, on the one hand, and the in-plane displacements and the computed membrane stress resultants on the other, is quite striking. This serves as another example of the static-geometric analogy first mentioned in section 5.4.2.

Once we have evaluated the in-plane displacements, we may compute the rotations from equation (5.48) with $\gamma_\phi = \gamma_\theta = 0$:

$$D_{\phi\theta} = -\frac{1}{R_\phi} (D_{n,\phi} - D_\phi) \tag{6.37a}$$

$$D_{\theta\phi} = -\frac{1}{R_0} (D_{n,\theta} - \sin\phi D_\theta) \tag{6.37b}$$

We now find it convenient to again consider the axisymmetric ($j = 0$) and nonsymmetric ($j > 0$) cases separately.

6.3.2.2 Axisymmetric Displacements. We drop the θ-dependent terms from equations (6.36) and (6.37) and integrate the former equations. We find, following section 4.3.2.1 and using equation (6.34b)

$$\bar{\xi} = \frac{D_\phi}{\sin\phi}$$

$$= \frac{1}{E} \int \frac{1}{h\sin\phi} [(R_\phi + \mu R_\theta)N_\phi - (R_\theta + \mu R_\phi)N_\theta]\, d\phi \tag{6.38a}$$

$$\bar{\psi} = \frac{D_\theta}{R_0} = 2\frac{(1+\mu)}{E} \int \frac{R_\phi}{hR_0} S\, d\phi \tag{6.38b}$$

$$D_n = \frac{R_\theta}{Eh} (N_\theta - \mu N_\phi) - \cot\phi D_\phi \tag{6.38c}$$

$$D_{\phi\theta} = -\frac{1}{R_\phi} (D_{n,\phi} - D_\phi) \tag{6.38d}$$

$$D_{\theta\phi} = \frac{D_\theta}{R_\theta} \tag{6.38e}$$

The indefinite integrals in equations (6.38a) and (6.38b) may be written in the alternate form

$$\bar{\xi} = \frac{D_\phi}{\sin\phi} = \bar{\xi}(\phi''') + \frac{1}{E} \int_{\phi'''}^{\phi} \frac{1}{h\sin\phi}$$

$$\times [(R_\phi + \mu R_\theta)N_\phi - (R_\theta + \mu R_\phi)N_\theta]\, d\phi \tag{6.39a}$$

$$\bar{\psi} = \frac{D_\theta}{R_0} = \bar{\psi}(\phi'''') + \frac{2(1+\mu)}{E} \int_{\phi''''}^{\phi} \frac{R_\phi}{hR_0} S\, d\phi \tag{6.39b}$$

where ϕ''' and ϕ'''' are the boundaries at which D_ϕ and D_θ are specified. We have used ϕ''' and ϕ'''' to emphasize that these boundaries are *not* the same boundaries where the corresponding stress resultants are specified, ϕ' and ϕ'', as defined in equations (4.10) and (4.11). Recalling the arguments of section 6.2.2, we cannot specify N_ϕ and D_ϕ, nor S and D_θ on the same boundary. Moreover, since we have presumably chosen the boundary values $N_\phi(\phi')$ and $S(\phi'')$ in the equilibrium solution, D_ϕ and D_θ must be specified at the other boundary in each case. Thus, we have little latitude in the choice for ϕ''' and ϕ''''.

For a dome, we again encounter indeterminate forms for D_ϕ and D_θ at the pole. It is easily shown using L'Hospital's rule that $D_\phi(0) = D_\theta(0) = 0$. Then, equations (6.38a–c) gives

$$D_n(0) = \frac{R_\theta(0)}{Eh} \left[N_\theta(0) - \mu N_\phi(0) \right] \tag{6.40}$$

As an example, we investigate a spherical dome under dead load, using the solution for stresses derived in section 4.3.2.2, and referring to figure 4–2 with $\phi_t = 0$. We wish to compute the displacement normal to the middle surface D_n at the lower boundary $\phi = \phi_b$.

Examining equation (6.38c) we observe that since $D_\phi(\phi_b) = 0$ as dictated by the requirement to fully develop $N_\phi(\phi_b)$, we need only the values of N_ϕ and N_θ at $\phi = \phi_b$. From equations (4.20) and (4.21), we have

$$N_\phi(\phi_b) = - \frac{qa}{1 + \cos \phi_b} \tag{6.41}$$

and

$$N_\theta(\phi_b) = qa \left(\frac{1}{1 + \cos \phi_b} - \cos \phi_b \right) \tag{6.42}$$

We substitute these values into equation (6.38c) with $R_\theta = a$, to get

$$D_n(\phi_b) = \frac{qa^2}{Eh} \left(\frac{1 + \mu}{1 + \cos \phi_b} - \cos \phi_b \right) \tag{6.43}$$

At the pole, the normal displacement is found using equations (4.22) and (6.40):

$$D_n(0) = \frac{-qa^2}{Eh} (1 + \mu) \tag{6.44}$$

As a further example, we consider the displacements of a hyperboloidal shell under self-dead load, previously studied in section 4.3.2.3. Referring to equations (6.38a) and (6.38b), since $S = 0$, $\bar{\psi}$ is 0. Then, $\bar{\xi}$ is evaluated from equation (6.39a), with equations (4.40) and (4.41) substituted for N_ϕ and N_θ and ϕ''' taken as ϕ_b. The resulting integral is quite complicated, but it is easily evaluated using a numerical algorithm, such as the trapezoidal method. Finally, D_n is found from equation (6.38c).

A study of the displacements for a parametric range of hyperboloidal shell dimensions, similar to that shown on figure 4–8 for the stress resul-

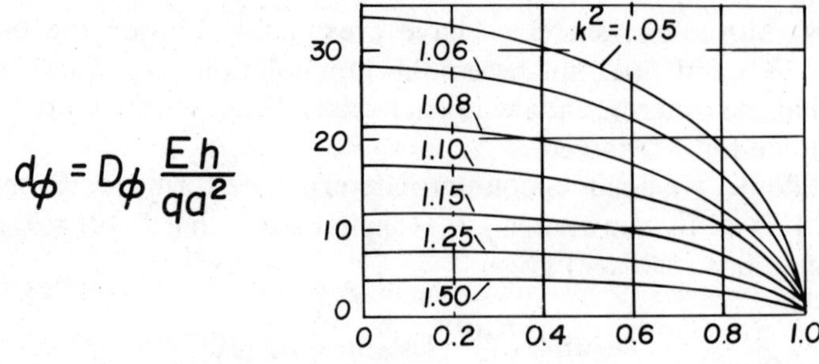

$$d_\phi = D_\phi \frac{Eh}{qa^2}$$

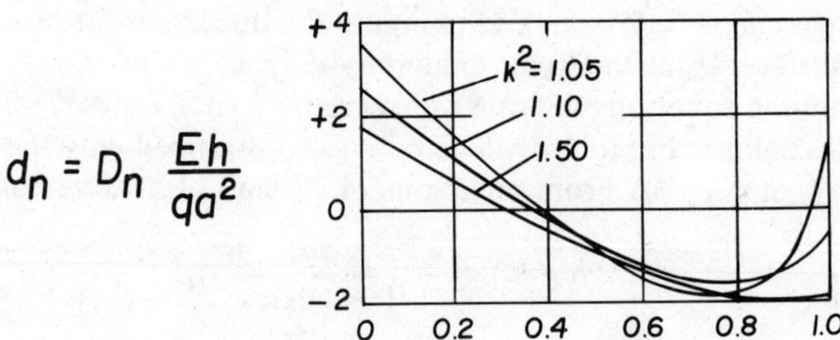

$$d_n = D_n \frac{Eh}{qa^2}$$

Source: P. L. Gould and S. L. Lee, "Hyperbolic Cooling Towers under Seismic Design Load," *Journal of the Structural Division, ASCE* 93, no. ST5 (October 1967): 95.

Figure 6–6. Nondimensional Displacements for a Hyperboloidal Shell with $a/t = 0.90$ and $a/s = 0.55$ (Reprinted with permission of American Society of Civil Engineers)

tants, is given in figure 6–6. The nondimensionalizing parameters are indicated on the figure. Also a complete tabulation is available in Gould.[13]

6.3.3.3 Nonsymmetric Loading. To complete the analogous treatment of the equations of equilibrium and compatibility, we apply to equations (6.36a) and (6.36b), the procedure initially used to obtain equation (4.83) from equations (4.6a) and (4.6b) in section 4.3.5.1. The result is

$$\frac{1}{R_\phi R_\theta \sin \phi} \left[\frac{R_\theta^2 \sin \phi}{R_\phi} \bar{\psi},_\phi \right]_{,\phi} + \frac{1}{R_\phi \sin^2 \phi} \bar{\psi},_{\theta\theta} = \frac{1}{R_\phi R_\theta \sin \phi \, Eh}$$

$$\times \left\{ 2(1 + \mu)(R_\theta S),_\phi + \left(\frac{1}{\sin \phi} \left[(R_\phi + \mu R_\theta)N_\phi - (R_\theta + \mu R_\phi)N_\theta \right] \right) \right\}$$

$$(6.45)$$

which is in the same form as equation (4.83).

All strains and displacements have been previously defined in harmonic form by equations (5.50) and (5.51), and the corresponding strain-displacement relations have been written as equations (5.52). With the stress resultants previously expanded in Fourier series in equation (4.84b), equations (6.34a–c) are easily written in separated form. Further, we take the variables $\bar{\psi}$ and $\bar{\xi}$ as

$$\begin{Bmatrix} \bar{\psi} \\ \bar{\xi} \end{Bmatrix} = \sum_{j=0}^{\infty} \begin{Bmatrix} \bar{\psi}^j \cos j\theta \\ \bar{\xi}^j \sin j\theta \end{Bmatrix} \tag{6.46}$$

and the separated form of equation (6.45), analogous to equation (4.86), is

$$\frac{1}{R_\phi R_\theta \sin \phi} \left[\frac{R_\theta^2 \sin\phi}{R_\phi} \bar{\psi}^j,_\phi \right],_\phi - \frac{j^2}{R_\phi \sin^2 \phi} \bar{\psi}^j = \frac{1}{R_\phi R_\theta \sin \phi \; Eh}$$

$$\times \left\{ 2(1 + \mu)(R_\theta S^j),_\phi - j\left(\frac{1}{\sin \phi} \left[(R_\phi + \mu R_\theta)N_\phi^j - (R_\theta + \mu R_\phi)N_\theta^j \right] \right) \right\}$$

$$\tag{6.47}$$

The final major equation for the displacements corresponds to the antisymmetrical case, $j = 1$, earlier discussed in section 4.3.6. Following the same steps used in section 4.3.6.1, we introduce the variable

$$\bar{\gamma}(\phi) = \bar{\psi}^1(\phi)R_\theta \sin \phi \tag{6.48}$$

into equation (6.47) with j taken as 1. The equation reduces to

$$\left(\frac{1}{R_\phi \sin \phi} \gamma,_\phi \right),_\phi = \frac{1}{R_\theta \sin \phi \; Eh}$$

$$\times \left\{ 2(1 + \mu)(R_\theta S^1),_\phi - \left(\frac{1}{\sin \phi} \left[(R_\phi + \mu R_\theta)N_\phi^1 - (R_\theta + \mu R_\phi)N_\theta^1 \right] \right) \right\}$$

$$\tag{6.49}$$

which may be solved as indicated in section 4.3.6.

We now turn to the asymmetric case, $j > 1$, which is described by equation (6.47). For spherical and cylindrical shells, it is possible to arrive at solutions for the asymmetric displacements that are analogous to the stress solutions given in section 4.3.7.2 and section 4.3.7.4, respectively.[14] For more complicated geometries, the transformation applied in section 4.3.7.5 may be of some value; however, little work has apparently been done in this area in terms of analytical solutions. Rather, for the most part, numerical techniques have been used to evaluate displacements for asymmetrically loaded shells of revolution.[15]

Now we have completed, in some detail, the displacement analysis for shells of revolution. Once equation (6.47) is solved for $\bar{\psi}^j$ to determine D_θ^j, $\bar{\xi}^j$ is computed from equation (6.36b) to get D_ϕ^j and then D_n^j is found from equation (6.34b).

It does not seem particularly useful to evaluate the membrane displacements for specific shell geometries in the same detail that is devoted to stress analysis, since the remaining treatment is self-evident based on the analogy established. However, it is of general interest to summarize some results of various studies that can be consulted by the interested reader.

First, for closed shells as shown in figure 4–1, certain relationships must exist among the displacements at the pole to avoid singularities in the solution. These have been derived in Brombolich and Gould,[16] and are summarized in table 6–1. The derivation of these relationships is discussed further in section 9.3.4.4.

Next, since hyperboloidal shells under seismic loading were considered in much detail, note that the nondimensionalized displacements for the M distribution are tabulated in Gould and Lee.[17]

Last, we should discuss the general influences of the various harmonic components of displacement $j = 0$, $j = 1$, and $j > 1$ on a rotational shell of the tower type. It is convenient to do this with respect to the R–Z Cartesian coordinates system shown on figure 2–11.

For the $j = 0$ case, the resultant displacement will be principally an elongation or shortening of the shell in the Z direction, with any R deformation being an expansion or contraction of the parallel circles, labeled as D^0 on figure 6–7(a).

For the $j = 1$ case, the resultant displacement will mainly consist of a lateral movement in the R direction, with comparatively little movement in the Z direction. This is illustrated in figure 6–7(b). The maximum lateral displacement D^1 must be checked to verify the applicability of the small deformation theory (assumption [2], table 1–1).

As discussed in chapter 4, both the $j = 0$ and $j = 1$ cases may be solved from the elementary theories of axial deformation and beam bending. Consequently, there is no distortion of the cross section. However, for the $j > 1$ case, the displacements will vary around the circumference in proportion to $\sin j\theta$ and $\cos j\theta$, and the cross section will distort as shown in figure 6–7(c). In problems with considerable $j > 1$ participation, the limits of the small deformation theory are frequently taxed by the circumferential distortions, where the relative displacements represented by the difference in the peak amplitudes should be checked. This difference is labeled as $2D^j$ on the figure.

For an actual loading that may consist of contributions from more than one harmonic, the maximum relative circumferential displacement, after summing all participating harmonics, should be checked against the limits of small deformation theory.

Table 6–1
Displacement Constraints for Closed Shells

Harmonic Number j	Continuous Dome @ $\phi = 0$ (Figure 4–1a)	Discontinuous Meridian @ $\phi = \phi_t$ (Figure 4–1b)
0	$D_\phi = D_\theta = D_{\phi\theta} = D_{\theta\phi} = 0$	$D_\theta = D_{\phi\theta} = D_{\theta\phi} = 0;$ $D_\theta = -\tan\phi_t\, D_n$
1	$D_n = 0$ $D_\phi + D_\theta = 0;\ D_{\phi\theta} + D_{\theta\phi} = 0$	$D_{\phi\theta} = D_{\theta\phi} = 0;\ D_\phi = -\cos\bar\phi\, D_\theta;$ $D_n = -\sin\phi_t\, D_\theta$
>1	$D_\phi = D_\theta = D_n = D_{\phi\theta} = D_{\theta\phi} = 0$	

6.3.4 Shells of Translation

6.3.4.1 Cylindrical Shells. We consider the general strain-displacement relationship, equations (5.46a–c), and the strain-stress resultant relationships, equations (6.33a–c), with the curvilinear coordinates adopted for the cylindrical geometry in section 4.4.1. Thus, we take $\alpha = X, \beta = \theta; A = 1, B = a;$ and $R_\alpha = \infty, R_\beta = a$ to get

$$\epsilon_X = D_{X,X} = \frac{1}{Eh}\left[(N_X - \mu N_\theta) + (1 - \mu)N_T\right] \tag{6.50a}$$

$$a\epsilon_\theta = D_{\theta,\theta} + D_n = \frac{a}{Eh}\left[(N_\theta - \mu N_X) + (1 - \mu)N_T\right] \tag{6.50b}$$

$$a\omega = aD_{\theta,X} + D_{X,\theta} = \frac{2(1 + \mu)a}{Eh}\,S \tag{6.50c}$$

As explained in section 6.3.3.1, the thermal terms are conveniently absorbed into the stress resultants and are not carried forth explicitly.

We first integrate equation (6.50a) to get

$$D_X = \frac{1}{Eh}\int (N_X - \mu N_\theta)\,dx + f_3(\theta) \tag{6.51}$$

From equation (6.50c), we write

$$D_\theta = -\frac{1}{a}\int D_{X,\theta}\,dX + \frac{1(1 + \mu)}{Eh}\int S\,dX + f_4(\theta) \tag{6.52}$$

and then from equation (6.50b), we find

$$D_n = \frac{a}{Eh}(N_\theta - \mu N_X) - D_{\theta,\theta} \tag{6.53}$$

The membrane rotations are found from equation (5.43), with $\gamma_\alpha = \gamma_\beta = 0$:

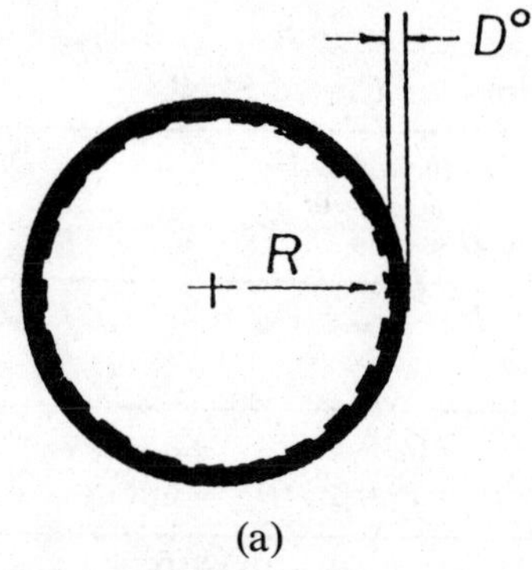

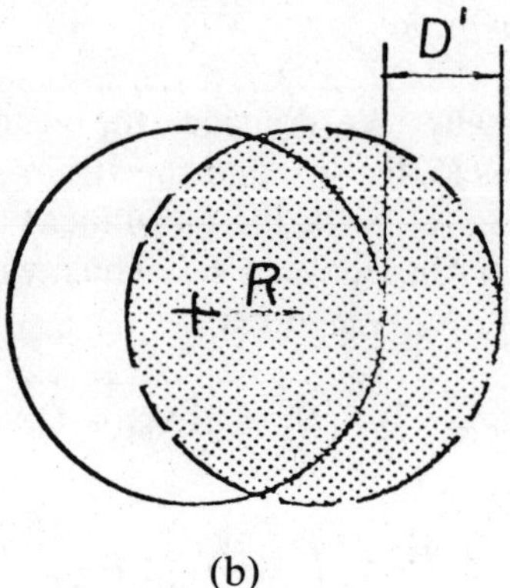

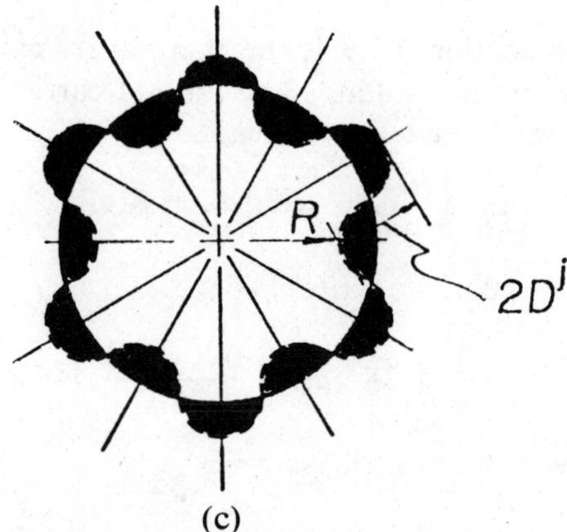

Figure 6–7. Harmonic Components of the Displacement of a Shell of Revolution

$$D_{X\theta} = -D_{n,X} \tag{6.54}$$

and

$$D_{\theta X} = -\frac{1}{a}(D_{n,\theta} - D_{\theta}) \tag{6.55}$$

A ready example is the simply supported cylindrical shell subject to the first harmonic of the Fourier series expansion for the dead load. The stress analysis for this loading was carried out in section 4.4.1, and the stress resultants are given by equation (4.169). The boundary conditions must be stated with respect to the displacements. The longitudinal symmetry dictates that $D_X(L/2) = 0$. Since the boundary corresponds to the *simply supported* condition discussed in section 6.2.2, we also have $D_\theta(0) = 0$. These boundary conditions yield $f_3(\theta) = f_4(\theta) = 0$.

Proceeding, we find by substituting equations (4.169) into equations (6.51)–(6.55), and carrying out the integrations and substitutions,

$$D_X = \frac{4q_d}{\pi Eh}\left(\frac{L}{\pi}\right)\left[\frac{2L^2}{\pi^2 a} - \mu a\right]\cos\frac{\pi X}{L}\cos\theta \tag{6.56a}$$

$$D_\theta = \frac{4q_d}{\pi Eh}\left(\frac{L^2}{\pi^2}\right)\left[\frac{2L^2}{\mu^2 a^2} + 4 + 3\mu\right]\sin\frac{\pi X}{L}\sin\theta \tag{6.56b}$$

$$D_n = -\frac{4q_d}{\pi Eh}\left[(4 + \mu)\frac{L^2}{\pi^2} + \frac{2L^4}{\pi^4 a^2} + a^2\right]\sin\frac{\pi X}{L}\cos\theta \tag{6.56c}$$

$$D_{X\theta} = \frac{4q_d}{\pi Eh}\left(\frac{\pi}{L}\right)\left[(4 + \mu)\frac{L^2}{\pi^2} + \frac{2L^4}{\pi^4 a^2} + a^2\right]\cos\frac{\pi X}{L}\cos\theta \tag{6.56d}$$

$$D_{\theta X} = -\frac{4q_d}{\pi Eh}\left[a - \frac{2\mu L^2}{\pi^2 a}\right]\sin\frac{\pi X}{L}\sin\theta \tag{6.56e}$$

The preceding expressions are useful in the bending analysis of open cylindrical shells, where the membrane theory solution serves as the particular solution. The boundary displacements, as computed from equations (6.56a–c), are corrected by edge forces and moments to satisfy prescribed compatibility conditions on the longitudinal boundaries.[18] The procedure is quite analogous to the classical flexibility method of structural analysis, as discussed briefly in section 6.3.1 and is explored in more detail in chapter 9.

We may at this point consider an example that represents an exception to the general association of *membrane theory analysis* with *statically determinate systems*. Occasionally, boundary conditions are encountered that do not grossly violate the requirements discussed in section 4.2, and yet do not permit the *a priori* determination of the stress resultants before considering the displacements. We might call this a *statically indeterminate membrane theory* problem.

As an illustration, reconsider the previous example with an axial constraint condition

$$D_X(0) = D_X(L) = 0 \tag{6.57a}$$

replacing the condition

$$N_X(0) = N_X(L) = 0 \tag{6.57b}$$

The static condition enabled $f_1(\theta)$ and $f_2(\theta)$ in equations (4.168a) and (4.168b) to be set equal to 0 and, subsequently, the explicit expressions for S and N_X to be written as equations (4.169a) and (4.169b). We will retain the other static boundary conditions $S(L/2) = 0$, since the symmetry is not altered; thus, $f_1(\theta)$ is still 0. Proceeding, the expressions for the stress resultants in the modified problem presently being considered are written from equations (4.168) as

$$S = \frac{8q_d L}{\pi^2} \cos \frac{\pi X}{L} \sin \theta \tag{6.58a}$$

$$N_X = \frac{8q_d L^2}{\pi^3 a} \sin \frac{\pi X}{L} \cos \theta + f_2(\theta) \tag{6.58b}$$

$$N_\theta = -\frac{4q_d a}{\pi} \sin \frac{\pi X}{L} \cos \theta \tag{6.58c}$$

We note from equation (6.58b) that the function $f_2(\theta)$ remains to be determined, obviously by consideration of the displacements. If we substitute equations (6.58a) and (6.58b) into equations (6.51) and (6.52), we have

$$D_X = D_{XS} + \frac{1}{Eh} f_2(\theta) X + f_3(\theta) \tag{6.59a}$$

$$D_\theta = D_{\theta S} - \frac{1}{2aEh} f_2(\theta),_\theta X^2 - \frac{1}{a} f_3(\theta),_\theta X + f_4(\theta) \tag{6.59b}$$

where D_{XS} and $D_{\theta S}$ represent the corresponding displacements from the simply supported cases, equations (6.56a) and (6.56b), respectively. We now have three functions of integration to be evaluated from the boundary conditions

$$D_X(0) = D_X(L) = 0 \tag{6.60a}$$

$$D_\theta(0) = D_\theta(L) = 0 \tag{6.60b}$$

$$D_X\left(\frac{L}{2}\right) = 0 \tag{6.60c}$$

These equations form only three independent conditions because of the symmetry of the problem. Since $D_{\theta S}$ and D_{XS} automatically satisfy equations (6.60b) and (6.60c), we find from evaluating $D_X(0)$, $D_\theta(0)$, and $D_X(L/2)$ that

$$D_{XS}(0,\theta) + f_3(\theta) = 0 \tag{6.61a}$$

$$f_4(\theta) = 0 \tag{6.61b}$$

$$\frac{1}{Eh} f_2(\theta) \frac{L}{2} + f_3(\theta) = 0 \tag{6.61c}$$

from which

$$f_2(\theta) = \frac{2Eh}{L} D_{XS}(0,\ \theta) \tag{6.62a}$$

$$f_3(\theta) = -D_{XS}(0,\ \theta) \tag{6.62b}$$

$$f_4(\theta) = 0 \tag{6.62c}$$

It is easily verified that the conditions on $D_X(L)$ and $D_\theta(L)$ are similarly satisfied by equations (6.62a–c).

To complete the analysis, we substitute equations (6.62a–c) into equations (6.58) and (6.59) and take D_{XS} and $D_{\theta S}$ from equations (6.56a) and (6.56b). The modified expression for the stress resultants N_X is

$$N_X = -\frac{8q_d}{\pi^2} \left\{ \frac{L^2}{\pi a} \sin \frac{\pi X}{L} - \left[\frac{2L^2}{\pi^2 a} - \mu a \right] \right\} \cos \theta \tag{6.63}$$

with S and N_θ given by equations (6.58a) and (6.58c), respectively. The corresponding displacements are

$$D_X = \frac{4q_d}{\pi Eh} \left(\frac{L}{\pi} \right) \left[\frac{2L^2}{\pi^2 a} - \mu a \right] \left[\cos \frac{\pi X}{L} + \frac{2X}{L} - 1 \right] \cos \theta \tag{6.64a}$$

$$D_\theta = \frac{4q_d}{\pi Eh} \left(\frac{L}{\pi} \right) \left[\frac{L}{\pi} \left(\frac{2L^2}{\pi^2 a^2} + 4 + 3\mu \right) \sin \frac{\pi X}{L} \right.$$

$$\left. + \left(\frac{2L^2}{\pi^2 a} - \mu a \right) \left(\frac{X^2}{aL} - \frac{X}{a} \right) \right] \sin \theta \tag{6.64b}$$

The remaining displacements D_n, $D_{X\theta}$, and $D_{\theta X}$ are easily evaluated from equations (6.53)–(6.55).

We should remember that the applications of this statically indeterminate membrane analysis are somewhat restricted, since the admissible boundary conditions may not grossly violate the membrane theory requirements.

6.3.4.2 Shells with Double Curvature. For doubly curved shells of the form considered in section 4.4.2, the computation of displacements due to the membrane stresses is not widely treated in the literature. This is probably due to two main reasons:

(a) One of the principal uses for the membrane displacements is to incorporate them into a flexibility type general solution, as described in the previous section. Although this approach is quite applicable for shells of revolution and for cylindrical shells, it is not particularly suited for doubly curved translational shells.

(b) The integration of the stress resultant-displacement relations generally must be carried out numerically, even for the shell of revolution geometry, although the stress resultants may have been evaluated analytically. Rather than deal with partially analytical, partially numerical solutions, it is often expedient to employ a strictly numerical approach. Such a technique for translational shells is described in Hedgren and Billington.[19] Also, the general techniques of finite differences[20] and finite elements[21] have been applied to this class of problem.

Notes

1. R. D Lowrey and P. L. Gould, "Thermal Analysis of Orthotropic Layered Shells of Revolution by the Finite Element Method," Proc. of the IASS Symposium on Shell Structures and Climatic Influences, University of Calgary, Alberta, Canada, July 1972, pp. 315–325.

2. S. B. Dong and F. K. W. Tso, "On a Laminated Orthotropic Shell Theory Including Transverse Shear Deformations," *Journal of Applied Mechanics, Trans. ASME* 39, series E, no. 4 (December 1972): 1091–1097.

3. K. P. Buchert, *Buckling of Shell and Shell-Like Structures* (Columbia, Mo.: K. P. Bucher and Associates, 1973), pp. 5–10, 31–34.

4. S. Timoshenko and S. Woinowsky-Krieger, *Theory of Plates and Shells*, 2nd ed. (New York: McGraw-Hill, 1959), pp. 368–369.

5. Buchert, *Buckling of Shell and Shell-Like Structures*.

6. E. Reissner, "Some Aspects of the Theory of Thin Elastic Shells," *Journal, Boston Society of Engineers*, Boston, Mass., 42, no. 2, (April 1956): 100–133.

7. J. J. Huffington, Jr., "Theoretical Determination of Rigidity Properties of Orthogonally Stiffened Plates," *Journal of Applied Mathematics, ASME*, paper no. 55–A–12 (March 1956): 15–20.

8. G. Kirchhoff, "Vorlesungen über Mathematische Physik," vol. 1. *Mechanik*, 1877, p. 450.

9. Lord Kelvin and P. G. Tait, *Treatise on Natural Philosophy*, vol. 1, pt. 2, 1883, p. 188.

10. S. Timoshenko and J. N. Goodier, *Theory of Elasticity*, 2nd ed. (New York: McGraw-Hill, 1951), p. 33.

11. P. G. Glockner, "Symmetry in Structural Mechanics," *Journal of the Structural Division, ASCE* 99, no. ST1 (January 1973): 71–89.

12. V. V. Novozhilov, *Thin Shell Theory* [translated from 2nd Russian ed. by P. G. Lowe (Groningen, The Netherlands: Noordhoff, 1964), p. 118].

13. P. L. Gould and S. L. Lee, "Hyperbolic Cooling Towers under Seismic Design Load," *Journal of the Structural Division, ASCE* 93, no. ST3, (June 1967): 87–109.

14. W. Flügge, *Stresses in Shells*, 2nd ed. (Berlin: Springer-Verlag, 1973), pp. 85–87, 121–124.

15. D. W. Martin, J. S. Maddock, and W. E. Scriven, "Membrane Displacements in Hyperbolic Cooling Towers Due to Wind- and Dead-Loading," *Proc. Institution of Civil Engineers* 28, 1964, pp. 327–337.

16. L. J. Brombolich and P. L. Gould, "Finite Element Analysis of Shells of Revolution by Minimization of the Potential Energy Functional," Proc. Conference on Applications of the Finite Element Method in Civil Engineering, Vanderbilt University, Nashville, Tenn., November 1969, pp. 279–307; L. J. Brombolich and P. L. Gould, "A High-Precision Curved Shell Finite Element," *Synoptic, AIAA Journal* 10, no. 6 (June 1972): 727–728.

17. Gould and Lee, "Hyperbolic Cooling Towers."

18. "Design of Cylindrical Concrete Roofs," *ASCE Manual of Engineering Practice* 31 (New York: American Society of Civil Engineers, 1952).

19. A. W. Hedgren and D. P. Billington, "Numerical Analysis of Translational Shell Roofs," *Journal of the Structural Division, ASCE* 92, no. ST1 (February 1966): 223–244.

20. M. Soare, *Application of Finite Difference Equations to Shell Analysis* (Oxford: Pergamon Press, 1967).

21. R. W. Clough and C. P. Johnson, "Finite Element Analysis of Arbitrary Thin Shells," *Concrete Thin Shells*, ACI publication SP–28 (Detroit: American Concrete Institute, 1971), pp. 333–363.

7

Energy and Approximate Methods

7.1 General

An energy formulation is often viewed as an alternative to the differential equation approach in solid mechanics. There is, however, a direct connection between the energy and the differential equation approaches through the Principle of Virtual Displacements and through various extremum principles. We will not pursue this connection, since it is beyond our immediate scope.

Our interest in introducing energy considerations here is twofold. First, the energy approach is the basis for many of the powerful numerical techniques for solving shell and plate problems. Second, energy solutions are an important resource even in the classical formulations that we stress in this book. We illustrate the latter application in some detail in chapter 9, and the materials in this chapter may be deferred until the appropriate sections are encountered without loss in continuity.

In this chapter, we highlight some of the more important aspects of energy methods. We primarily stress those items essential to support the ensuing applications. Also, we introduce some approximations that form the basis of many contemporary numerical techniques.

7.2 Strain Energy

The strain energy for an elastic body U_ϵ may be written in terms of the strain energy density dU_ϵ as

$$U_\epsilon = \int_V dU_\epsilon \tag{7.1}$$

where V represents the volume of the continuum.

In terms of the orthogonal curvilinear coordinates, as defined in chapter 2, and the differential thickness $d\zeta$, as shown on figure 3–1(a), the strain energy density follows from the linear theory of elasticity:

$$
\begin{aligned}
dU_\epsilon = [&\sigma_{\alpha\alpha}(\alpha, \beta, \zeta)\epsilon_\alpha(\alpha, \beta, \zeta) + \sigma_{\beta\beta}(\alpha, \beta, \zeta)\epsilon_\beta(\alpha, \beta, \zeta) \\
&+ \sigma_{nn}(\alpha, \beta, \zeta)\epsilon_n(\alpha, \beta, \zeta) + \sigma_{\alpha\beta}(\alpha, \beta, \zeta)\omega(\alpha, \beta\ \zeta) \\
&+ \sigma_{\alpha n}(\alpha, \beta)\gamma_\alpha(\alpha, \beta) + \sigma_{\beta n}(\alpha, \beta)\gamma_\beta(\alpha, \beta)]\, dV
\end{aligned}
\tag{7.2}
$$

256

The stress terms are introduced in section 3.1, while the strains are defined in section 5.3.

For a shell or plate, the differential volume dV may be expressed in terms of the differential area dS and thickness $d\zeta$ as

$$dV = dSd\zeta$$

$$= A(\zeta)\,d\alpha B(\zeta)\,d\beta d\zeta$$

$$= A\left(1 + \frac{\zeta}{R_\alpha}\right)d\alpha B\left(1 + \frac{\zeta}{R_\beta}\right)d\beta d\zeta \tag{7.3a}$$

or, after dropping terms of $O(h/R){:}1$,

$$dV = AB\,d\alpha d\beta d\zeta \tag{7.3b}$$

It is generally preferable to write the strain energy in terms of one type of basic quantity; i.e., stresses, strains, or displacements. To retain generality, we choose strains, which are connected to each of the others by a single set of equations, (6.1) and (5.46), respectively.

We proceed by noting that $\sigma_{nn} = 0$ (assumption [4], table 1–1) and by rewriting equation (7.1) in the matrix notation introduced in equation (6.2):

$$U_\epsilon = \frac{1}{2}\int_V \lfloor\sigma\rfloor\{\epsilon\}\,dV \tag{7.4}$$

Now, substituting equation (6.2) into (7.4), we get

$$U_\epsilon = \frac{1}{2}\int_V \lfloor[\mathbf{H}]\{\epsilon\} - \{\sigma_T\}\rfloor\{\epsilon\}\,dV \tag{7.5}$$

Keeping in mind that the elements of $\{\sigma_T\}$ may be regarded as known, equation (7.5) is an expression of the strain energy in terms of strain-type quantities alone. We may now multiply out the terms in equation (7.5). Since the arithmetic is quite lengthy and straightforward, it is omitted here. After integration through the thickness and some rearrangement, the resulting expression is

$$U_\epsilon = \frac{E}{2(1-\mu^2)}\int_S \left\{h\left[\epsilon_\alpha + \epsilon_\beta)^2 - 2(1-\mu)\left(\epsilon_\alpha\epsilon_\beta - \frac{\omega^2 + \gamma_\alpha^2 + \gamma_\beta^2}{4}\right)\right]\right.$$

$$+ \frac{h^2}{4}\left[(\epsilon_\alpha + \epsilon_\beta)(\kappa_\alpha + \kappa_\beta) - (1-\mu)(\epsilon_\alpha\kappa_\beta + \epsilon_\beta\kappa_\alpha - 2\omega\tau)\right]$$

$$+ \frac{h^3}{12}\left[(\kappa_\alpha + \kappa_\beta)^2 - 2(1-\mu)\kappa_\alpha\kappa_\beta - \tau^2)\right]$$

$$- \bar{\alpha}(1 + \mu)(\epsilon_\alpha + \epsilon_\beta) \int_{-h/2}^{h/2} T(\zeta)\,d\zeta$$

$$- \bar{\alpha}(1 + \mu)(\kappa_\alpha + \kappa_\beta) \int_{-h/2}^{h/2} T(\zeta)\zeta\,d\zeta \bigg\} AB\,d\alpha d\beta \qquad (7.6)$$

Modified expressions for U_ϵ which correspond to other stress-strain laws may be written in the same manner using the appropriate [C], as discussed in chapter 6.

If for the moment we neglect the temperature dependent terms, we may group equation (7.6) as

$$U_\epsilon = \frac{E}{2(1 - \mu^2)} \int_S \left\{ h[\mathrm{I}] + \frac{h^2}{4}\,[\mathrm{II}] + \frac{h^3}{12}\,[\mathrm{III}] \right\} AB\,d\alpha d\beta \qquad (7.7)$$

It is convenient to interpret equation (7.7) with respect to the three component expressions [I, II, III], which are, respectively, linear, quadratic, and cubic in the thickness h. First, the linear component $Eh/[2(1 - \mu^2)]\,[\mathrm{I}]$ represents extensional and shearing energy, which is predominant in shells that behave primarily in accordance with the membrane theory. The quadratic component $Eh^2/[8(1 - \mu^2)]\,[\mathrm{II}]$ is a coupling between extensional and shearing, and twisting and bending terms, and enters into some geometrically nonlinear formulations. Finally, the cubic component $Eh^3/[24(1 - \mu^2)]\,[\mathrm{III}]$ contains both bending and twisting energy which are the primary resistance mechanisms in the flexure of thin plates. In many applications based on energy methods, only one or two of the three components are required.

7.3 Potential Energy of the Applied Loads

In equation (3.11), the surface loading vector $\mathbf{q}$ is defined in terms of loading intensities per unit area of middle surface $q_\alpha(\alpha, \beta)$, $q_\beta(\alpha, \beta)$, and $q_n(\alpha, \beta)$. Here, we may also admit loads which act along a coordinate line. These *line loads* are defined in the same fashion as the surface loading vectors

$$\hat{\mathbf{q}}(\alpha, \bar{\beta}) = \hat{q}_\alpha(\alpha, \bar{\beta})\mathbf{t}_\alpha + \hat{q}_\beta(\alpha, \bar{\beta})\mathbf{t}_\beta + \hat{q}_n(\alpha, \bar{\beta})\mathbf{t}_n \qquad (7.8a)$$

for a load acting along the s_β coordinate line corresponding to $\beta = \bar{\beta}$, and

$$\hat{\mathbf{q}}(\bar{\alpha}, \beta) = \hat{q}(\bar{\alpha}, \beta)\mathbf{t}_\alpha + \hat{q}(\bar{\alpha}, \beta)\mathbf{t}_\beta + \hat{q}_n(\bar{\alpha}, \beta)\mathbf{t}_n \qquad (7.8b)$$

for a load acting along the s_α coordinate line corresponding to $\alpha = \bar{\alpha}$. Boundary reactions are frequently represented by such line loads. Furthermore, we may accommodate concentrated loads defined by

258

$$\tilde{\mathbf{q}}(\bar{\alpha}, \bar{\beta}) = \tilde{q}_\alpha(\bar{\alpha}, \bar{\beta})\mathbf{t}_\alpha + \tilde{q}_\beta(\bar{\alpha}, \bar{\beta})\mathbf{t}_\beta + \tilde{q}_n(\bar{\alpha}, \bar{\beta})\mathbf{t}_n \qquad (7.9)$$

for a load acting at $(\bar{\alpha}, \bar{\beta})$. No body forces are included in development. For a thin continuum, it is generally sufficient to refer such forces to the middle surface as we did in chapter 3 for gravity loading.

Corresponding to each of the loading terms are the components of the middle surface displacement vector $\mathbf{\Delta}(\alpha, \beta)$. These components have been defined in equation (5.3a) as $D_\alpha(\alpha, \beta)$, $D_\beta(\alpha, \beta)$, and $D_n(\alpha, \beta)$.

It is convenient to define the *change* in potential energy of the applied loading U_q as the product of each loading component and the corresponding displacement component. Further, when a positive displacement occurs, U_q *decreases*, so that the product of the correspondents is given a *negative* sign. Thus we have

$$U_q = -\left\{ \int\!\!\int_S \mathbf{q}(\alpha, \beta) \cdot \mathbf{\Delta}(\alpha, \beta) A(\alpha, \beta) B(\alpha, \beta)\, d\alpha d\beta \right.$$

$$+ \sum_S \left[\int_{s_\alpha} \hat{\mathbf{q}}(\alpha, \bar{\beta}) \cdot \mathbf{\Delta}(\alpha, \bar{\beta}) A(\alpha, \bar{\beta})\, d\alpha \right.$$

$$+ \left. \int_{s_\beta} \hat{\mathbf{q}}(\bar{\alpha}, \beta) \cdot \mathbf{\Delta}(\bar{\alpha}, \beta) B(\bar{\alpha}, \beta)\, d\beta \right]$$

$$\left. + \sum_S \tilde{\mathbf{q}}(\bar{\alpha}, \bar{\beta}) \cdot \mathbf{\Delta}(\bar{\alpha}, \bar{\beta}) \right\} \qquad (7.10a)$$

or, in terms of the components,

$$U_q = -\left\{ \int\!\!\int_S q_i(\alpha, \beta) D_i(\alpha, \beta) A(\alpha, \beta) B(\alpha, \beta)\, d\alpha d\beta \right.$$

$$+ \sum_S \left[\int\!\!\int_{s_\alpha} \hat{q}_i(\alpha, \bar{\beta}) D_i(\alpha, \bar{\beta}) A(\alpha, \bar{\beta})\, d\alpha \right.$$

$$+ \left. \int_{s_\beta} \hat{q}_i(\bar{\alpha}, \beta) D_i(\bar{\alpha}, \beta) B(\bar{\alpha}, \beta)\, d\beta \right]$$

$$\left. + \sum_S \tilde{q}_i(\bar{\alpha}, \bar{\beta}) D_i(\bar{\alpha}, \bar{\beta}) \right\} \qquad (i = \alpha, \beta, n) \qquad (7.10b)$$

In equations (7.10a) and (7.10b), the second term signifies that the load potentials for each line load acting on the shell surface are summed, and the third term indicates that load potentials for each concentrated load are similarly summed.

Note that equation (7.8) can be generalized to include distributed, line, and concentrated couples corresponding to the rotations $D_{\alpha\beta}$ and $D_{\beta\alpha}$.

7.4 Energy Principles and Rayleigh–Ritz Method

7.4.1 Principle of Virtual Displacements

Consider an elastic body subject to a set of surface tractions composed of the distributed load vectors $\mathbf{q}(\alpha, \beta)$, as defined in equation (3.11e), as well as line load vectors $\hat{\mathbf{q}}(\alpha, \bar{\beta})$ and $\hat{\mathbf{q}}(\bar{\alpha}, \beta)$ and concentrated load vectors $\tilde{\mathbf{q}}(\bar{\alpha}, \bar{\beta})$, as defined in the previous section. If we assume that there are no heat or inertial effects, then the Law of Conservation of Energy requires that the work done by the surface tractions be equal to the strain energy stored in the material.

Now, a change is imposed on the displacement field $\mathbf{\Delta}$ to a new position $\mathbf{\Delta} + \delta\mathbf{\Delta}$, where $\delta(\ )$ is the variational operator. For our purposes, $\delta\mathbf{\Delta}$ may be regarded as a *linear increment* of the vector $\mathbf{\Delta}$. Also, $\delta\mathbf{\Delta}$ is assumed to be compatible with the constraints of the system, although this restriction is not necessary but only convenient.[1] The source of $\delta\mathbf{\Delta}$ is *not* specified; i.e., it does not necessarily result from any particular loading system. Hence, $\delta\mathbf{\Delta}$ is called a *virtual* displacement.

Next, we write the energy balance corresponding to the virtual displacement. The virtual work performed by the surface tractions is given by

$$
\begin{aligned}
\delta \bar{U}_q = {} & \int_S \mathbf{q}(\alpha, \beta) \cdot \delta\mathbf{\Delta}(\alpha, \beta) A(\alpha, \beta) B(\alpha, \beta)\, d\alpha\, d\beta \\[6pt]
& + \sum_S \left[\int_{s_\alpha} \hat{\mathbf{q}}(\alpha, \bar{\beta}) \cdot \delta\mathbf{\Delta}(\alpha, \bar{\beta}) A(\alpha, \bar{\beta})\, d\alpha \right. \\[6pt]
& \left. + \int_{s_\beta} \hat{\mathbf{q}}(\bar{\alpha}, \beta) \cdot \delta\mathbf{\Delta}(\bar{\alpha}, \beta) B(\bar{\alpha}, \beta)\, d\beta \right] \\[6pt]
& + \sum_S \tilde{\mathbf{q}}(\bar{\alpha}, \bar{\beta}) \cdot \delta\mathbf{\Delta}(\bar{\alpha}, \bar{\beta})
\end{aligned}
\tag{7.11}
$$

while the strain energy is changed by δU_ϵ, which, from equation (7.1), is written as

$$
\delta U_\epsilon = \delta \int_V dU_\epsilon
\tag{7.12}
$$

The r.h.s. of equation (7.12)

$$
\delta \int_V dU_\epsilon
$$

may be evaluated from equation (7.5); however, since equation (7.5) as yet is not written as an explicit function of the displacement Δ, and since variations or increments of Δ are being considered, we choose to leave the r.h.s. of equation (7.12) in the general form.

Then, the energy balance is written by equating equations (7.11) and (7.12):

$$\delta \bar{U}_q = \delta \int_V dU_\epsilon \qquad (7.13)$$

which is a statement of the Principle of Virtual Displacement. This principle is a special form of the Principle of Virtual Work, which is regarded by many contemporary mechanicians as the cornerstone of solid mechanics. (For a complete discussion of the interrelation of the various energy principles of solid mechanics, the interested reader is referred to Washizu.[2])

The Principle of Virtual Displacement, as stated in this section, can be taken as an alternate statement of the condition of static equilibrium as derived in chapter 3. This interpretation is used freely in the ensuing sections.

7.4.2 Principle of Minimum Total Potential Energy

It is convenient to make some further assumptions at this point. First, we assume that the volume V does not change during the virtual displacement, and that the surface tractions also do not vary. Then the variational operator $\delta(\ )$ is restricted to variations or increments in *displacement* only. We now recognize that $\delta \bar{U}_q$, as defined in equation (7.11), is simply $-\delta U_q$, where the load potential U_q was defined in equation (7.10a). Hence, we may rewrite equation (7.13) as

$$\delta U_\tau = \delta U_\epsilon + \delta U_q = \delta(U_\epsilon + U_q) = 0 \qquad (7.14)$$

where U_τ is known as the total potential energy. Bear in mind that U_τ itself is an integral function of various algebraic functions of the dependent variables, as demonstrated by equations (7.6) and (7.10). Such a function is called a *functional* in the terminology of the Calculus of Variations. Equation (7.14) states that *the total potential energy of an elastic system in equilibrium must be stationary with respect to all displacements satisfying the boundary conditions*. Technically, the term *stationary* corresponds to the first variation of the functional U_τ, δU_τ, vanishing, and may indicate an absolute or relative maximum or minumum; however, for our applications here, the stationary condition may be regarded as an absolute or relative minimum.

From the standpoint of solution strategy, an alternate interpretation of the principle is helpful: we seek a displacement field $\Delta(\alpha, \beta)$ which satisfies the kinematic boundary conditions and makes U_τ a minimum; then, the stress resultants and stress couples computed from $\boldsymbol{\Delta}$ via the strain-displacement and constitutive laws will satisfy the equilibrium considerations.

The classical solution for the problem of finding the minimum of a functional is the fundamental problem of the Calculus of Variations and is not within our scope here. The interested reader is referred to Forray[3] for an introductory treatment of this subject. For our purposes, we wish to pursue the application of one of the so-called *direct methods* of the Calculus of Variations to the solution of equation (7.14). Of the several possibilities available, the Rayleigh–Ritz method is perhaps the best known and simplest to understand, and thus we consider this technique first.

However, before proceeding, we might reflect on the first sentence of this chapter, where we noted that an energy formulation may be regarded as an alternative to a differential equation approach. Historically, the latter was called the method of *effective causes*, whereas the former was known as the method of *final causes*. The renowned mathematician Euler, who was the inventor of the Calculus of Variations, observed in an argument which concisely blends science and theology: "Since the fabric of the universe is most perfect, and is the work of a most wise Creator, nothing whatsoever takes place in the universe in which some relation of maximum and minimum does not appear. Wherefore there is absolutely no doubt that every effect in the universe can be explained as satisfactorily from final causes, by the aid of the method of maxima and minima, as it can from the effective causes themselves. . . ."[4] The resounding versatility and success of the contemporary energy-based finite element method appears to leave even Euler's lofty observations understated.

7.4.3 Rayleigh–Ritz Method

Consider the displacement vector $\boldsymbol{\Delta}(\alpha, \beta)$, which was defined originally in equation (5.3a) to include D_α, D_β, and D_n. Because we may have additional generalized displacements, such as $D_{\alpha\beta}$ and $D_{\beta\alpha}$ as defined in equations (5.22a) and (5.22b), and because in some cases not all of the displacement components are included, we take $\boldsymbol{\Delta}$ in the general form

$$\{\boldsymbol{\Delta}\} = \{\Delta_1 \Delta_2 \ldots \Delta_k \ldots \Delta_m\} \tag{7.15}$$

where the index m can be set according to the problem at hand.

The next step is to assume that each $\Delta_k (k = 1, m)$ in equation (7.15) is of the form

$$\Delta_k = c_{k1}\phi_{k1} + c_{k2}\phi_{k2} + \ldots c_{kl}\phi_{kl} + \ldots c_{kn}\phi_{kn} \quad (k = 1, m)$$

$$= \sum_{l=1}^{n} c_{kl}\phi_{kl} \tag{7.16}$$

In equation (7.16), each ϕ_k represents a linearly independent coordinate function, and each c_{kl} is an as yet undetermined *constant* coefficient. It is not necessary that each coordinate function satisfies all of the kinematic boundary conditions, but only that the sequence Δ_k meets this requirement. The truncation index for the sequence n may be set at a different value for each Δ_k, and it is anticipated that improved results will be obtained by increasing n, which means including more terms in the sequence. Commonly, polynomials or basic triginometric functions are selected for the coordinate functions ϕ_{kl}, but a wide variety of functions may be suitable according to the problem.

Now, the approximations for Δ_k are substituted into the strain energy and the change in potential energy expressions to evaluate equation (7.14). In the strain energy U_ϵ as given by equation (7.6), we first must evaluate the corresponding strains by using the appropriate strain-displacement relationships. After the integrations indicated in equation (7.6) and (7.10) are carried out, U_τ becomes an algebraic function rather than an integral function. Therefore, the extremum problem of the Calculus of Variations is transformed into the maximum–minimum problem of differential calculus. The maximum–minimum problem, in turn, is solved by satisfying

$$
\frac{\partial U_\tau}{\partial c_{kl}} = 0 \qquad \begin{pmatrix} k = 1, m \\ l = 1, n \end{pmatrix} \tag{7.17}
$$

which produces a set of simultaneous algebraic equations for the coefficients c_{kl}. Following the solution of these equations, the displacement functions Δ_k can be readily used to evaluate strains and curvatures, from which stress resultants and couples may be computed.

The convergence of this method in a rigorous mathematical sense is considered to be a difficult theoretical question.[5] In practice, improved results are obtained in two ways: first, by selecting coordinate functions which closely resemble the actual displacement field—the value of physical intuition in this regard is obvious; second, by using an increasing number of terms in each sequence. Of course, this increases the number of simultaneous equations and the computational effort required.

The careful reader may have noted that no specific reference was made as to the chosen coordinate functions *a priori* satisfying the internal compatibility or strain-displacement conditions. This illustrates one of the advantages of a displacement formulation. For reasonably chosen coordinate functions, the ability to compute the strains from the strain-displacement conditions in the course of evaluating U_ϵ represents, in fact, the satisfaction of the compatibility requirements, so that little difficulty is encountered in this regard.

Another point of general interest is to mention the limitations of the method. Obviously, the requirement of the displacement function sequences satisfying the boundary conditions can be quite restrictive in the case of irregularly shaped continua. The application of the Rayleigh–Ritz method in a *piecewise* fashion to subdivisions or finite elements of the entire medium, with the undetermined coefficients chosen to enforce continuity requirements along interelement boundaries as well as to satisfy the external boundary conditions, has become one of the most powerful techniques in applied mechanics: the *finite element method*. A wide variety of applications of this technique are discussed in Zienkiewicz.[6]

Even before the appearance of the finite element method in the form presently familiar to engineers, Timoshenko suggested that the Rayleigh–Ritz technique—originated by Nobel laureate Lord Rayleigh in his epic treatise "Theory of Sound" and elaborated by W. Ritz, who considered applications to the analysis of thin plates—has spurred more research in the strength of materials and the theory of elasticity than any other single mathematical tool.[7] The vast literature accompanying the finite element method resoundingly reinforces this claim.

Also, we should mention that the Rayleigh–Ritz method may be generalized by taking each of the c_{kl} coefficients in equation (7.16) as an unknown *function* of one of the independent variables, rather than as simply a *constant*. In this case, equation (7.16) is generalized to

$$\Delta_k = \phi_{k0}(\alpha, \beta) + \sum_{l=1}^{n} c_{kl}(\alpha \text{ or } \beta)\, \phi_{kl}(\alpha, \beta) \tag{7.18}$$

It is usually convenient to choose ϕ_{k0} to satisfy all of the boundary conditions; then, each term of the sequence $c_{kl}\phi_{kl}(l = 1, n)$ must vanish on the boundaries.

This extension of the basic Rayleigh–Ritz technique is known as the *Kantorovich method*. If applied skillfully, it can reduce the numerical work considerably for some problems.[8]

7.5 Galerkin Method

The Galerkin method is quite similar in execution to the Rayleigh–Ritz method, in that the solution is postulated to be represented by a sequence of coordinate functions, such as equation (7.16). The most apparent operational difference is that in the Galerkin technique, these coordinate functions are substituted into the *governing differential equations* rather than into an energy expression.

As an illustration, assume that the governing equations have been reduced to a single equation of the form

$$\mathbf{B}(\Delta_k) = f_k(\alpha, \beta, n) \tag{7.19}$$

where Δ_k is one element of the general displacement vector $\{\Delta\}$, equation (7.15), and $\mathbf{B}$ is a linear differential operator. For example, the Laplacian operator in Cartesian coordinates, obtained by setting

$$\mathbf{B}(\ ) = (\ )_{,xx} + (\ )_{,yy} \tag{7.20}$$

is frequently encountered in plate problems. The r.h.s. of equation (7.19) represents the known elements of the system; e.g., the applied surface loading. When equation (7.18) is substituted into equation (7.19), we have

$$\mathbf{B}\left(\sum_{l=1}^{n} C_{kl}\phi_{kl}\right) - f_k(\alpha, \beta, n) = R_k(\alpha, \beta, n) \tag{7.21}$$

where R_k is a residual error function. Recall that Δ_k has been postulated to satisfy the boundary conditions of the problem. The essence of the Galerkin method is to make Δ_k also satisfy the governing equation as closely as possible. This means that the residual error R_k should be minimized.

The condition for minimizing R_k is to require that R_k be orthogonal to *each* ϕ_{kl} over the domain of the problem;[9] That is,

$$\int_V R_k\phi_{kl}\,dV = 0 \tag{7.22a}$$

or

$$\int_V \left[\mathbf{B}\left(\sum_{l=1}^{n} c_{kl}\phi_{kl}\right) - f_k(\alpha, \beta, n)\right]\phi_{kl}\,dV = 0 \quad (l = 1, n) \tag{7.22b}$$

This produces a system of n linear simultaneous algebraic equations in the n unknown coefficients c_{kl}.

Obviously, this procedure becomes increasingly involved when there are multiple equations and unknowns in the system, instead of just a single equation in Δ_k as we have considered here. But, for a large class of second-order ordinary differential equations, the Galerkin method leads to an identical set of coefficients as produced by the Rayleigh–Ritz method,[10] and may be preferable if it is more convenient to work with the governing equations rather than the energy functional. Moreover, there are problems for which no satisfactory variational principle has been determined, but for which a set of governing differential equations is available. This suggests that the Galerkin method may be broader in applicability than even the Rayleigh–Ritz method.

Notes

1. F. L. DiMaggio, "Principle of Virtual Work in Structural Analysis," *Journal of the Structural Division, ASCE* 86, no. ST11 (November 1960):65–77.

2. K. Washizu, *Variational Methods in Elasticity and Plasticity* (New York: Pergamon Press, 1968).

3. M. J. Forray, *Variational Calculus in Science and Engineering* (New York: McGraw-Hill, 1968), chaps. 1 and 2.

4. S. Timoshenko, *History of Strength of Materials* (New York: McGraw-Hill, 1953), p. 31. Copyright, 1953 by McGraw-Hill, Inc., and used with permission of McGraw-Hill Book Co.

5. R. Courant and D. Hilbert, *Methods of Mathematical Physics*, vol. 1 (New York: Interscience Publishers, 1966), pp. 175–176.

6. O. C. Zienkiewicz, *The Finite Element Method in Engineering Science* (London: McGraw-Hill, 1971).

7. Timoshenko, *History of Strength of Materials*, pp. 338–339, 399–401.

8. Forray, "Variational Calculus," pp. 199–201.

9. Ibid.

10. Ibid., pp. 189–199.

8

Bending of Plates

8.1 Governing Equations

8.1.1 General Formulation

The equilibrium equations for initially flat plates are stated as equations (3.25a–e); the strain-displacement relations are given as equations (5.54a–h), or with transverse shearing strains suppressed, as equations (5.55a–c). Taken together with the stress resultant-strain relationships in the form of matrix 6–2, the requisite boundary conditions, as discussed in section 6.2, and specifically the Kirchhoff conditions, equations (6.25) and (6.27), the elements of a quite general plate theory are available and substantiated.

In this chapter, the highlights of such a theory are presented. The comprehensive volume by Szilard[1] is a source for extensions of the theory, solution techniques, and applications. Also, the classical work of Timoshenko and Woinowsky-Krieger[2] is rich in solution procedures and practical aspects of plate analysis. Our purpose here is only to deduce the theory of plates from a general surface structure formulation—a somewhat unique approach—and to leave the application to specialized works.

It is convenient to consider the development of the theory of plates in several stages, initially neglecting transverse shearing strains and in-plane stress resultants, and also assuming isotropic material properties. Then the theory may be generalized as required.

When we neglect the in-plane forces, equations (3.25a) and (3.25b) are eliminated, leaving

$$[(BQ_\alpha)_{,\alpha} + (AQ_\beta)_{,\beta}] + q_n AB = 0 \qquad (8.1a)$$

$$(BM_{\alpha\beta})_{,\alpha} + (AM_\beta)_{,\beta} + B_{,\alpha}M_{\beta\alpha} - A_{,\beta}M_\alpha - Q_\beta AB = 0 \qquad (8.1b)$$

$$(BM_\alpha)_{,\alpha} + (AM_{\beta\alpha})_{,\beta} + A_{,\beta}M_{\alpha\beta} - B_{,\alpha}M_\beta - Q_\alpha AB = 0 \qquad (8.1c)$$

Along with the elimination of the in-plane forces, we also disregard D_α, D_β, and the extensional strains and in-plane shearing strains on the middle plate, ϵ_α, ϵ_β, and ω. With transverse shearing strains also suppressed, we have our complete set of compatibility equations given by equations (5.55a–c). Of course, we still have extensional and in-plane shearing strains away from the middle plane, as computed from equations (5.36), (5.37), and (5.45):

$$\epsilon_\alpha(\zeta) = \zeta\kappa_\alpha \tag{8.2a}$$

$$\epsilon_\beta(\zeta) = \zeta\kappa_\beta \tag{8.2b}$$

$$\omega(\zeta) = 2\zeta\tau \tag{8.2c}$$

where κ_α and κ_β are the curvatures and τ is the twist of the middle plane. On substituting equations (8.2a–c) into equations (5.55a–c), we find

$$\kappa_\alpha(\zeta) = -\left[\frac{1}{A}\left(\frac{1}{A}\,D_{n,\alpha}\right)_{,\alpha} + \frac{A_{,\beta}}{AB^2}\,D_{n,\beta}\right] \tag{8.3a}$$

$$\kappa_\beta(\zeta) = -\left[\frac{1}{B}\left(\frac{1}{B}\,D_{n,\beta}\right)_{,\beta} + \frac{B_{,\alpha}}{A^2B}\,D_{n,\alpha}\right] \tag{8.3b}$$

$$\tau(\zeta) = -\frac{1}{2}\left\{\frac{1}{A}\left(\frac{1}{B}\,D_{n,\beta}\right)_{,\alpha} + \frac{1}{B}\left(\frac{1}{A}\,D_{n,\alpha}\right)_{,\beta}\right.$$

$$\left. - \frac{1}{AB}\left[\frac{A_{,\beta}}{A}\,D_{n,\alpha} + \frac{B_{,\alpha}}{B}\,D_{n,\beta}\right]\right\} \tag{8.3c}$$

These terms are seen to be functions of the normal displacement D_n only.

The stress couples are easily expressed in terms of D_n by substitution of equations (8.3a–c) into the appropriate constitutive laws. For isotropic materials, we have, from matrix 6–2 and equations (8.3a–c)

$$M_\alpha = -D\left\{\frac{1}{A}\left(\frac{1}{A}\,D_{n,\alpha}\right)_{,\alpha} + \frac{A_{,\beta}}{AB^2}D_{n,\beta}\right.$$

$$\left. + \mu\left[\frac{1}{B}\left(\frac{1}{B}D_{n,\beta}\right)_{,\beta} + \frac{B_{,\alpha}}{A^2B}\,D_{n,\alpha}\right]\right\} \tag{8.4a}$$

$$M_\beta = -D\left\{\frac{1}{B}\left(\frac{1}{B}\,D_{n,\beta}\right)_{,\beta} + \frac{B_{,\alpha}}{A^2B}\,D_{n,\alpha}\right.$$

$$\left. + \mu\left[\frac{1}{A}\left(\frac{1}{A}\,D_{n,\alpha}\right)_{,\alpha} + \frac{A_{,\beta}}{AB^2}\,D_{n,\beta}\right]\right\} \tag{8.4b}$$

$$M_{\alpha\beta} = -D\frac{(1-\mu)}{2}\left\{\frac{1}{A}\left(\frac{1}{B}\,D_{n,\beta}\right)_{,\alpha} + \frac{1}{B}\left(\frac{1}{A}\,D_{n,\alpha}\right)_{,\beta}\right.$$

$$\left. - \frac{1}{AB}\left[\frac{A_{,\beta}}{A}\,D_{n,\alpha} + \frac{B_{,\alpha}}{B}\,D_{n,\beta}\right]\right\} \tag{8.4c}$$

in which D is the flexural rigidity of the plate given by

$$D = \frac{Eh^3}{12(1 - \mu^2)} \tag{8.4d}$$

Assume that $M_{\alpha\beta} = M_{\beta\alpha}$, with the justification following the same argument offered in taking $N_{\alpha\beta} = N_{\beta\alpha}$ in section 4.1. Note that when the transverse shearing strains are excluded, the transverse shear stress resultants can no longer be evaluated from the corresponding constitutive relationship (matrix 6–2). Rather, equations (8.1b) and (8.1c) are solved for

$$Q_\beta = \frac{1}{AB} \left[(BM_{\alpha\beta})_{,\alpha} + (AM_\beta)_{,\beta} + B_{,\alpha}M_{\beta\alpha} - A_{,\beta}M_\alpha \right] \tag{8.5a}$$

and

$$Q_\alpha = \frac{1}{AB} \left[(BM_\alpha)_{,\alpha} + (AM_{\beta\alpha})_{,\beta} + A_{,\beta}M_{\alpha\beta} - B_{,\alpha}M_\beta \right] \tag{8.5b}$$

so that Q_α and Q_β can be computed once the stress couples have been determined. This is, of course, analogous to the elementary theory of beams, where the shear force is evaluated from equilibrium considerations rather than from a constitutive law. Only transverse shearing *strains* can be suppressed; the corresponding *forces* are required for equilibrium.

The formulation of the plate bending problem is completed as follows:

1. Introduce equations (8.4a–d) into (8.5a) and (8.5b).
2. Differentiate (8.5a) and (8.5b) by β and α, respectively.
3. Substitute into equation (8.1a).

The resulting equation expresses *equilibrium* in the normal direction in terms of the single *displacement* D_n, and constitutes the governing equation of the system. This is a classical displacement formulation, and, once D_n is evaluated, the stress couples may be computed from the differentiation of equation (8.4a–d) and the transverse shears from equation (8.5a) and (8.5b).

We also have the effective transverse shears, $\bar{Q}_\alpha$ and $\bar{Q}_\beta$,

$$\bar{Q}_\alpha = Q_\alpha + \frac{M_{\alpha\beta,\beta}}{B} \tag{8.6a}$$

and

$$\bar{Q}_\beta = Q_\beta + \frac{M_{\beta\alpha,\alpha}}{A} \tag{8.6b}$$

which we obtain from equations (6.25) and (6.27). Also, the rotations $D_{\alpha\beta}$ and $D_{\beta\alpha}$ are obtained from equation (5.54g) and (5.54h) with $\gamma_\alpha = \gamma_\beta = 0$:

$$D_{\alpha\beta} = \psi_\alpha = -\frac{1}{A} D_{n,\alpha} \tag{8.7a}$$

$$D_{\beta\alpha} = \psi_\beta = -\frac{1}{B} D_{n,\beta} \tag{8.7b}$$

It is evident from the complexity of the foregoing expressions that the algebra involved in carrying out the steps outlined in the previous paragraph becomes quite involved. For our purposes, it is sufficient at this point to specialize the further development for the most common case encountered in practice, Cartesian coordinates.

8.1.2 Cartesian Coordinates

For Cartesian coordinates, as shown on figure 8–1, $\alpha = X$, $\beta = Y$, $n = Z$, and $A = B = 1$. Then equations (8.4) become

$$M_X = -D(D_{Z,XX} + \mu D_{Z,YY}) \tag{8.8a}$$

$$M_Y = -D(D_{Z,YY} + \mu D_{Z,XX}) \tag{8.8b}$$

$$M_{XY} = M_{YX} = -D(1 - \mu)D_{Z,XY} \tag{8.8c}$$

Similarly, equations (8.5a) and (8.5b) reduce to

$$Q_Y = [M_{XY,X} + M_{Y,Y}]$$
$$= -D[D_{Z,YYY} + \mu D_{Z,XXY} + (1 - \mu)D_{Z,XXY}]$$
$$= -D[D_{Z,YYY} + D_{Z,XXY}] \tag{8.9a}$$
$$Q_X = [M_{X,X} + M_{YX,Y}]$$
$$= -D[D_{Z,XXX} + \mu D_{Z,XYY} + (1 - \mu)D_{Z,XYY}]$$
$$= -D[D_{Z,XXX} + D_{Z,XYY}] \tag{8.9b}$$

Next we take $\partial/\partial Y$ of equation (8.9a), $\partial/\partial X$ of equation (8.9b), and substitute into equation (8.1a) to get

$$Q_{X,X} + Q_{Y,Y} + q_Z = 0 \tag{8.10}$$

or

$$-D[D_{Z,XXXX} + 2D_{Z,XXYY} + D_{Z,YYYY}] + q_Z = 0 \tag{8.11}$$

Equation (8.11) may be written concisely as

$$\nabla^2(\nabla^2 D_Z) = \nabla^4 D_Z = \frac{q_Z}{D} \tag{8.12}$$

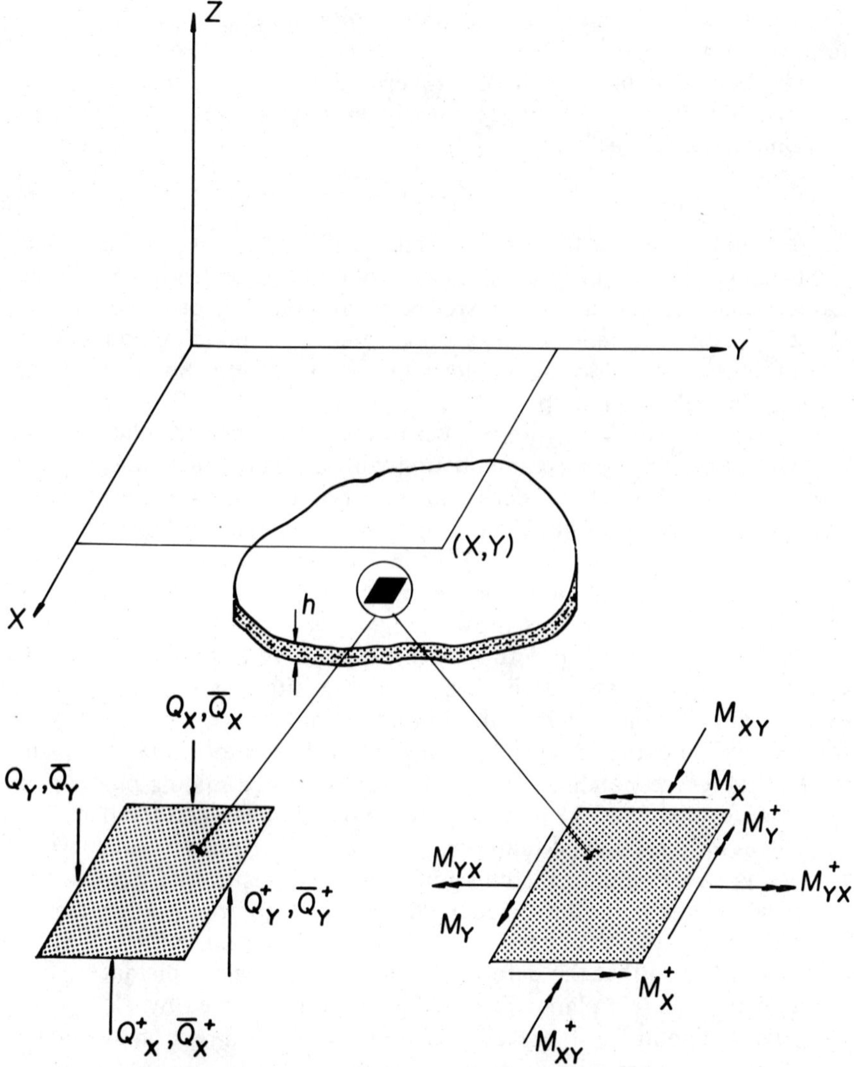

Figure 8–1. Plate in Cartesian Coordinates

where

$$\nabla^2(\ ) = (\ )_{,XX} + (\ )_{,YY} \tag{8.13a}$$

and

$$\nabla^4(\ ) = \nabla^2[\nabla^2(\ )] = (\ )_{,XXXX} + 2(\ )_{,XXYY} + (\ )_{,YYYY} \tag{8.13b}$$

$\nabla^2(\ )$ and $\nabla^4(\ )$ are commonly called the *Laplacian* and *biharmonic* operators, respectively.

The homogeneous part of the governing equation of the engineering theory of medium-thin plates, commonly termed the *plate equation*, is the *biharmonic equation*

$$\nabla^4 D_Z = 0 \tag{8.14}$$

The recognition of this form is important in at least two ways. First, there is a considerable reservoir of mathematical knowledge dealing with biharmonic equations that can be transformed to our plate theory.[3] Second, because $\nabla^4(\ )$ is invarient, we may obtain the corresponding equation for any other system of orthogonal curvilinear coordinates by simply transforming the biharmonic operator.

Of particular interest with respect to the preceding formulation is the reduction of the system to an equation with a single dependent variable. Stepping back through the derivation, we find that the key step was the expression of the extensional and shear strains in terms of the normal displacement D_Z only, equation (8.3). To interpret this from a physical standpoint, we examine a segment of the middle plane along a normal section traced by the X–Z plane, as shown in figure 8–2.

We show the segment deformed in the positive Z direction. Both the displacement D_Z and the derivative $D_{Z,X}$ are positive as shown. We focus on an arbitrary point *1* which lies initially at a distance ζ from the undeformed middle plane. Since all geometry is referred to the middle plane, we find the corresponding point on the underformed middle plane, point *2*. The deformation of the plate displaces point *2* a distance D_Z, parallel to the *original Z* direction (assumption [2], table 1–1) to point *3*. Since we are neglecting shearing deformations, the deformed position of point *1* must lie on the normal to the deformed middle plane $D_{Z,X}$ at point *3* (assumption [3]). Similarly, the thickness remains constant during deformation (assumption [4]), so that the deformed point must remain a distance ζ from the deformed middle plane. The final position is denoted by *4*.

Note that point *1* also has a displacement in the X direction denoted by D_X. Since D_X is negative corresponding to the positive sense of $D_{Z,X}$ as shown on figure 8–2

$$D_X = -\zeta D_{Z,X} \tag{8.15}$$

Because the Lamé parameters are constant, the strain in the X direction is simply, from equation (5.50a),

$$\epsilon_X(\zeta) = D_{X,X}$$

$$= -\zeta D_{Z,XX} \tag{8.16}$$

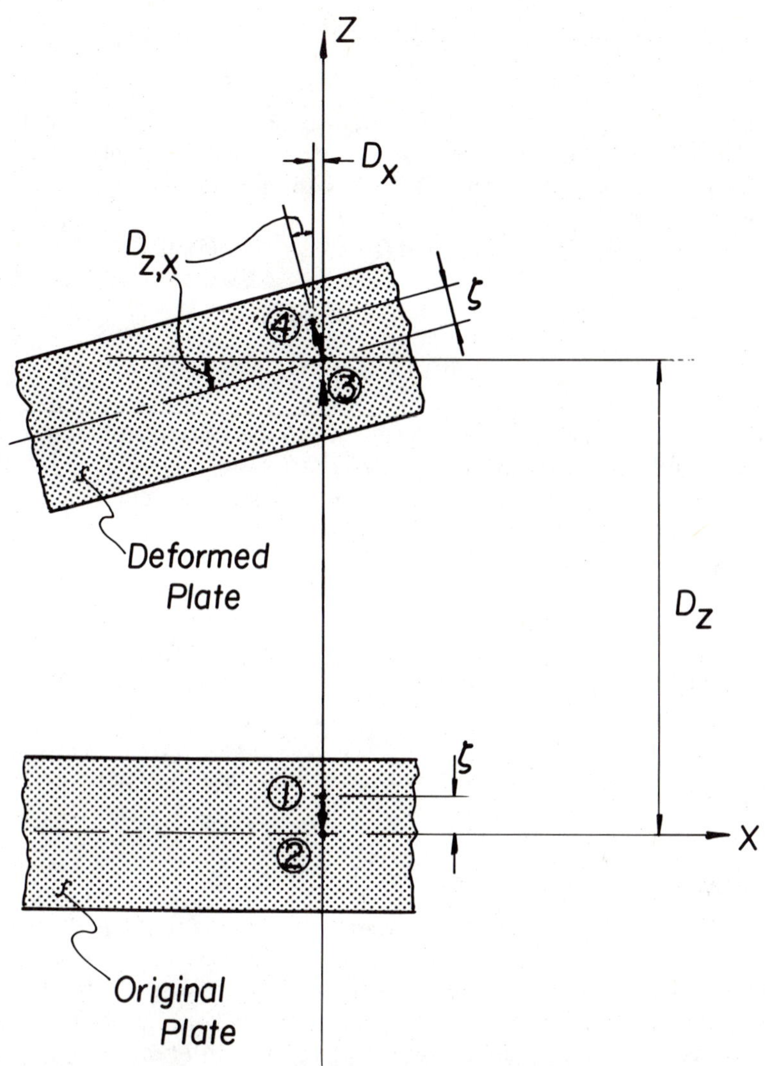

Figure 8–2. Displacements for a Plate

which corresponds to equation (8.3a) specialized for Cartesian coordinates.

An identical argument referred to a normal section traced by the *YZ* plane gives

$$D_Y = -\zeta D_{Z,Y} \qquad (8.17)$$

and

$$\epsilon_Y(\zeta) = -zD_{Z,YY} \tag{8.18}$$

corresponding to equation (8.3b), suitably specialized. Equations (8.15) and (8.17) clearly illustrate the coupling of the *in-plane* displacements to the *normal* displacement, which is the key to the relatively simple form of the plate equation.

The shearing strain, specialized from equation (5.50c), is

$$\omega = \frac{1}{2}(D_{X,Y} + D_{Y,X})$$

$$= -\zeta D_{Z,XY} \tag{8.19}$$

With the strains expressed in terms of D_n, the subsequent steps of the derivation follow directly as previously shown.

To make the formulation complete we also list the effective shears $\bar{Q}_X$ and $\bar{Q}_Y$. From equations (8.6a) and (8.6b),

$$\bar{Q}_X = Q_X + M_{XY,Y}$$

$$= -D[D_{Z,XXX} + (2 - \mu)D_{Z,XYY}] \tag{8.20a}$$

$$\bar{Q}_Y = Q_Y + M_{XY,X}$$

$$= -D[D_{Z,YYY} + (2 - \mu)D_{Z,XXY}] \tag{8.20b}$$

The rotations D_{XY} and D_{YX} are found from equation (8.7) as

$$D_{XY} = \psi_X = -D_{Z,X} \tag{8.21a}$$

$$D_{YX} = \psi_Y = -D_{Z,Y} \tag{8.21b}$$

Finally, in terms of general remarks, it is obvious that equation (8.11) is the two-dimensional counterpart of the familiar Bernoulli–Euler equation, which is the cornerstone of beam theory. Taking $D_Z = D_Z(X)$ or $D_Z(Y)$ only in equation (8.11) and neglecting Poisson's ratio in the constant D, we can immediately derive the governing differential equation for a beam of unit width and depth h.

8.1.3 Boundary Conditions

The boundary conditions corresponding to the theory of plates without transverse shearing strains are easily synthesized from section 6.2.2 and figure 6–3.

We consider boundaries $\alpha = \bar{\alpha}$ and $\beta = \bar{\beta}$ and list the most common idealized boundary conditions in figure 8–3. These conditions are given in

Condition and Symbol		Boundary	
		$\alpha = \bar{\alpha}$	$\beta = \bar{\beta}$
Fixed or Clamped		$D_Z = D_{\alpha\beta} = 0$ or $D_Z = D_{Z,\alpha} = 0$	$D_Z = D_{\beta\alpha} = 0$ or $D_Z = D_{Z,\beta} = 0$
Hinged or Simply Supported		$D_Z = 0;\ M_\alpha = 0$ or $D_Z = D_{Z,\alpha\alpha} = 0$	$D_Z = 0;\ M_\beta = 0$ or $D_Z = D_{Z,\beta\beta} = 0$
Free		$\bar{Q}_\alpha = M_\alpha = 0$	$\bar{Q}_\beta = M_\beta = 0$

Figure 8–3. Boundary Conditions for Plates

homogeneous form, but may be generalized as described at the conclusion of this section.

The roller condition, as defined in chapter 6, cannot be accommodated unless the plate can sustain in-plane forces that, for the moment, are not considered.

With respect to the *fixed* condition, often called a *clamped* boundary, note from equation (8.7) that $D_{\alpha\beta} = 0$ and $D_{\beta\alpha} = 0$ imply $D_{Z,\alpha} = 0$ and $D_{Z,\beta} = 0$, respectively.

The *hinged* condition, also termed a *simply-supported* boundary, may be written in an alternate form for Cartesian coordinates, where the boundaries are coincident with the coordinate lines. We consider first a boundary defined by $X = \bar{X}$, for example, in figure 8–1. The condition of simple support states that $D_Z(\bar{X},\ Y) = 0$. Further, it is implied that the boundary is straight and therefore $D_{Z,YY}(\bar{X},\ Y) = 0$. A similar argument with respect to a boundary $Y = \bar{Y}$ leads to $D_{Z,XX}(X,\ \bar{Y}) = 0$. Equations (8.8a) and (8.8b) for simply supported boundaries become

$$M_X(\bar{X},\ Y) = -DD_{Z,XX}(\bar{X},\ Y) = 0 \tag{8.22}$$

and

$$M_Y(X,\ \bar{Y}) = -DD_{Z,YY}(X,\ \bar{Y}) = 0 \tag{8.23}$$

which, in view of the preceding discussion, generalize to

$$M_X(\bar{X},\ Y) = -D\nabla^2 D_Z(\bar{X},\ Y) = 0 \tag{8.24}$$

and

$$M_Y(X, \bar{Y}) = -D\nabla^2 D_Z(X, \bar{Y}) = 0 \qquad (8.25)$$

Thus we have

$$M_X = 0 \rightarrow D_{Z,XX} = 0 \rightarrow \nabla^2 D_Z = 0$$

and

$$M_Y = 0 \rightarrow D_{Z,YY} = 0 \rightarrow \nabla^2 D_Z = 0$$

The statement of the simply supported boundary condition as $\nabla^2 D_Z = 0$ is predicated on all edges being straight as well as undeflecting.

We now consider the *free*-edge condition. In section 6.2.3, we presented in some detail the contraction of the general boundary conditions into the Kirchhoff conditions as a result of the supression of transverse shearing strains. For the plate problem, equations (8.6a) and (8.6b) indicate that the transverse shear and the twisting stress couple are combined into an effective transverse shear which is prescribed to vanish on a free edge.

It should also be mentioned that the homogeneous boundary conditions discussed in this section may be generalized to include prescribed edge displacements and forces. For example, within the definition of fixed and hinged conditions, a nonzero value of the transverse displacement D_Z may be accommodated. This may be of interest in support settlement problems. Also a specified edge moment may be inserted in place of M_α or $M_\beta = 0$ in the hinged and free conditions. Depending on the solution routine, known edge moments and edge forces may be grouped with the applied loading terms, but also should be regarded as force boundary conditions.

8.2 Strain Energy for Plates in Flexure

Refer to the general expression for U_ϵ as given by equation (7.6). We are presently considering no extensional strains, and, in accordance with our previous development, we choose to neglect transverse shearing strains. Therefore, we have only the component III in equation (7.7) remaining along with the bending thermal term:

$$U_\epsilon = \frac{D}{2} \int_S \left\{ \left[(\kappa_\alpha + \kappa_\beta)^2 - 2(1 - \mu)(\kappa_\alpha \kappa_\beta - \tau^2) \right] \right.$$

$$\left. - \bar{\alpha}(1 + \mu)(\kappa_\alpha + \kappa_\beta) \int_{-h/2}^{h/2} T(\zeta)\zeta \, d\zeta \right\} AB \, d\alpha d\beta \qquad (8.26)$$

Since we are interested in a displacement formulation, we may substitute the strain-displacement relationships, as given by equation (5.55), into equation (8.26). This may be specialized for Cartesian coordinates by taking $\alpha = X$, $\beta = Y$, $n = Z$ and $A = B = 1$ in equation (5.55) to get

$$\kappa_X = -D_{Z,XX} \tag{8.27a}$$

$$\kappa_Y = -D_{Z,YY} \tag{8.27b}$$

$$\tau = -D_{Z,XY} \tag{8.27c}$$

Then, substituting equations (8.27a–c) into equation (8.26), we have

$$U_\epsilon = \frac{D}{2} \int_S \left\{ \left[(D_{Z,XX} + D_{Z,YY})^2 - 2(1 - \mu)(D_{Z,XX}D_{Z,YY} - D_{Z,XY}^2) \right] \right.$$
$$\left. + \bar{\alpha}(1 + \mu)(D_{Z,XX} + D_{Z,YY}) \int_{-h/2}^{h/2} T(\zeta)\zeta \, d\zeta \right\} dY \, dX \tag{8.28}$$

We may write equation (8.28) in the absence of thermal effects as

$$U_\epsilon = U_{\epsilon 1} + U_{\epsilon 2}$$

$$= \frac{D}{2} \int_S \{ (\nabla^2 D_Z)^2 - 2(1 - \mu)[D_{Z,XX}D_{Z,YY} - (D_{Z,XY})^2] \} \, dY \, dX \tag{8.29}$$

8.3 In-Plane Forces and Finite Deformations

8.3.1 Modification of Equilibrium Equations

We are interested in the analysis of plates that are subjected to forces acting in the middle plane along with the transverse loading. If the forces acting in the middle plane are compressive in at least one direction, plate instability is a possibility. Further, if there are no transverse forces but only in-plane compressive forces acting, we can formulate the two-dimensional analogue of the classical Euler column buckling problem. To accomplish this, we must relax our basic assumption [2], which enabled the equilibrium equations to be formulated with respect to the *undeformed* middle surface, and now include the effect of the in-plane forces on the equilibrium in the normal direction.

Refer to figure 3–2(a), and consider a section of the middle surface along an α coordinate line of length ds_α. We show two such sections on figure 8–4 both before and after deformation. Figure 8–4(a) is for N_α, while figure 8–4(b) pertains to $N_{\beta\alpha}$. The surface rotation $D_{\alpha\beta}$ is denoted in accordance with figure 5–2. Recall that if transverse shearing strains are neglected, $D_{\alpha\beta}$ can be expressed in terms of $D_{n,\alpha}$ by equation (8.7a). Note

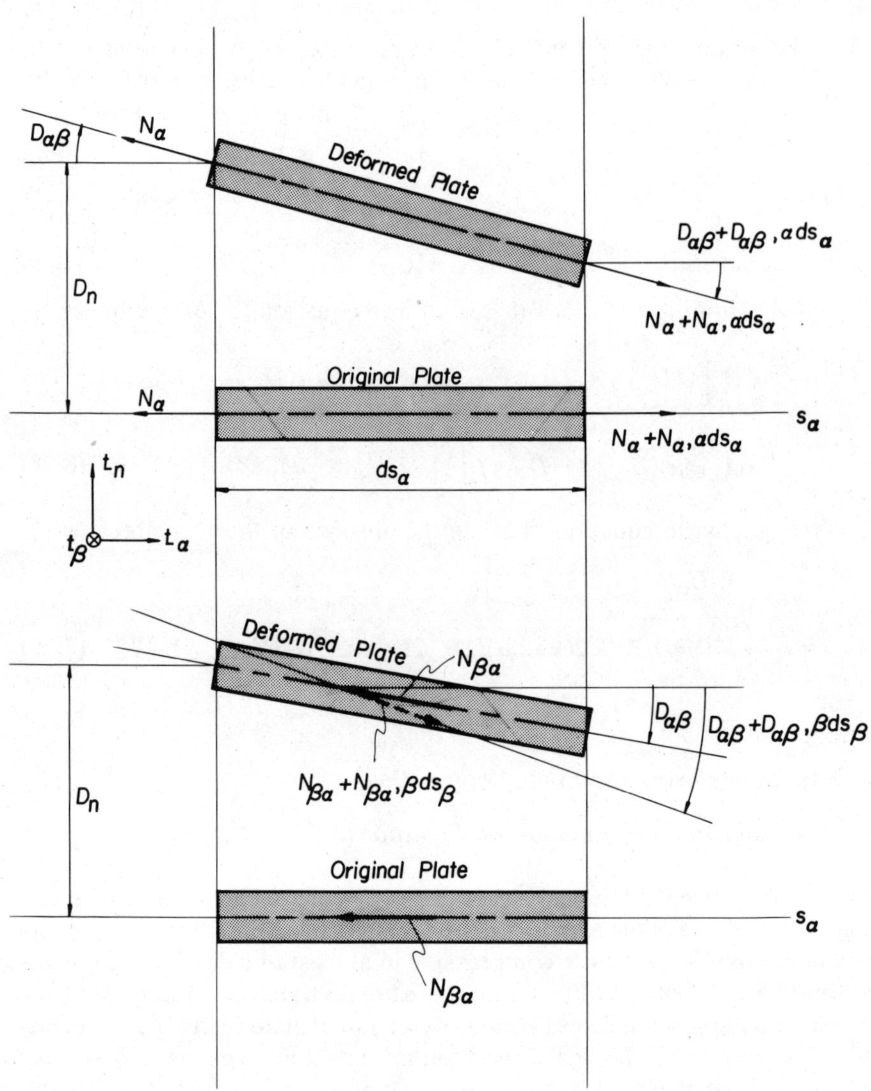

Figure 8–4. In-Plane Forces Acting on a Plate

that figure 8–4 corresponds to the positive sense of $D_{\alpha\beta}$ and the negative sense of $D_{n,\alpha}$, since the $(+)\mathbf{t}_\alpha$ direction is opposite to that shown on figure 5–2(a).

Consider the stress resultant vectors $\mathbf{F}_\alpha$ and $\mathbf{F}_\beta$ as defined in equations (3.11a) and (3.11b). First, from figure 8–4(a), we see that the stress resultant N_α has a normal component $-N_\alpha \sin D_{\alpha\beta} \simeq -N_\alpha D_{\alpha\beta}$, which adds a force $-N_\alpha D_{\alpha\beta} \mathbf{t}_n \, ds_\beta$ to $\mathbf{F}_\alpha$ in equation (3.11a). The negative sign is in accordance with the sign convention defined on figure 3–2(a). Next, from figure 8–4(b) we have the stress resultant $N_{\beta\alpha}$ with a normal component $-N_{\beta\alpha} \cdot D_{\alpha\beta}$, which contributes a force $-N_{\beta\alpha} D_{\alpha\beta} \mathbf{t}_n \, ds_\alpha$ to $\mathbf{F}_\beta$ in equation (3.11b). Similar sections along a β coordinate line show the forces $-N_{\alpha\beta} D_{\beta\alpha} \mathbf{t}_n \, ds_\beta$ adding to $\mathbf{F}_\alpha$ in equation (3.11a), and $-N_\beta D_{\beta\alpha} \mathbf{t}_n \, ds_\alpha$ going to $\mathbf{F}_\beta$ in equation (3.11b). Since these additional normal forces associate only with Q_α and Q_β, respectively, as coefficients of $\mathbf{t}_n$ in the two equations, they are easily traced through to equations (3.25a–e), the scalar force equilibrium equations for a plate. The in-plane equilibrium equations are unaffected, while the generalization of equation (3.25c) is

$$[B(Q_\alpha - N_\alpha D_{\alpha\beta} - N_{\alpha\beta} D_{\beta\alpha})]_{,\alpha}$$

$$+ [A(Q_\beta - N_{\beta\alpha} D_{\alpha\beta} - N_\beta D_{\beta\alpha})]_{,\beta} + q_n AB = 0 \qquad (8.30)$$

It is convenient to expand equation (8.30) and eliminate certain terms by introducing the in-plane equilibrium equations, (3.25a) and (3.25b). We will continue for Cartesian coordinates, $\alpha = X$, $\beta = Y$, $n = Z$, and $A = B = 1$. Then equations (3.25a) and (3.25b) become

$$N_{X,X} + N_{YX,Y} + q_X = 0 \qquad \text{(a)}$$
$$N_{XY,X} + N_{Y,Y} + q_Y = 0 \qquad \text{(b)}$$
$$(8.31)$$

while equation (8.30) is written as

$$Q_{X,X} + Q_{Y,Y} - D_{XY}(N_{X,X} + N_{YX,Y}) - D_{YX}(N_{XY,X} + N_{Y,Y})$$

$$- N_X D_{XY,X} - N_Y D_{YX,Y} - N_{YX} D_{XY,Y} - N_{XY} D_{YX,X} + q_Z = 0 \qquad (8.32)$$

Substituting equations (8.31a) and (8.31b) into (8.32), and setting $N_{XY} = N_{YX}$, we have

$$Q_{X,X} + Q_{Y,Y} - N_X D_{XY,X} - N_Y D_{YX,Y} - N_{XY}(D_{YX,X} + D_{XY,Y})$$

$$+ D_{XY} q_X + D_{YX} q_Y + q_Z = 0 \qquad (8.33)$$

A more familiar form of equation (8.33), applicable for transverse loading only, is found by letting $q_X = q_Y = 0$ and suppressing transverse shearing strains, whereupon, from equation (8.7), we find

$$D_{XY} = -D_{Z,X}; \quad D_{YX} = -D_{Z,Y} \qquad (8.34)$$

Also, introducing equations (8.10)–(8.12) into (8.33), we find

$$D\nabla^4 D_Z - N_X D_{Z,XX} - N_Y D_{Z,YY} - 2N_{XY} D_{Z,XY} = q_Z \qquad (8.35)$$

as the governing equation for the deflection of a plate in the presence of lateral forces.

8.3.2 Modification of Strain Energy

The expression for strain energy, as given by equations (7.6) and (7.7), must be supplemented when coupling of the normal displacements and the in-plane strains is included, since nonlinear strain terms are required. We shall refer back to the basic description of deformation as discussed in sections 5.2 and 5.3. However, we will not attempt to develop a complete nonlinear theory, but only to retain those higher order terms containing D_n. This corresponds to a modified finite deformation theory in which the in-plane displacements remain small, but the rotations are regarded as moderate.

We first consider equations (5.13) and (5.14) for A' and B'. Noting from equations (5.10c) and (5.12c), respectively, that ψ_α and ψ_β contain D_n terms, we have

$$A' = A\left[(1 + \epsilon_\alpha)^2 + \frac{1}{A^2}\, D_{n,\alpha}^2 \right]^{1/2} \tag{8.36}$$

and

$$B' = B\left[(1 + \epsilon_\beta)^2 + \frac{1}{B^2}\, D_{n,\beta}^2 \right]^{1/2} \tag{8.37}$$

Using the binomial theorem, we may simplify these expressions to

$$A' = A\left(1 + \epsilon_\alpha + \frac{1}{2A^2}\, D_{n,\alpha}^2\right) \tag{8.38}$$

and

$$B' = B\left(1 + \epsilon_\beta + \frac{1}{2B^2} D_{n,\beta}^2\right) \tag{8.39}$$

We now return to the definition of the middle surface strains in section 5.3.1. We denote the modified strains as $\bar{\epsilon}_\alpha$, $\bar{\epsilon}_\beta$, and $\bar{\omega}$, respectively, and retain the basic definitions of these components of strain. Then, equation (5.25), in view of equation (8.38), becomes

$$\bar{\epsilon}_\alpha = \epsilon_\alpha + \frac{1}{2}\left(\frac{D_{n,\alpha}}{A}\right)^2 \tag{8.40}$$

Correspondingly, equation (8.39) generalizes to

$$\bar{\epsilon}_\beta = \epsilon_\beta + \frac{1}{2}\left(\frac{D_{n,\beta}}{B}\right)^2 \tag{8.41}$$

For the shearing strain, we note the product $\psi_\alpha\psi_\beta$ in equation (5.27), which previously had been neglected. Referring to equations (5.10c) and (5.12c), and keeping only the D_n terms

$$\psi_\alpha \psi_\beta \simeq \left(\frac{D_{n,\alpha}}{A}\right)\left(\frac{D_{n,\beta}}{B}\right)$$

Therefore, from equation (5.27), we find

$$\bar{\omega} = \omega + \left(\frac{D_{n,\alpha}}{A}\right)\left(\frac{D_{n,\beta}}{B}\right) \tag{8.42}$$

To obtain a modified expression for the strain energy, refer to the basic expression, equation (7.4). Assume that the strain energy due to bending is not changed by the axial forces, and thus remains the cubic component of equation (7.7), $D/2$ [III]. Also assume that the in-plane stress resultants are due entirely to applied edge loading in the plane of the plate, in which case the *in-plane* stress resultants are unchanged during bending. This implies that the external and internal work done by these constant stress resultants acting through the corresponding external and internal *in-plane displacements* will cancel in the energy balance when a virtual transverse displacement is introduced. This may be formally substantiated by rather involved arguments,[4] which are not repeated here. Therefore, the *additional* strain energy $U_{\bar{\epsilon}}$ is due entirely to the straining of the middle surface as a result of the bending. With these assumptions, we can write

$$U_{\bar{\epsilon}} = \int_S \left[N_\alpha(\bar{\epsilon}_\alpha - \epsilon_\alpha) + N_\beta(\bar{\epsilon}_\beta - \epsilon_\beta) + N_{\alpha\beta}(\bar{\omega} - \omega) \right] AB\, d\alpha\, d\beta \tag{8.43}$$

Note that there is no 1/2 coefficient in equation (8.43), since the in-plane stress resultants are already acting when the additional middle surface strains occur.

Substituting equations (8.40)–(8.42) into equation (8.43) gives

$$U_{\bar{\epsilon}} = \frac{1}{2} \int_S \left[N_\alpha\left(\frac{D_{n,\alpha}}{A}\right)^2 + N_\beta\left(\frac{D_{n,\beta}}{B}\right)^2 \right.$$

$$\left. + 2N_{\alpha\beta}\left(\frac{D_{n,\alpha}}{A}\right)\left(\frac{D_{n,\beta}}{B}\right) \right] AB\, d\alpha\, d\beta \tag{8.44}$$

and the total strain energy for plate bending in the presence of constant in-plane forces is

$$U_\epsilon + U_{\bar{\epsilon}} = U_{\epsilon+\bar{\epsilon}} = \frac{D}{2}\,[\text{III}] + U_{\bar{\epsilon}} \tag{8.45}$$

where III is defined in equations (7.6) and (7.7).

For Cartesian coordinates, equation (8.44) becomes

$$U_{\bar{\epsilon}} = \frac{1}{2} \int_S \, [N_X D_{Z,X}^2 + N_Y D_{Z,Y}^2 + 2N_{XY} D_{Z,X} D_{Z,Y}] \, dX \, dY \tag{8.46}$$

and U_ϵ is given by equation (8.29).

8.3.3 Finite Deformations of Plates

If we consider the modified expressions for the middle surface strains, equations (8.40)–(8.42) and restrict ourselves to Cartesian coordinates, we have, in view of equations (5.54a–c),

$$\bar{\epsilon}_X = D_{X,X} + \frac{1}{2}\,(D_{Z,X})^2 \tag{8.47a}$$

$$\bar{\epsilon}_Y = D_{Y,Y} + \frac{1}{2}\,(D_{Z,Y})^2 \tag{8.47b}$$

$$\bar{\omega} = D_{Y,X} + D_{X,Y} + D_{Z,X} D_{Z,Y} \tag{8.47c}$$

Equations (8.47) may be combined into a single compatibility equation in terms of D_Z and the strains.

We form certain second partials assuming sufficient continuity so that the order of differentiation can be altered:

$$\bar{\epsilon}_{X,YY} = D_{X,XYY} + (D_{Z,XY})^2 + D_{Z,X} D_{Z,XYY} \tag{8.48a}$$

$$\bar{\epsilon}_{Y,XX} = D_{Y,XXY} + (D_{Z,XY})^2 + D_{Z,Y} D_{Z,XXY} \tag{8.48b}$$

$$\bar{\omega}_{,XY} = D_{Y,XXY} + D_{X,XYY} + (D_{Z,XY})^2$$
$$+ D_{Z,XX} D_{Z,YY} + D_{Z,X} D_{Z,XYY}$$
$$+ D_{Z,Y} D_{Z,XXY} \tag{8.48c}$$

Now, adding equations (8.48a) and (8.48b) and subtracting (8.48c), we have

$$\bar{\epsilon}_{X,YY} + \bar{\epsilon}_{Y,XX} - \bar{\omega}_{,XY} = (D_{Z,XY})^2 - D_{Z,XX} D_{Z,YY} \tag{8.49}$$

We now eliminate the strains in favor of the stress resultants. From matrix 6–2, in the absence of thermal terms,

$$N_X = \frac{Eh}{1 - \mu^2}\,(\epsilon_X + \mu\epsilon_Y) \tag{8.50a}$$

$$N_Y = \frac{Eh}{1 - \mu^2}\,(\epsilon_Y + \mu\epsilon_X) \tag{8.50b}$$

$$N_{XY} = \frac{Eh}{2(1 + \mu)}\omega \tag{8.50c}$$

which may be inverted to

$$\epsilon_X = \frac{1}{Eh}(N_X - \mu N_Y) \tag{8.51a}$$

$$\epsilon_Y = \frac{1}{Eh}(N_Y - \mu N_X) \tag{8.51b}$$

$$\omega = \frac{2(1 + \mu)}{Eh}N_{XY} \tag{8.51c}$$

Now, replacing ϵ_X, ϵ_Y, and ω by $\bar{\epsilon}_X$, $\bar{\epsilon}_Y$, and $\bar{\omega}$ in equation (8.51) and introducing this equation into equation (8.49), we have

$$\frac{1}{Eh}\,[N_{X,YY} + N_{Y,XX} - \mu(N_{X,XX} + N_{Y,YY}) - 2(1 + \mu)N_{XY}]$$

$$= (D_{Z,XY})^2 - D_{Z,XX}D_{Z,YY} \tag{8.52}$$

which, along with equation (8.35), constitute the compatibility and equilibrium equations, respectively.

There are still four unknowns remaining in the two equations, so that further refinement is necessary.

If a stress function F is defined, such that

$$N_X = F_{,YY} \tag{8.53a}$$

$$N_Y = F_{,XX} \tag{8.53b}$$

$$N_{XY} = -F_{,XY} \tag{8.53c}$$

and equations (8.53a–c) are introduced into equations (8.52) and (8.35), we obtain

$$\nabla^4 F(X,\ Y) = Eh[\,(D_{Z,XY})^2 - D_{Z,XX}D_{Z,YY}] \tag{8.54}$$

and

$$\nabla^4 D_Z(X,\ Y) = \frac{1}{D}\,[q_Z + F_{,YY}D_{Z,XX} + F_{,XX}D_{Z,YY} - 2F_{,XY}D_{Z,XY}] \tag{8.55}$$

which are known as the von Kármán equations for the large deflection of plates, after the famous contemporary mechanician.

The von Kármán equations are coupled and nonlinear. The nonlinearity arises from the relaxation of assumption [2], and the enforcement of this assumption immediately reduces equation (8.55) to the equation of the linear theory, equation (8.12). Also, the equations are written in invariant form, and thus may be readily transformed to other coordinate systems.

Notes

1. R. Szilard, *Theory and Analysis of Plates* (Englewood Cliffs, N.J.: Prentice-Hall, 1974).

2. S. Timoshenko and S. Woinowksy-Krieger, *Theory of Plates and Shells*, 2nd ed. (New York: McGraw-Hill, 1959).

3. M. Filonenko-Borodich, Theory of Elasticity [translated from Russian by M. Konayeva (New York: Dover, 1965) pp. 260–268].

4. Timoshenko and Woinowsky-Krieger, *Theory of Plates and Shells*, pp. 380–387.

9 Shell Bending and Instability

9.1 General

We have derived in earlier chapters the equilibrium, strain-displacement, and constitutive equations, and stated the required boundary conditions for the bending theory of shells, referred to a system of orthogonal curvilinear coordinates. Also, we have derived strain energy and potential energy expressions that can be incorporated into an energy formulation of the shell theory. In this chapter, these equations are specialized for various classes of shells, as we have done for the membrane theory equations in chapter 4.

Before proceeding, again note that many shells may achieve equilibrium through membrane action alone, provided the requisite conditions for membrane behavior are closely approached by the actual shell. For such shells, the bending is a secondary phenomenon often confined to narrow regions near boundaries, geometric discontinuities, and concentrated loads. On the other hand, there are shells for which the membrane theory solution is grossly violated by the physical situation. The bending behavior may alter the stress pattern from that computed by the membrane theory in two ways. First and most obvious, significant transverse shearing forces and bending and twisting moments can develop. Also, the pattern of in-plane stress resultants may be altered markedly by the bending deformations. It is this latter possibility, whereby the transverse loading is resisted by a *combination* of in-plane forces and transverse shearing forces, which distinguishes the bending theory of plates. Although a shell may seriously violate the membrane theory requirements, there still remains the mechanism of resisting *transverse loading* primarily with *in-plane forces*, which is the basic initial attraction of this structural form.

In the study of shell bending, cylindrical and conical shells are considered apart from rotational and translational shells, although they technically may fall into one or both of these classes. The bending behavior of shells with zero Gaussian curvature is quite distinct, however, in that these shells cannot develop any membrane forces in the direction where the radius of curvature is infinite. As we have seen in chapter 4, the membrane theory solutions for such shells generally consist of the entire transverse loading being sustained in the curved direction, a one-way resistance pattern. When the membrane boundary conditions in the curved direction are violated, drastic alteration of the stress pattern is often the result because the zero curvature direction is of no help except in a bending mode.

Another important reason for studying cylindrical shells especially as a distinct class is that fairly simple analytical solutions may be found, whereas for more complex geometries, such solutions are relatively scarce. Moreover, many of the techniques for solving the governing equations for other geometries involve various approximations and simplifications that cast the equations into a form similar to that of the cylindrical shell. Thus the cylindrical shell solutions assume added importance from the standpoint of generality.

Finally, we note that cylindrical shells are probably the most frequently encountered shell form and certainly merit careful attention.

9.2 Circular Cylindrical Shells

9.2.1 Specialization of Equations

In the discussion of the membrane theory of circular cylindrical shells (section 4.3.3.2), we chose $\alpha = Z$, $\beta = \theta$, $A = 1$, and $B = a$, as shown in figure 4–12(a). For this geometry, $R_\alpha = \infty$ and $R_\beta = a$. Then in section 4.4.1, the axial coordinate was taken as X in deference to its widespread usage in current literature. We should feel comfortable with either choice. For vertical vessels in which the axis of rotation coincides with the global Z axis, $\alpha = Z$ is the logical choice, whereas for horizontal shells, $\alpha = X$ seems equally suitable. We use $\alpha = X$ in this chapter.

The force equilibrium equations, (3.17a–c), and moment equilibrium equations, (3.18a) and (3.18b), become

$$aN_{X,X} + N_{\theta X,\theta} + q_X a = 0 \tag{9.1a}$$

$$aN_{X\theta,X} + N_{\theta,\theta} + Q_\theta + q_\theta a = 0 \tag{9.1b}$$

$$aQ_{X,X} + Q_{\theta,\theta} - N_\theta + q_n a = 0 \tag{9.1c}$$

$$-aM_{X\theta,X} - M_{\theta,\theta} + Q_\theta a = 0 \tag{9.1d}$$

$$aM_{X,X} + M_{\theta X,\theta} - Q_X a = 0 \tag{9.1e}$$

The strain-displacement relations, equations (5.46a–h), are

$$\epsilon_X = D_{X,X} \tag{9.2a}$$

$$\epsilon_\theta = \frac{1}{a}(D_{\theta,\theta} + D_n) \tag{9.2b}$$

$$\omega = D_{\theta,X} + \frac{1}{a} D_{X,\theta} \tag{9.2c}$$

$$\kappa_X = D_{X\theta,X} \tag{9.2d}$$

$$\kappa_\theta = \frac{1}{a} D_{\theta X,\theta} \tag{9.2e}$$

$$\tau = \frac{1}{2} \left[D_{\theta X,X} + \frac{1}{a} D_{X\theta,\theta} \right] \tag{9.2f}$$

$$\gamma_X = -(D_{n,X} + D_{X\theta}) \tag{9.2g}$$

$$\gamma_\theta = -\frac{1}{a} (D_{n,\theta} + D_\theta) - D_{\theta X} \tag{9.2h}$$

If transverse shearing strains are neglected, equations (5.47a–c) replace equations (9.2d–h):

$$\kappa_X = -D_{n,XX} \tag{9.3a}$$

$$\kappa_\theta = \frac{1}{a^2} (D_{n,\theta\theta} - D_{\theta,\theta}) \tag{9.3b}$$

$$\tau = \frac{1}{a} \left(\frac{1}{2} D_{\theta,X} - D_{n,X\theta} \right) \tag{9.3c}$$

Also, we have the effective shearing forces evaluated from equations (6.24)–(6.27):

$$\bar{N}_{X\theta} = N_{X\theta} + \frac{M_{X\theta}}{a} \tag{9.3d}$$

$$\bar{Q}_X = Q_X + \frac{M_{X\theta,\theta}}{a} \tag{9.3e}$$

$$\bar{N}_{\theta X} = N_{\theta X} \tag{9.3f}$$

$$\bar{Q}_\theta = Q_\theta + M_{\theta X,X} \tag{9.3g}$$

The constitutive relations are given by matrix 6–2, with α and β taken as X and θ, respectively. These are restated here only in the matrix form of equation (6.10)

$$\{N\}_{X\theta} = [D]\{\epsilon\}_{X\theta} - \{N_T\}_{X\theta} \tag{9.4}$$

with the subscripts serving to remind us of our choice of coordinates.

It is easily verified that the membrane theory equations are recoverable from the preceding expressions. With the bending terms neglected and $N_{\theta X} = N_{X\theta} = S$, equations (9.1a–c) reduce to equations (4.160) and equations (9.2a–c), and the corresponding parts of equation (9.4) are identical to equations (6.50).

9.2.2 Axisymmetrical Loading

9.2.2.1 Displacement Formulation and Solution. Axisymmetrically loaded circular cylindrical shells are employed as pressure vessels and tanks in many industrial applications. Some examples are shown in figures 2–8(r) and (u). For axisymmetrical loading, the terms q_θ, $N_{\theta X}$, $N_{X\theta}$, $M_{\theta X}$, $M_{X\theta}$, Q_θ, ω, κ_θ, τ, γ_θ, D_θ, $D_{\theta X}$, and all terms differentiated with respect to θ drop out. This leaves the three equilibrium equations

$$N_{X,X} + q_X = 0 \tag{9.5a}$$

$$Q_{X,X} - \frac{N_\theta}{a} + q_n = 0 \tag{9.5b}$$

$$M_{X,X} - Q_X = 0 \tag{9.5c}$$

and the strain-displacement relationships

$$\epsilon_X = D_{X,X} \tag{9.6a}$$

$$\epsilon_\theta = \frac{D_n}{a} \tag{9.6b}$$

$$\kappa_X = D_{X\theta,X} \tag{9.6c}$$

$$\gamma_X = -D_{n,X} - D_{X\theta} \tag{9.6d}$$

or, in the absence of transverse shearing strains

$$\kappa_X = -D_{n,XX} \tag{9.7}$$

We also have the pertinent constitutive relationships from equation (9.4). These are written explicitly from matrix 6–2 as

$$N_X = \frac{Eh}{1 - \mu^2}(\epsilon_X + \mu\epsilon_\theta) - N_{XT} \tag{9.8a}$$

$$N_\theta = \frac{Eh}{1 - \mu^2}(\mu\epsilon_X + \epsilon_\theta) - N_{\theta T} \tag{9.8b}$$

$$M_X = D\kappa_X - M_{XT} \tag{9.8c}$$

$$M_\theta = \mu D\kappa_X - M_{\theta T} \tag{9.8d}$$

$$Q_X = \frac{\lambda Eh}{2(1+\mu)} \gamma_X \tag{9.8e}$$

When transverse shearing strains are suppressed, equation (9.8e) is discarded and Q_X is evaluated from equation (9.5c).

We now express the equilibrium equations in terms of the displacements in the classical fashion of a displacement formulation.

First, we substitute equations (9.6a–d) into equations (9.8a–e) to get the stress resultants and couples in terms of the displacements. Also, in view of equation (6.9), we set $N_{XT} = N_{\theta T} = N_T$ and $M_{XT} = M_{\theta T} = M_T$. Then we have

$$N_X = \frac{Eh}{1-\mu^2}\left(D_{X,X} + \mu\frac{D_n}{a}\right) - N_T \tag{9.9a}$$

$$N_\theta = \frac{Eh}{1-\mu^2}\left(\mu D_{X,X} + \frac{D_n}{a}\right) - N_T \tag{9.9b}$$

$$M_X = DD_{X\theta,X} - M_T \tag{9.9c}$$

$$M_\theta = DD_{X\theta,X} - M_T \tag{9.9d}$$

$$Q_X = -\frac{\lambda Eh}{2(1+\mu)}(D_{n,X} + D_{X\theta}) \tag{9.9e}$$

In the absence of transverse shearing strain, we use equation (9.7) for κ_X and find

$$M_X = -DD_{n,XX} - M_T \tag{9.10a}$$

$$M_\theta = -\mu DD_{n,XX} - M_T \tag{9.10b}$$

and, from equations (9.5c) and (9.10a),

$$Q_X = -DD_{n,XXX} - M_{T,X} \tag{9.10c}$$

Also, in equation (9.3e), with $M_{X\theta} = 0$, $\bar{Q}_X = Q_X$.

Now, substituting equations (9.9a–e) into (9.5a–c), we get

$$\frac{Eh}{1-\mu^2}\left(D_{X,X} + \frac{\mu}{a}D_n\right)_{,X} = -q_X + N_{T,X} \tag{9.11a}$$

$$\frac{-Eh}{1+\mu}\left[\frac{\lambda}{2}(D_{n,X} + D_{X\theta})_{,X} + \frac{1}{a(1-\mu)}\left(\mu D_{X,X} + \frac{D_n}{a}\right)\right] = -q_n - \frac{N_T}{a} \tag{9.11b}$$

$$DD_{X\theta,XX} + \frac{\lambda Eh}{2(1+\mu)}(D_{n,X} + D_{X\theta}) = M_{T,X} \tag{9.11c}$$

The loading and thermal terms are presumed as known and transposed to the r.h.s. The resulting set of three equations in the three unknowns D_X, D_n, and $D_{X\theta}$ constitutes the displacement formulation. We will not reduce these equations further, but concentrate on the elementary theory in which transverse shearing strains are neglected.

With $\gamma_X = 0$, $D_{X\theta}$ is expressed in terms of D_n by equation (9.6d). We need retain only the first of equations (9.11a–c), along with a new equation found by substituting equations (9.9b) and (9.10c) into equation (9.5b):

$$\frac{Eh}{1 - \mu^2}\left(D_{X,X} + \frac{\mu}{a}D_n\right)_{,X} = -q_X + N_{T,X} \tag{9.12a}$$

$$-DD_{n,XXXX} - \frac{Eh}{a(1 - \mu^2)}\left(\mu D_{X,X} + \frac{D_n}{a}\right) = -q_n - \frac{N_T}{a} + M_{T,XX} \tag{9.12b}$$

This set may be reduced to a single equation. We first integrate both sides of equation (9.12a), which gives

$$\frac{Eh}{1 - \mu^2}\left(D_{X,X} + \frac{\mu}{a}D_n\right) = -\int q_X\,dX + N_T + C \tag{9.13a}$$

where C is an integration constant.

Comparing equation (9.13a) with equation (9.9a), we find

$$N_X = -\int q_X\,dX + C \tag{9.13b}$$

We may rewrite the integral in equation (9.13a) in the alternate form introduced in chapter 4 by choosing $C = N_X(0)$. Then we have

$$N_X(X) = -\int_0^X q_X\,dX + N_X(0) \tag{9.13c}$$

Next, we solve equation (9.13a) for

$$D_{X,X} = \frac{1 - \mu^2}{Eh}\left[-\int_0^X q_X\,dX + N_T + N_X(0)\right] - \frac{\mu}{a}D_n \tag{9.13d}$$

which we substitute into equation (9.12b) to find

$$-DD_{n,XXXX} - \frac{Eh}{a(1 - \mu^2)}\left\{\mu\left(\frac{1 - \mu^2}{Eh}\left[-\int_0^X q_X\,dX + N_T + N_X(0)\right]\right.\right.$$

$$\left.\left. - \frac{\mu}{a}D_n\right) + \frac{D_n}{a}\right\} = -q_n - \frac{N_T}{a} + M_{T,XX} \tag{9.14}$$

which consolidates to

$$DD_{n,XXXX} + \frac{Eh}{a^2}D_n = q_n + \left(\frac{1-\mu}{a}\right)N_T - M_{T,XX}$$

$$+ \frac{\mu}{a}\left[\int_0^X q_X\,dX - N_X(0)\right] \tag{9.15}$$

Equation (9.15) is often written in the form

$$D_{n,XXXX} + 4k^4 D_n =$$

$$\frac{1}{D}\left(q_n + \frac{(1-\mu)}{a}N_T - M_{T,XX} + \frac{\mu}{a}\left[\int_0^X q_X\,dX - N_X(0)\right]\right) \tag{9.16a}$$

where

$$k^4 = \frac{3(1-\mu^2)}{a^2 h^2} = \frac{Eh}{4Da^2} \tag{9.16b}$$

Equation (9.16) may be recognized as the governing equation for a well-documented problem in mechanics, the bending of a prismatic beam on an elastic foundation.[1] Just as an elastic foundation takes a portion of any transverse loading applied to a beam, the shell parameter Eh/a^2 which appears explicitly in equation (9.15), supplements the flexural rigidity D of the cylindrical shell in resisting the transverse loading q_n. Once the governing equation is solved for D_n, the in-plane stress resultants are computed from equations (9.9a) and (9.9b), and the stress couples and transverse shear resultant from equations (9.10a–c).

It is instructive to consider a somewhat specialized case, in which the thermal terms are dropped and the axial load $q_X = 0$. Then, from equation (9.13c), $N_X = $ constant. If one boundary—e.g., $X = 0$—is unconstrained against axial deformation, $N_X(0) = 0$ and, therefore, $N_X = 0$ throughout the shell. With these simplifications, equation (9.9a) gives

$$D_{X,X} = -\mu\frac{D_n}{a} \tag{9.17}$$

and equation (9.9b) becomes

$$N_\theta = \frac{Eh}{a}D_n \tag{9.18}$$

which is a remarkably simple *algebraic* relationship between the major stress resultant and the primary displacement, considering the complexity of the basic formulation with which we began.

The displacement formulation of the axisymmetrically loaded cylindrical shell may be generalized to accommodate shells in which the thickness $h = h(X)$ in a straightforward matter.

292

The homogeneous solution to equation (9.16) is written from the auxiliary equation

$$m^4 + 4k^4 = 0 \qquad\qquad (9.19a)$$

with roots

$$m = \pm k(1 + i),\ \pm k(1 - i) \qquad\qquad (9.19b)$$

as[2]

$$D_{nh} = e^{-kX}(C_1 \cos kX + C_2 \sin kX)$$
$$+ e^{kX}(C_3 \cos kX + C_4 \sin kX) \qquad\qquad (9.20)$$

where C_1–C_4 are integration constants.

The particular solution is, of course, dependent on the loading, and the constants C_1–C_4 are found from applying the appropriate boundary conditions, as we see in several examples later.

9.2.2.2 Semi-Infinite Cylindrical Shells. The complete solution of equation (9.16) involves four integration constants, as shown in equation (9.20). This constitutes a two-point boundary value problem. Within the class of axisymmetrically loaded cylindrical shells, however, there are many cases for which the forces at one boundary do not materially affect those at the other boundary. Such shells are termed *semi-infinite*.

We may demonstrate this behavior by referring to equation (9.20). If X is measured from one boundary, the factor e^{kX} will grow very large as X increases unless $C_3 = C_4 = 0$, whereas the factor e^{-kX} will cause the other terms to attenuate. Thus the solution simplifies to

$$D_{nh} = e^{-kX}(C_1 \cos kX + C_2 \sin kX) \qquad\qquad (9.21)$$

with C_1 and C_2 evaluated from the boundary conditions at $X = 0$.

The semi-infinite approach simplifies the ensuing arithmetic considerably. The range of shell parameters for which this assumption is valid may be determined by considering a shell with an arbitrary edge loading and observing the behavior of the solution as the distance from the loaded boundary increases. As shown in figure 9–1, a transverse shearing force Q_0 and bending moment M_0 are applied uniformly around the circumference at $X = 0$. The positive signs are chosen in accordance with figure 3–2. We have, from equations (9.10a) and (9.10c),

$$M_0 = M_X(0) = -DD_{n,XX}(0) \qquad\qquad (9.22a)$$

$$Q_0 = Q_X(0) = -DD_{n,XXX}(0) \qquad\qquad (9.22b)$$

from which

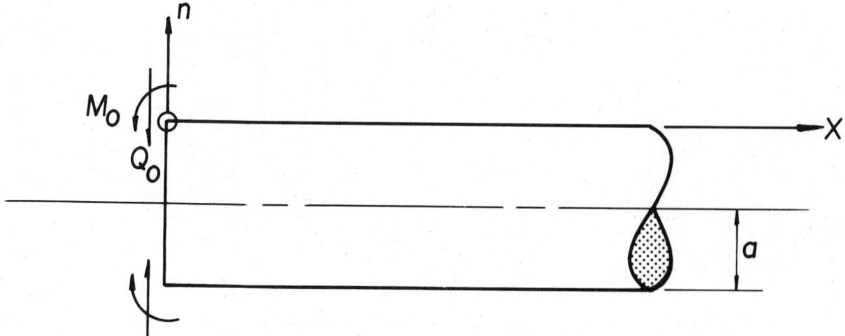

Figure 9–1. Semi-Infinite Edge-Loaded Cylindrical Shell

$$C_1 = -\frac{1}{2k^3D}(Q_0 + M_0) \tag{9.23a}$$

and

$$C_2 = \frac{M_0}{2k^2D} \tag{9.23b}$$

Before we carry the solution further, it is convenient to define the functions

$$F_1(kX) = e^{-kX}(\cos kX + \sin kX) \tag{9.24a}$$

$$F_2(kX) = e^{-kX}(\cos kX - \sin kX) \tag{9.24b}$$

$$F_3(kX) = e^{-kX}\cos kX = \frac{1}{2}(F_1 + F_2) \tag{9.24c}$$

$$F_4(kX) = e^{-kX}\sin kX = \frac{1}{2}(F_1 - F_2) \tag{9.24d}$$

Now we may express D_n and the various derivatives required to obtain other components of the solution by matrix 9–1.

Since $F_1(0) = F_2(0) = 1$ and decrease as X increases, and since F_3 and F_4 are linear combinations of F_1 and F_2, the behavior of $F_1(X)$ and $F_2(X)$ is indicative of the propagation of the effects of the boundary loads. These functions are plotted on figure 9–2 and are tabulated in Timoshenko and Woinowski-Krieger and Tsui.[3] It is apparent from the figure that the edge loads will produce insignificant effects at about $kX = 3$. In the literature, it is common to find $L \geq \Pi/k$ designated as the length for which a given shell is considered to be a long shell and, therefore, treatable by the semi-infinite approach.

Matrix 9–1

$$
\begin{Bmatrix} D_n \\ D_{n,x} \\ D_{n,xx} \\ D_{n,xxx} \end{Bmatrix}
= \frac{1}{D}
\begin{bmatrix}
0 & -\dfrac{M_0}{2k^2} & -\dfrac{Q_0}{2k^3} & 0 \\[2ex]
\dfrac{Q_0}{2k^2} & 0 & \dfrac{M_0}{k} & 0 \\[2ex]
-M_0 & 0 & 0 & -\dfrac{Q_0}{k} \\[2ex]
0 & -Q_0 & 0 & 2kM_0
\end{bmatrix}
\begin{Bmatrix} F_1(kX) \\ F_2 kX) \\ F_3(kX) \\ F_4(kX) \end{Bmatrix}
\begin{matrix} (a) \\ (b) \\ (c) \\ (d) \end{matrix}
$$

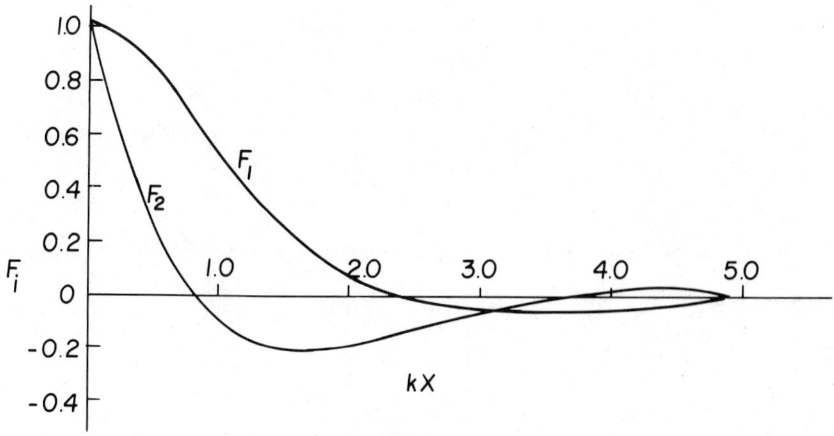

Figure 9–2. Solution Functions for Semi-Infinite Shells

Note that the edge effects that have been discussed here, Q_0 and M_0, are *self-equilibrating* with respect to the overall equilibrium of the shell. The corresponding dissipation of these edge disturbances may be regarded as a demonstration of St. Venant's principle, which we have mentioned in connection with the derivation of the Kirchhoff boundary conditions in section 6.2.3. Recall that there may be another type of edge effect which is *not* self-equilibrated and readily penetrates to the opposite boundary. An example is the axial line load $\bar{N}$ shown in figure 4–12(b) and discussed in section 4.3.3.2. We should be careful to reserve the semi-infinite simplification for cases with self-equilibrating edge loads, such as Q_0 and M_0.

Finally, with respect to the problem illustrated on figure 9–1, we may obtain explicit expressions for the stress resultants from equations (9.10a–c), (9.18), and matrix 9–1:

$$M_X(X) = -DD_{n,XX} = M_0 F_1(kX) + \frac{Q_0}{k} F_4(kX) \tag{9.25a}$$

$$M_\theta(X) = \mu M_X \tag{9.25b}$$

$$Q_X(X) = -DD_{n,XXX} = Q_0 F_2(kX) - 2kM_0 F_4(kX) \tag{9.25c}$$

$$N_\theta(X) = \frac{Eh}{a} D_n(X) = \frac{Eh}{2ak^2 D} \left[M_0 \, F_2(kX) + \frac{Q_0}{k} F_3(kX) \right] \tag{9.25d}$$

This solution is utilized frequently in the ensuing examples.

9.2.2.3 Circumferential Line Loading. Consider a shell subjected to a circumferential line load P (force/length) as shown in figure 9–3(a), and as-

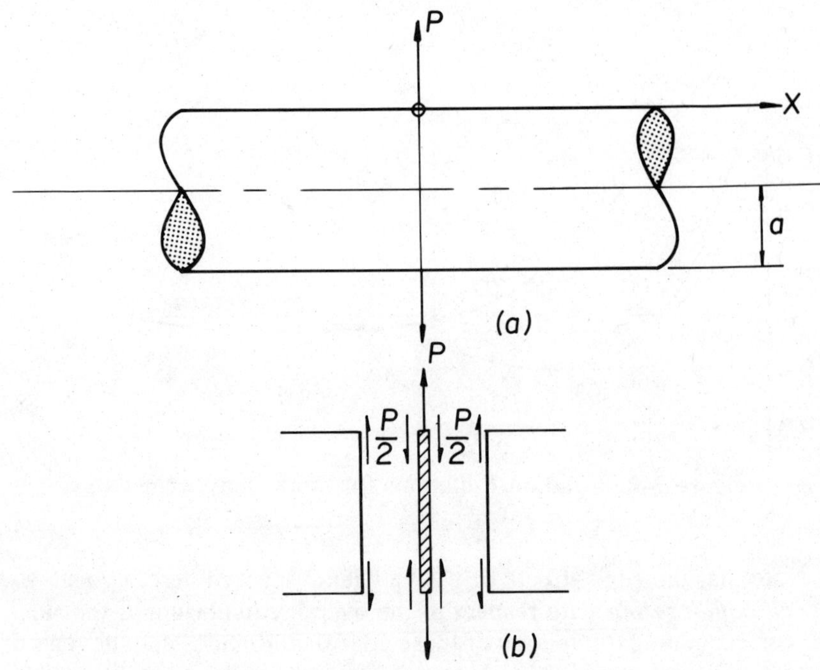

Figure 9–3. Symmetric Line Loading on a Cylindrical Shell

sume that the shell extends a distance of at least Π/k from the point of application of P in each direction. A free body diagram of a narrow ring under the load, figure 9–3(b), reveals that the problem reduces to the case of an edge-loaded semi-infinite shell, similar to that treated in the previous section with boundary conditions

$$Q_0 = Q_x(0) = -\frac{P}{2} \tag{9.26a}$$

$$D_{x\theta}(0) = -D_{n,x}(0) = 0 \tag{9.26b}$$

From matrix 9–1(b) and equations (9.26a) and (9.26b), we have

$$M_0 = -\frac{Q_0}{2k} = \frac{P}{4k} \tag{9.27}$$

The displacement function and stress resultants are written by substituting equations (9.26a) and (9.27) into matrix 9–1 and equations (9.25a–d) and simplifying the ensuing expressions, using equations (9.24c) and (9.24d):

$$D_n(X) = \frac{-P}{4k^3D}\left[\frac{1}{2}F_2(kX) - F_3(kX)\right] = \frac{P}{8k^3D}F_1(kX) \quad (9.28a)$$

$$M_X(X) = \frac{P}{2k}\left[\frac{1}{2}F_1(kX) - F_4(kX)\right] = \frac{P}{4k}F_2(kX) \quad (9.28b)$$

$$M_\theta(X) = \mu M_X(X) \quad (9.28c)$$

$$Q_X(X) = -\frac{P}{2}[F_2(kX) + F_4(kX)] = -\frac{P}{2}F_3(kX) \quad (9.28d)$$

$$N_\theta(X) = \frac{Eh}{a}D_n(X) \quad (9.28e)$$

Under the load at $X = 0$, we find the maximum values

$$D_n(0) = \frac{P}{8k^3D} \quad (9.29a)$$

$$M_X(0) = \frac{P}{4k} \quad (9.29b)$$

$$Q_X(0) = -\frac{P}{2} \quad (9.29c)$$

$$N_\theta(0) = \frac{PEh}{8ak^3D} = \frac{Pak}{2} \quad (9.29d)$$

These effects die out with increasing X, as previously indicated.

This solution is of interest in the case of a cylindrical shell with a circular ring stiffener, which we consider later, and also in generating solutions for loading distributed in the X direction by using the Green's function approach.

9.2.2.4 Axially Distributed Loading. We now consider a radial loading uniformly distributed around the circumference but arbitrarily distributed along the X axis, as shown in figure 9–4. To treat this problem from a general standpoint, we choose the Green's function technique. In this approach, the response of the shell to a single concentrated load acting at a *point of application C*, which is arbitrarily located within the loaded region, is studied first. The influence of this load is evaluated at a specified *point of observation*. Then, the effect of the entire distributed load at the point of observation is computed by integrating over the loaded region. In this general example, we must treat two separate cases: (a) point of observation outside the loaded region of the shell, such as point A in figure 9–4;

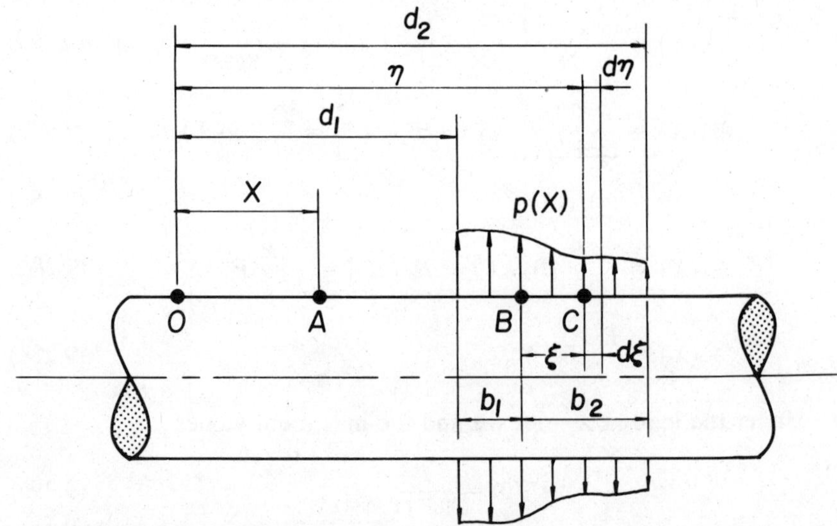

Figure 9–4. Distributed Symmetrical Loading on a Cylindrical Shell

and (b) point of observation within the loaded region of the shell, such as point B in figure 9–4.

We first select the point of observation as A, outside the loaded region at a distance X from a convenient origin O. We then evaluate the concentrated load produced by $p(\eta)$ acting on the differential element $d\eta$ at the point of application C as $p(\eta)\,d\eta$. The displacement at A due to this loading is given by equation (9.28a), with the argument of $F_1(kX)$ taken as the distance between A and C, $\eta - X$:

$$D_n(A, C) = D_n(X, \eta) = \frac{p(\eta)\,d\eta}{8k^3D}\,F_1(k[\eta - X]) \tag{9.30}$$

Then the displacement due to the entire distributed loading is found as

$$D_n(A) = D_n(X) = \int_{\eta=d_1}^{\eta=d_2} D_n(X, \eta)\,d\eta \tag{9.31}$$

which is easily evaluated once $p(X)[$ or $p(\eta)]$ is specified.

Second, locate the point of observation at B, within the loaded region. Point B is a distance $X = d_1 + b_1$ from the origin O, and is a distance ξ from the point of application C. The displacement at B due to the concentrated load $P(\xi)\,d\xi$ is

$$D_n(B, C) = D_n(X, \xi) = \frac{p(\xi)\,d\xi}{8k^3D}\,F_1(k\xi) \tag{9.32}$$

and the displacement of B due to the entire load is found to be

$$D_n(B) = D_n(X) = \int_{\xi=0}^{\xi=b_1} D_n(X, \xi)\, d\xi + \int_{\xi=0}^{\xi=b_2} D_n(X, \xi)\, d\xi \qquad (9.33)$$

where $b_2 = d_2 - d_1 - b_1$. For both integrals in the above equation, the coordinate ξ is positive as measured from point B.

For a point of observation at the edge of the loaded region, A and B are coincident, and equations (9.31) and (9.33) should be identical. With $b_1 = 0$, $b_2 = d_2 - d_1$ in equation (9.33), and the identity is confirmed.

The stress resultants and couples for specific loadings $p(X)$ may be evaluated by similiar integrations or by subsequent differentiations of the computed $D_n(X)$, as indicated in equations (9.25).

9.2.2.5 Built-In Shell Under Internal Pressure and Temperature Gradiant.

Next, we examine the cylindrical shell shown in figure 9–5(a) that is subject to a uniform internal pressure p and a linear temperature gradiant—T_1 on the outside and T_2 on the inside. Both the pressure and the thermal gradiant are taken as constant along the length of the shell, and the ends are fixed against translation and rotation. We again assume that L is sufficiently large so that semi-infinite analysis is valid. A similar problem is considered by Kraus[4] using the more general short-shell solution, but only for a uniform temperature change through the thickness.

Proceeding, we have the homogeneous solution from equation (9.21)

$$D_{nh} = e^{-kX}(C_1 \cos kX + C_2 \sin kX) = C_1 F_3(kX) + C_2 F_4(kX) \qquad (9.34)$$

We now examine equation (9.16a) with respect to the construction of a particular solution, and consider each term on the r.h.s. individually.

First, for the pressure term $q_n = p$, we select

$$D_{np1} = \frac{p}{4k^4 D} = \frac{pa^2}{Eh} \qquad (9.35)$$

To find the contribution of the temperature gradiant, we first evaluate N_T and M_T from equations (6.9a) and (6.9b). It is helpful to resolve the linear gradiant into symmetric and antisymmetric components, as shown in figure 9–5(b):

$$T(\zeta) = T_s + \zeta T_a \qquad (9.36a)$$

where

$$T_s = \frac{T_1 + T_2}{2} \qquad (9.36b)$$

$$T_a = \frac{T_1 - T_2}{2} \qquad (9.36c)$$

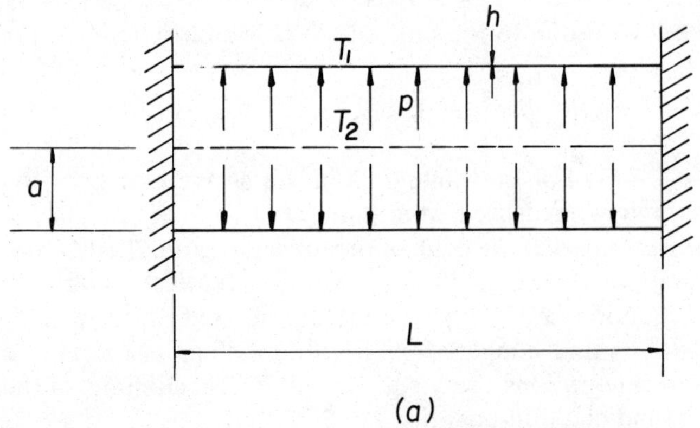

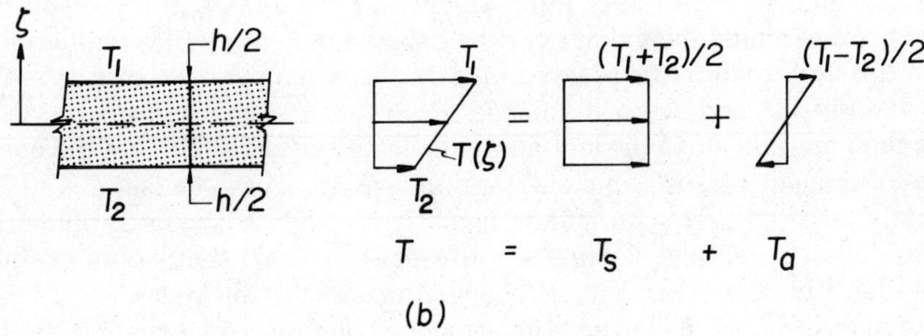

Figure 9–5. Built-In Cylindrical Shell under Pressure and Thermal Loadings

and ζ = an auxiliary normal coordinate measured from the middle surface. Then, from equation (6.9a) and (6.9b),

$$N_T = \frac{E\bar{\alpha}}{1 - \mu} \int_{-h/2}^{h/2} T(\zeta)\, d\zeta = \frac{E\bar{\alpha}h}{1 - \mu} T_s \qquad (9.37a)$$

$$M_T = \frac{E\bar{\alpha}}{1 - \mu} \int_{-h/2}^{h/2} T(\zeta)\zeta\, d\zeta = \frac{E\bar{\alpha}h^2}{6(1 - \mu)} T_a \qquad (9.37b)$$

Now, in equation (9.16a), we set

$$4k^4 D_{np2} = \frac{(1 - \mu)}{aD} \cdot \frac{E\bar{\alpha}h}{(1 - \mu)} T_s \qquad (9.38)$$

from which

$$D_{np2} = \bar{\alpha}a T_s \qquad (9.39)$$

Since $M_{T,xx} = q_x = 0$ in this case, no particular solutions are required for these terms.

Finally, we have the axial stress resultant term $N_x(0)$, which gives

$$4k^4 D_{np3} = -\frac{\mu}{aD} N_x(0) \tag{9.40a}$$

or

$$D_{np3} = -\frac{\mu}{4k^4 aD} N_x(0) = -\frac{\mu a}{Eh} N_x(0) \tag{9.40b}$$

Collecting all of the particular solution contributions, we have

$$D_{np} = D_{np1} + D_{np2} + D_{np3} = \frac{pa^2}{Eh} + \bar{\alpha} a T_s - \frac{\mu a}{Eh} N_x(0) \tag{9.41}$$

We may make an interesting observation concerning the particular solution by referring back to our treatment of membrane theory displacements for cylindrical shells in section 6.3.4.1. Specifically, we consider the relevant part of equation (6.53) for the membrane theory normal displacement

$$D_n = \frac{a}{Eh} (N_\theta - \mu N_x) \tag{9.42a}$$

Now we substitute the *membrane theory* stress resultant N_θ, as given by equation (4.62b),

$$N_\theta = pa \tag{9.42b}$$

along with

$$N_x = 0 \tag{9.42c}$$

into equation (9.43a), which gives

$$D_n = D_{nm1} = \frac{pa^2}{Eh} \tag{9.42d}$$

which is the same as equation (9.35). For the thermal term, referring to equations (6.9a) and (6.9b), we introduce

$$N_x = N_\theta = N_T = \frac{E\bar{\alpha}h}{1 - \mu} T_s \tag{9.43a}$$

into equation (9.42a). Then,

$$D_n = D_{nm2} = \frac{a}{Eh} \cdot \frac{E\bar{\alpha}h}{1 - \mu} T_s(1 - \mu) = \bar{\alpha} a T_s \tag{9.43b}$$

which is identical to equation (9.39). Similarly, with $N_X = N_X(0)$ in equation (9.42a),

$$D_n = D_{nm3} = -\frac{\mu a}{Eh} N_X(0) \tag{9.43c}$$

which matches equation (9.40b). We thus have quantified the earlier assertion that the membrane theory solution frequently serves as a particular solution to the bending theory equations.

We now proceed with the general solution, which is the sum of equations (9.34) and (9.41), by enforcing the boundary conditions

$$D_n(0) = 0 \tag{9.44a}$$

and

$$D_{X\theta}(0) = -D_{n,X}(0) = 0 \tag{9.44b}$$

The first condition gives

$$C_1 + \frac{pa^2}{Eh} + \bar{\alpha}aT_s - \frac{\mu a}{Eh} N_X(0) = 0 \tag{9.45a}$$

while the second yields

$$k(-C_1 + C_2) = 0 \tag{9.45b}$$

whereupon

$$C_1 = -\left[\frac{pa^2}{Eh} + \bar{\alpha}aT_s - \frac{\mu a}{Eh} N_X(0)\right] \tag{9.46a}$$

$$C_2 = C_1 \tag{9.46b}$$

To evaluate the term $N_X(0)$ explicitly, we integrate equation (9.13d), giving

$$D_X = \int_0^X D_{X,X}\, dX + D_{X,X}(0) \tag{9.47a}$$

with

$$D_{X,X}(0) = \frac{1 - \mu^2}{Eh} [N_T + N_X(0)] \tag{9.47b}$$

in view of equation (9.44a). With the end of the shell restrained at $X = 0$,

$$D_X(0) = 0 \tag{9.47c}$$

and

$$N_X(0) = -N_T = -\frac{E\bar{\alpha}h}{1 - \mu}T_s \tag{9.47d}$$

Then, substituting equation (9.47d) into equations (9.46a) and (9.46b),

$$C_1 = -\left[\frac{pa^2}{Eh} + \left(\frac{1}{1 - \mu}\right)\bar{\alpha}aT_s\right] \tag{9.48a}$$

$$C_2 = C_1 \tag{9.48b}$$

Correspondingly, equation (9.41) can be consolidated into

$$D_{np} = \frac{pa^2}{Eh} + \left(\frac{1}{1 - \mu}\right)\bar{\alpha}aT_s \tag{9.49}$$

and together with equation (9.34),

$$D_{nh} = C_1F_3(kX) + C_2F_4(kX) \tag{9.50}$$

constitutes the general solution. We also need $D_{X,X}$, which is found from equation (9.13d):

$$D_{X,X} = -\frac{\mu}{a}D_n(X) \tag{9.51}$$

Explicit expressions for the stress resultants and couples may be written by applying equations (9.9) and (9.10) to this solution. These are routine and are omitted for brevity. However, we may surmise from our previous studies of the semi-infinite cylindrical shell that the homogeneous part of the solution D_{nh} will be most influential near the ends and will diminish as X increases, whereas the particular part D_{np}, which is also the membrane theory solution, will predominate away from the ends. Also, note from equation (9.10) that there will be bending in the shell—even after the effects of D_{nh} diminish—because of the constant thermal moment M_T.

9.2.2.6 Short Cylindrical Shells. When the distance between boundary points, L, is such that $L < \Pi/k$, the semi-infinite assumption is no longer valid, and the general equation, (9.20), should be used. We first consider the extension of the edge-loaded cylinder problem, shown in figure 9–1, to the case where line moments and transverse shearing forces are applied to both ends (figure 9–6).

We rewrite equation (9.20) as

$$D_{nh} = C_1F_5(kX) + C_2F_6(kX) + C_3F_7(kX) + C_4F_8(kX) \tag{9.52a}$$

where

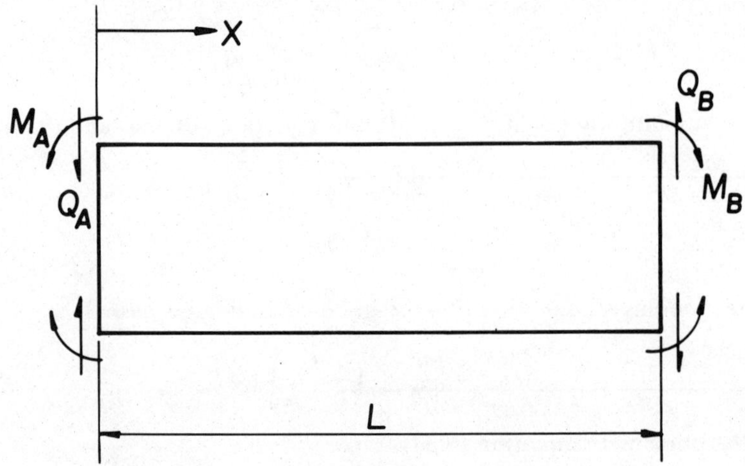

Figure 9–6. Edge-Loaded Cylindrical Shell

$$F_5(kX) = e^{-kX} \cos kX$$
$$F_6(kX) = e^{-kX} \sin kX$$
$$F_7(kX) = e^{kX} \cos kX$$
$$F_8(kX) = e^{kX} \sin kX \qquad (9.52b)$$

The boundary conditions for the loading in figure 9–6 are

$$M_X(0) = -DD_{n,XX}(0) = M_A$$
$$Q_X(0) = -DD_{n,XXX}(0) = Q_A$$
$$M_X(L) = -DD_{n,XX}(kL) = M_B$$
$$Q_X(L) = -DD_{n,XXX}(kL) = Q_B \qquad (9.53)$$

We now turn to the tedious calculation of the derivatives of D_{nh}. We first evaluate

$$D_{nh,X} = -C_1 k[F_5(kX) + F_6(kX)] + C_2 k[F_5(kX) - F_6(kX)]$$
$$+ C_7 k[F_7(kX) - F_8(kX)] + C_8 k[F_7(kX) + F_8(kX)] \qquad (9.54)$$

and then continue with the differentiation in operator notation, e.g. $\mathbf{D}^n = d^n(\ )/dX^n$

$$F_5(kX) = F_5$$
$$\mathbf{D}^1 F_5 = -kF_5 - kF_6$$
$$\mathbf{D}^2 F_5 = \mathbf{D}^1(\mathbf{D}^1 F_5) = -k\mathbf{D}^1 F_5 - k\mathbf{D}^1 F_6$$

$$F_6(kX) = F_6$$

$$\mathbf{D}^1 F_6 = -kF_6 + kF_5$$

$$\mathbf{D}^2 F_6 = \mathbf{D}^1(\mathbf{D}^1 F_6) = -k\mathbf{D}^1 F_6 + k\mathbf{D}^1 F_5$$

$$\mathbf{D}^3 F_6 = \mathbf{D}^1(\mathbf{D}^2 F_6) = -k\mathbf{D}^2 F_6 + k\mathbf{D}^2 F_5$$

$$F_7(kX) = F_7$$

$$\mathbf{D}^1 F_7 = kF_7 - kF_8$$

$$\mathbf{D}^2 F_7 = \mathbf{D}^1(\mathbf{D}^1 F_7) = k\mathbf{D}^1 F_7 - k\mathbf{D}^1 F_8$$

$$\mathbf{D}^3 F_7 = \mathbf{D}^1(\mathbf{D}^2 F_7) = k\mathbf{D}^2 F_7 - k\mathbf{D}^2 F_8$$

$$F_8(kX) = F_8$$

$$\mathbf{D}^1 F_8 = kF_8 + kF_7$$

$$\mathbf{D}^2 F_8 = \mathbf{D}^1(\mathbf{D}^1 F_8) = k\mathbf{D}^1 F_8 + k\mathbf{D}^1 F_7$$

$$\mathbf{D}^3 F_8 = \mathbf{D}^1(\mathbf{D}^2 F_8) = k\mathbf{D}^2 F_8 + k\mathbf{D}^2 F_7$$

By repeated substitution, the various derivatives can be written in terms of F_5–F_8:

$$\mathbf{D}^1 F_5 = -k(F_5 + F_6)$$

$$\mathbf{D}^2 F_5 = 2k^2 F_6$$

$$\mathbf{D}^3 F_5 = 2k^3(F_5 - F_6)$$

$$\mathbf{D}^1 F_6 = k(F_5 - F_6)$$

$$\mathbf{D}^2 F_6 = -2k^2 F_5$$

$$\mathbf{D}^3 F_6 = 2k^3(F_5 + F_6)$$

$$\mathbf{D}^1 F_7 = k(F_7 - F_8)$$

$$\mathbf{D}^2 F_7 = -2k^2 F_8$$

$$\mathbf{D}^3 F_7 = -2k^3(F_7 + F_8)$$

$$\mathbf{D}^1 F_8 = k(F_7 + F_8)$$

$$\mathbf{D}^2 F_8 = 2k^2 F_7$$

$$\mathbf{D}^2 F_8 = 2k^3(F_7 - F_8)$$

Now we insert the appropriate derivatives into equation (9.53), noting that $F_5(0) = F_7(0) = 1$ and $F_6(0) = F_8(0) = 0$, to get matrix 9–2, or

Matrix 9-2

$$
\begin{bmatrix}
0 & -1 & 0 & 1 \\
k & k & -k & k \\
F_6(kL) & -F_5(kL) & -F_8(kL) & F_7(kL) \\
k[F_5(kL) & k[F_5(kL) & -k[F_7(kL) & k[F_7(kL) \\
-F_6(kL)] & +F_6(kL)] & +F_8(kL)] & -F_8(kL)]
\end{bmatrix}
\begin{Bmatrix}
C_1 \\
C_2 \\
C_3 \\
C_4
\end{Bmatrix}
= \frac{-1}{2k^2D}
\begin{Bmatrix}
M_A \\
Q_A \\
M_B \\
Q_B
\end{Bmatrix}
$$

$$[\mathbf{F}]\{\mathbf{C}\} = -\frac{1}{2k^2D}\{\mathbf{B}\} \tag{9.55a}$$

from which

$$\{\mathbf{C}\} = -\frac{1}{2k^2D}[\mathbf{F}^{-1}]\{\mathbf{B}\} \tag{9.55b}$$

Although $\{\mathbf{C}\}$ may be written explicitly in algebraic form, the resulting expressions are cumbersome; also, in the computer era, it is routine to evaluate $\{\mathbf{C}\}$ by specifying numerical values for the shell parameters and performing the inversion and multiplication. We will regard the problem as solved at this point and consider an application.

9.2.2.7 Ring-Stiffened Cylindrical Shells.

Ring-stiffened cylindrical tubes are commonly used in pressure vessels and submersible vehicles. An example is shown in figure 2–8(r), and an idealization is depicted in figure 9–7. The rings are often relatively close together, so that the semi-infinite simplification may not be valid.

As the cylinder deforms in the radial direction, the ring will retard the expansion or contraction. The resulting contact force between the cylinder and the stiffener is shown as P in figure 9–7. This problem is somewhat similar in concept to that shown in figure 4–4, where the ring beam of a dome is analyzed. In that case, the strain incompatibility at the interface of the shell and the beam could not be resolved because of the limitations of the membrane theory. Here, however, we are able to enforce the deformation compatibility between the ring stiffener and the shell.

We assume that the stiffener thickness $h_s << L$, so that the reaction can be considered to act on the shell as a circumferential line load. Also, we take the inside radius of the stiffener as $\bar{a} = a + h/2$, where a is the radius at the shell middle surface and h is the shell thickness; b is the width of the stiffener.

The radial deformation of the ring stiffener due to the line load P (force/length of circumference) is easily computed by considering the stiffener to be a short cylindrical shell with radius $= \bar{a} + b/2$, thickness $= b$, and length $= h_s$. Then we apply equation (6.53), which for the loading and geometry under consideration, reduces to

$$D_n = D_s = \frac{[\bar{a} + (b/2)]}{Eb} N_\theta \tag{9.56a}$$

Assuming the line load to be uniformly distributed over the length h_s, and using equation (4.62b),

$$N_\theta = \frac{P}{h_s}\left(\bar{a} + \frac{b}{2}\right) \tag{9.56b}$$

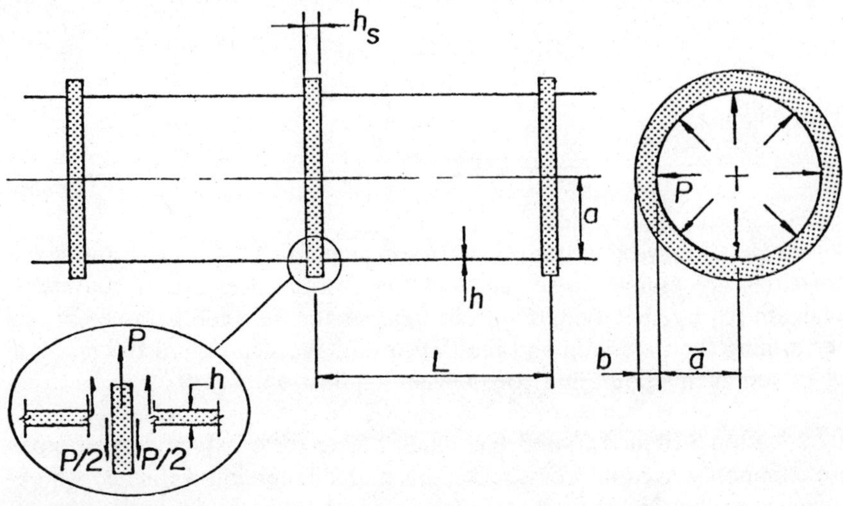

Figure 9–7. Ring-Stiffened Cylindrical Shell

so that the stiffener deformation in terms of the unknown contact force P is

$$D_s = \frac{P[\bar{a} + (b/2)]^2}{Ebh_s} \qquad (9.56c)$$

Now we examine the forces acting on the shell. We assume the loading is a uniform internal pressure p, so that the particular solution is given by equation (9.35):

$$D_{np} = D_{np1} = \frac{pa^2}{Eh} \qquad (9.57)$$

Also we have the force P applied to the shell by the ring stiffener. Assuming P is directed outward on the ring as depicted on the figure, an equal and opposite force reacts on the shell as shown in the inset. This is the same situation depicted in figure 9–3, but with the sense reversed. Thus we may find end conditions from equation (9.26), referring to figure 9–6 for the correct algebraic signs for $Q_x(0) = Q_A$ and $Q_x(L) = Q_B$:

$$Q_x 0) = \frac{P}{2} \qquad (9.58a)$$

$$D_{X\theta}(0) = -D_{n,x}(0) = 0 \qquad (9.58b)$$

and

$$Q_X(L) = -\frac{P}{2} \tag{9.58c}$$

$$D_{X\theta}(L) = -D_{n,x}(L) = 0 \tag{9.58d}$$

Substituting equations (9.58a) and (9.58c) into the second and fourth rows of matrix 9–2, and equations (9.58b) and (9.58d) into equation (9.54) yields four equations in the four unknown integration constants C_1–C_4 as well as the contact force P. We obtain an additional relationship by equating the radial displacements of the shell and the stiffener:

$$D_n(0) = D_{nh}(0) + D_{np}(0) = D_s \tag{9.59}$$

where D_{nh} is given by equation (9.52a), D_{np} by equation (9.57), and D_s by equation (9.56c).

As mentioned previously, the short-shell solution is best carried out by inserting numerical values pertaining to the case at hand. We may, however, easily proceed for the case where $L > \pi/k$ and the semi-infinite solution is valid. We need only the end conditions given by equations (9.58a) and (9.58b), and may immediately write from equation (9.29) with the appropriate sign change

$$D_{nh}(0) = -\frac{P}{8k^3D} \tag{9.60}$$

Correspondingly, equation (9.59) becomes

$$-\frac{P}{8k^3D} + \frac{pa^2}{Eh} = \frac{P[\bar{a} + (b/2)]^2}{bh_s} \tag{9.61}$$

from which P is found. Then, the remaining stress resultants and couples are routinely calculated.

A slight variation in this problem is noted in Timoshenko and Woinowsky-Krieger.[5] If the shell is closed at both ends, an axial stress resultant

$$N_X = \frac{p\pi a^2}{2\pi a} = \frac{pa}{2}$$

is present. This adds the term given by equation (9.40b) to the particular solution. Thus,

$$D_{np3} = -\frac{\mu a}{Eh} N_X(0) = -p\left(\frac{\mu a^2}{2Eh}\right) \tag{9.62}$$

so that in equation (9.57) and the ensuing equations,

$$D_{np} = \frac{pa^2}{Eh}(1 - \mu/2) \tag{9.63}$$

9.2.3 General Loading

9.2.3.1 Displacement Formulation and Solution. The analysis of cylindrical shells for nonaxisymmetrical surface loading commands a prominent place in the literature, and a comprehensive development of the subject is beyond our objectives in this text. We should mention, however, that the formulation and application of this theory has attracted some of the most prominent twentieth-century mechanicians and engineers, including L. H. Donnell, J. Kempner, H. Schorer, and N. J. Hoff in the United States; A. Aas Jakobsen in Norway; H. Reissner, W. Flügge, U. Finsterwalder, F. Dischinger, and W. Zerna in Germany; J. E. Gibson and R. S. Jenkins in Great Britain; E. Torroja in Spain; and V. V. Novozhilov, A. I. Lur'e, and V. Z. Vlasov in the U.S.S.R. Comprehensive historical reviews of these works are found in Flügge[6] from the western standpoint, and in Novozhilov[7] from the Soviet standpoint. Apparently, the Soviet work in this field did not have much impact in the west until the publication of the English language translation of Vlasov's monograph.[8]

We begin the displacement formulation with transverse shearing strains neglected by eliminating the transverse shear stress resultants from equations (9.1b) and (9.1c), using equations (9.1d) and (9.1e). We also take $N_{\theta X} = N_{X\theta}$ and $M_{\theta X} = M_{X\theta}$, leaving the three equilibrium equations

$$aN_{X,X} + N_{X\theta,\theta} + q_X a = 0 \tag{9.64a}$$

$$aN_{X\theta,X} + N_{\theta,\theta} + M_{X\theta,X} + \frac{1}{a}M_{\theta,\theta} + aq_\theta = 0 \tag{9.64b}$$

$$aM_{X,XX} + \frac{1}{a}M_{\theta,\theta} + 2M_{X\theta,X\theta} - N_\theta + q_n a = 0 \tag{9.64c}$$

Next, we substitute equations (9.2a–c) and (9.3a–c) into (9.4) to get the stress resultant-displacement equations

$$N_X = \frac{Eh}{1 - \mu^2}\left[D_{X,X} + \frac{\mu}{a}(D_{\theta,\theta} + D_n)\right] - N_{XT} \tag{9.65a}$$

$$N_\theta = \frac{Eh}{1 - \mu^2}\left[\mu D_{X,X} + \frac{1}{a}(D_{\theta,\theta} + D_n)\right] - N_{\theta T} \tag{9.65b}$$

$$N_{X\theta} = \frac{Eh}{2(1 + \mu)}\left[D_{\theta,X} + \frac{1}{a}D_{X,\theta}\right] \tag{9.65c}$$

$$M_X = -D\left[D_{n,XX} + \frac{\mu}{a^2}(D_{n,\theta\theta} - D_{\theta,\theta})\right] - M_{XT} \tag{9.65d}$$

$$M_\theta = -D\left[\mu D_{n,XX} + \frac{1}{a^2}(D_{n,\theta\theta} - D_{\theta,\theta})\right] - M_{\theta T} \tag{9.65e}$$

$$M_{X\theta} = D\left(\frac{1-\mu}{a}\right)\left[\frac{1}{2}D_{\theta,X} - D_{n,X\theta}\right] \tag{9.65f}$$

Then these expressions for the resultants and couples are inserted into equations (9.64a–c) to get the equilibrium equations in terms of the displacements. Following equation (6.11), we get again set $N_{XT} = N_{\theta T} = N_T$ and $M_{XT} = M_{\theta T} = M_T$:

$$D_{X,XX} + \left(\frac{1-\mu}{2a^2}\right)D_{X,\theta\theta} + \left(\frac{1+\mu}{2a}\right)D_{\theta,X\theta} + \left(\frac{\mu}{a}\right)D_{n,X}$$

$$= \left(\frac{1-\mu^2}{Eh}\right)(-q_X + N_{T,X}) \tag{9.66a}$$

$$\left(\frac{1-\mu}{2}\right)D_{\theta,XX} + \left(\frac{1}{a^2}\right)D_{\theta,\theta\theta} + \left(\frac{1+\mu}{2a}\right)D_{X,X\theta} + \left(\frac{1}{a^2}\right)D_{n,\theta}$$

$$+ \frac{h^2}{12a^2}\left[\left(\frac{1-\mu}{2}\right)D_{\theta,XX} + \left(\frac{1}{a^2}\right)D_{\theta,\theta\theta} - D_{n,XX\theta} - \left(\frac{1}{a^2}\right)D_{n,\theta\theta\theta}\right]$$

$$= \frac{1-\mu^2}{Eh}\left(-q_\theta + \frac{N_{T,\theta}}{a} - \frac{M_{T,\theta}}{a^2}\right) \tag{9.66b}$$

$$\frac{h^2}{12}\left[D_{n,XXXX} + \left(\frac{2}{a^2}\right)D_{n,XX\theta\theta} + \left(\frac{1}{a^4}\right)D_{n,\theta\theta\theta\theta}\right]$$

$$- \frac{h^2}{12a^2}\left[\frac{1}{a}D_{\theta,XX\theta} + \left(\frac{1}{a^2}\right)D_{\theta,\theta\theta\theta}\right] + \frac{1}{a^2}(\mu a D_{X,X} + D_{\theta,\theta} + D_n)$$

$$= \frac{1-\mu^2}{Eh}\left(q_n + \frac{N_T}{a} - M_{T,XX} - \frac{M_{T,\theta\theta}}{a^2}\right) \tag{9.66c}$$

We may simplify equations (9.66b) and (9.66c) by examining the terms multiplied by h^2/a^2. We will drop all such terms as being of $O(h^2/a^2){:}1$. At this point, this step involves some presumption. Whereas the first two terms of the $h^2/12a^2[\;\;]$ expression in equation (9.66b) have corresponding terms in the same equation that are of $O(1)$, the remaining terms and all terms contained in $h^2/12a^2[\;\;]$ in equation (9.66c) involve derivatives not found elsewhere in the equation. Nevertheless, this assumption, if not fully justified, has been found to give good results for a wide class of problems and greatly simplifies the remaining derivation and solution. A discussion of the possible errors introduced by this step is contained in Kraus.[9]

Insofar as equation (9.66b) is concerned, we may regard the suppression of the $O(h^2/a^2)$ terms as being equivalent to neglecting the influence of the stress couples on the in-plane equilibrium equation, (9.66b). Also, ignoring the $O(h^2/a^2)$ terms in equation (9.66c) is equivalent to replacing the displacement curvature relationships, equation (9.3), by the corresponding equations for a plate. The latter equations are given by equations (5.55), with $\alpha = X$ and $\beta = \theta$, as

$$\kappa_X = -D_{n,XX} \tag{9.67a}$$

$$\kappa_\theta = -\frac{1}{a^2}D_{n,\theta\theta} \tag{9.67b}$$

$$\tau = -\frac{1}{a}D_{n,X\theta} \tag{9.67c}$$

These interpretations give a reasonable physical basis to the elimination of the $O(h^2/a^2)$ terms.

We now collect the simplified equations as

$$D_{X,XX} + \left(\frac{1-\mu}{2a^2}\right)D_{X,\theta\theta} + \left(\frac{1+\mu}{2a}\right)D_{\theta,X\theta} + \left(\frac{\mu}{a}\right)D_{n,X} = p_X \tag{9.68a}$$

$$\left(\frac{1-\mu}{2}\right)D_{\theta,XX} + \left(\frac{1}{a^2}\right)D_{\theta,\theta\theta} + \left(\frac{1+\mu}{2a}\right)D_{X,X\theta} + \left(\frac{1}{a^2}\right)D_{n,\theta} = p_\theta \tag{9.68b}$$

$$\frac{h^2}{12}\nabla^4 D_n + \frac{1}{a^2}(\mu a D_{X,X} + D_{\theta,\theta} + D_n) = p_n \tag{9.68c}$$

where

$$p_X = \frac{1-\mu^2}{Eh}(-q_X + N_{T,X}) \tag{9.69a}$$

$$p_\theta = \frac{1-\mu^2}{Eh}\left(-q_\theta + \frac{N_{T,\theta}}{a} - \frac{M_{T,\theta}}{a^2}\right) \tag{9.69b}$$

$$p_n = \frac{1-\mu^2}{Eh}\left[q_n + \frac{N_T}{a} - \nabla^2(M_T)\right] \tag{9.69c}$$

$$\nabla^2(\ ) = (\ )_{,XX} + \frac{1}{a^2}(\ )_{,\theta\theta} \tag{9.69d}$$

is the harmonic operator in the X–θ cylindrical coordinate system. Kraus[10] has derived a similar set of equations, applicable for noncircular cylindrical shells as well, by neglecting the stress couples in equation (9.64b) and using equations (9.67) for the changes in curvatures of the outset. The only perceptible difference appears to be that the thermal moment gradient term in p_θ, $M_{T,\theta}/a^2$, is not present in his equations.

It is easily verified that for axisymmetric loading, equations (9.68a) and (9.68c) are identical to equations (9.12a), and (9.12b), which reduce ultimately to equations (9.16a) and (9.16b).

The formulation continues by forming a judicious set of mixed partial derivatives as suggested by Kraus:[11] (a) $\partial^2/\partial X^2$ [equation (9.68a)], (b) $\partial^2/\partial\theta^2$ [equation (9.68a), and (c) $\partial^2/\partial X\partial\theta$ [equation (9.68b); and solving (a) and (b) for the mixed partials of D_θ:

$$D_{\theta,XXX\theta} = \frac{2a}{1 + \mu}\left[p_{X,XX} - D_{X,XXXX} - \left(\frac{1 - \mu}{2a^2}\right)D_{X,XX\theta\theta} - \left(\frac{\mu}{a}\right)D_{n,XXX}\right]$$

$$(9.70a)$$

$$D_{\theta,X\theta\theta\theta} = \frac{2a}{1 + \mu}\left[p_{X,\theta\theta} - D_{X,XX\theta\theta} - \left(\frac{1 - \mu}{2a^2}\right)D_{X,\theta\theta\theta\theta} - \left(\frac{\mu}{a}\right)D_{n,X\theta\theta}\right]$$

$$(9.70b)$$

and

$$\left(\frac{1 - \mu}{2}\right)D_{\theta,XXX\theta} + \left(\frac{1}{a^2}\right)D_{\theta,X\theta\theta\theta} + \left(\frac{1 + \mu}{2a}\right)D_{X,XX\theta\theta}$$

$$+ \left(\frac{1}{a^2}\right)D_{n,X\theta\theta} - p_{\theta,X\theta} = 0 \qquad (9.70c)$$

We now eliminate the first two terms in equation (9.70c) by using equations (9.70a) and (9.70b), and consolidate the remaining terms over the common denominator $a(1 + \mu)$ to get

$$\nabla^4 D_X = -\left(\frac{\mu}{a}\right)D_{n,XXX} + \left(\frac{1}{a^3}\right)D_{n,X\theta\theta} + p_{X,XX}$$

$$+ \frac{2}{a^2(1 - \mu)}p_{X,\theta\theta} - \frac{(1 + \mu)}{a(1 - \mu)}p_{\theta,X\theta} \qquad (9.71)$$

Equation (9.71) becomes the governing equation for D_X once D_n has been determined.

We now derive a similar equation for D_θ by interchanging equations (9.68a) and (9.68b) in the aforementioned mixed partial operations (a), (b), and (c) and solving (a) and (b) for the mixed partials of D_X. This produces

$$D_{X,XXX\theta} = \frac{2a}{1 + \mu}\left[p_{\theta,XX} - \left(\frac{1 - \mu}{2}\right)D_{\theta,XXXX} - \left(\frac{1}{a^2}\right)D_{\theta,XX\theta\theta} - \left(\frac{1}{a^2}\right)D_{n,XX\theta}\right]$$

$$(9.72a)$$

314

$$D_{X,X\theta\theta} = \frac{2a}{1+\mu}\left[p_{\theta,\theta\theta} - \left(\frac{1-\mu}{2}\right)D_{\theta,XX\theta\theta} - \left(\frac{1}{a^2}\right)D_{\theta,\theta\theta\theta\theta} - \left(\frac{1}{a^2}\right)D_{n,\theta\theta\theta}\right]$$

$$(9.72b)$$

and

$$D_{X,XXX\theta} + \left(\frac{1-\mu}{2a^2}\right)D_{X,X\theta\theta\theta} + \left(\frac{1+\mu}{2a}\right)D_{\theta,XX\theta\theta} + \left(\frac{\mu}{a}\right)D_{n,XX\theta} - p_{X,\theta} = 0$$

$$(9.72c)$$

Eliminating the mixed partials of D_X from equation (9.72c) and clearing, we have

$$\nabla^4 D_\theta = -\left(\frac{2+\mu}{a^2}\right)D_{n,XX\theta} - \left(\frac{1}{a^4}\right)D_{n,\theta\theta\theta} - \frac{(1+\mu)}{a(1-\mu)}p_{X,X\theta}$$

$$+ \left(\frac{2}{1-\mu}\right)p_{\theta,XX} + \left(\frac{1}{a^2}\right)p_{\theta,\theta\theta} \tag{9.73}$$

which becomes the governing equation for D_θ once D_n has been determined.

Finally, we may isolate the normal displacement D_n. The procedure is: (a) $\partial/\partial X$ [equation (9.71)]; (b) $\partial/\partial\theta$ [equation (9.73)]; and (c) ∇^4 [equation (9.68c)]. These operations give

$$(\nabla^4 D_X)_{,X} = \nabla^4(D_{X,X}) = -\left(\frac{\mu}{a}\right)D_{n,XXXX} + \left(\frac{1}{a^3}\right)D_{n,XX\theta\theta}$$

$$+ p_{X,XXX} + \frac{2}{a^2(1-\mu)}p_{X,X\theta\theta} - \frac{(1+\mu)}{a(1-\mu)}p_{\theta,XX\theta} \tag{9.74a}$$

$$(\nabla^4 D_\theta)_{,\theta} = \nabla^4(D_{\theta,\theta})$$

$$= -\left(\frac{2+\mu}{a^2}\right)D_{n,XX\theta\theta} - \left(\frac{1}{a^4}\right)D_{n,\theta\theta\theta\theta}$$

$$- \frac{(1+\mu)}{a(1-\mu)}p_{X,X\theta\theta} + \left(\frac{2}{1-\mu}\right)p_{\theta,XX\theta} + \left(\frac{1}{a^2}\right)p_{\theta,\theta\theta\theta} \tag{9.74b}$$

and

$$\frac{h^2}{12}\nabla^8 D_n + \frac{1}{a^2}[\mu a\nabla^4(D_{X,X}) + \nabla^4(D_{\theta,\theta}) + \nabla^4 D_n] = \nabla^4 p_n \tag{9.74c}$$

After substituting equations (9.74a) and (9.74b) into equations (9.74c), we have the desired equations for D_n:

$$\frac{h^2}{12}\nabla^8 D_n + \left(\frac{1 - \mu^2}{a^2}\right)D_{n,XXXX} = \nabla^4 p_n$$

$$-\frac{1}{a}\left[p_{X,XXX} - \frac{1}{a^2}p_{X,X\theta\theta} + \frac{(2 + \mu)}{a}p_{\theta,XX\theta} + \left(\frac{1}{a^3}\right)p_{\theta,\theta\theta\theta}\right] \qquad (9.75)$$

Equations (9.71), (9.73), and (9.75) are among the most famous equations in the theory of thin shells and are commonly referred to as, alternately, Donnell's equations, Jenkins's equations, or Vlasov's equations. The eighth order system is uncoupled for the displacements, so that once D_n is determined from equation (9.75), D_X and D_θ may be found by solving equations (9.71) and (9.73), respectively. These equations are also used, with slight modification, to study the dynamic response of circular cylindrical shells.[12]

The appropriate boundary conditions for the foregoing equations are established from figure 6–3, with $\alpha = X$ and $\beta = \theta$. The relationships in which there are no transverse shearing strains included are applicable here. Note that the equations derived pertain to open, as well as closed circular cylindrical shells. For closed shells, the boundary conditions in the θ direction are replaced by continuity conditions of the form $f(0) = f(2\pi)$, where $f(\theta)$ is a function which is continuous at $\theta = 0$.

Before considering some specific applications, it is instructive to make some general comments on the analytical solution of the governing equations for this theory, equations (9.71), (9.73) and (9.75), or, alternately, the coupled equations (9.68a–c). Complete solutions are fairly complicated algebraically and are extensively documented in specialized works, such as Flügge.[13]

Briefly, the solutions of the homogeneous equations for edge loadings at $X = \bar{X} =$ constant are of interest for both open and closed shells, and are obtained by taking the Fourier expansions

$$\begin{Bmatrix} D_{Xh} \\ D_{\theta h} \\ D_{nh} \end{Bmatrix} = \sum_{j=0}^{\infty} \begin{Bmatrix} D^j_{Xh}(X)\cos j\theta \\ D^j_{\theta h}(X)\sin j\theta \\ D^j_{nh}(X)\cos j\theta \end{Bmatrix} \qquad (9.76)$$

Note that for closed shells, $D_{ih}(0) = D_{ih}(2\pi)(i = X, \theta, n)$, which serve as the continuity or periodicity conditions. Then, for a general harmonic j, the substitution of equations (9.76) into the governing equations gives an eighth order algebraic system. The solution ultimately takes the general form

316

$$
\begin{aligned}
R^j(X,\theta) = C_G\{ & C^j_1[e^{-g_1 x}(f^j_1 \cos g_3x - f^j_2 \sin g_3x)] \\
+ & C^j_2[e^{-g_1 x}(f^j_2 \cos g_3x + f^j_1 \sin g_3x)] \\
+ & C^j_3[e^{-g_2 x}(f^j_3 \cos g_4x - f^j_4 \sin g_4x)] \\
+ & C^j_4[e^{-g_2 x}(f^j_4 \cos g_4x + f^j_3 \sin g_4x)] \\
+ & C^j_5[e^{g_1 x}(f^j_1 \cos g_3x + f^j_2 \sin g_3x)] \\
+ & C^j_6[e^{g_1 x}(-f^j_2 \cos g_3x + f^j_1 \sin g_3x)] \\
+ & C^j_7[e^{g_2 x}(f^j_3 \cos g_4x + f^j_4 \sin g_4x)] \\
+ & C^j_8[e^{g_2 x}(-f^j_4 \cos g_4x + f^j_3 \sin g_3x)]\}\begin{pmatrix} \cos j\theta \\ \text{or} \\ \sin j\theta \end{pmatrix}
\end{aligned}
\tag{9.77}
$$

in which $R^j(X,\theta)$ = a typical displacement, stress resultant, or stress couple for harmonic j; $x = X/a$; g_1–g_4 are quantities dependent on j, the dimensions of the shell, and the material properties; f^j_1–f^j_4 are quantities dependent on j, the dimensions of shell, and the material properties for each individual R^j; C_G is a constant dependent on the shell geometry and material properties; and C^j_1–C^j_8 are constants of integration.

In the case of a general shell, all eight constants are needed, and hence eight boundary conditions in the X direction are required, whereas for a semi-infinite shell, those terms associated with C^j_5–C^j_8 are suppressed, and therefore only four boundary conditions are needed.

For an edge loading at $X = \bar{X}$ such as we have studied for axisymmetric loading, the integration constants are determined by substituting the displacements into the stress resultants-displacement relations, equations (9.65), and setting the expressions for the specified edge loadings equal to the boundary values, $N_X(\bar{X})$, $\bar{N}_{X\theta}(\bar{X})$, $\bar{Q}_X(\bar{X})$, and/or $M_X(\bar{X})$. Additional equations may be obtained by the specification of kinematic boundary conditions $D_X(\bar{X})$, $D_\theta(\bar{X})$, $D_n(\bar{X})$, and/or $D_{\theta X}(X)$. For a semi-infinite shell, $\bar{X}$ corresponds to $\bar{X} = 0$ whereas for a complete shell, $\bar{X}$ stands for $X = L$ as well.

Now we consider the homogeneous solutions for a shell with an edge loading applied at a boundary $\theta = \bar{\theta}$ = constant. Obviously, this case is pertinent only for open cylindrical shells, such as those discussed in section 4.4.1. Here, series solutions of the form

$$
\begin{Bmatrix} D_{Xh} \\ D_{\theta h} \\ D_{nh} \end{Bmatrix} = \sum_{k=0}^{\infty} \begin{Bmatrix} D^k_{xh}(\theta) \cos k\pi \dfrac{X}{L} \\ D^k_{\theta h}(\theta) \sin k\pi \dfrac{X}{L} \\ D^k_{nh}(\theta) \cos k\pi \dfrac{X}{L} \end{Bmatrix}
\tag{9.78}
$$

are widely used. Again, an eighth order algebraic system must be evaluated for each harmonic component k, from which the solution may be expressed in the same form as equation (9.77), with j replaced by k, and the coordinates X and θ interchanged:

$$
\begin{aligned}
R^k = \; C_G\{ & C_1^k[e^{-g_1\theta}(f_1^k \cos g_3\theta - f_2^k \sin g_3\theta) \\
& + C_2^k[e^{-g_1\theta}(f_2^k \cos g_3\theta + f_1^k \sin g_3\theta)] \\
& + C_3^k[e^{-g_2\theta}(f_3^k \cos g_4\theta - f_4^k \sin g_4\theta) \\
& + C_4^k[e^{-g_2\theta}(f_4^k \cos g_4\theta + f_3^k \sin g_4\theta)] \\
& + C_5^k[e^{g_1\theta}(f_1^k \cos g_3\theta + f_2^k \sin g_3\theta)] \\
& + C_6^k[e^{g_1\theta}(-f_2^k \cos g_3\theta + f_1^k \sin g_3\theta)] \\
& + C_7^k[e^{g_2\theta}(f_3^k \cos g_4\theta + f_4 \sin g_4\theta)] \\
& + C_8^k[e^{g_2\theta}(-f_4^k \cos g_4\theta + f_3^k \sin g_3\theta)]
\end{aligned}
\begin{pmatrix} \cos K\Pi\dfrac{X}{L} \\[4pt] \text{or} \\[4pt] \sin k\Pi\dfrac{X}{L} \end{pmatrix}
$$

$$(9.79)$$

Again, the effects at opposite boundaries may be uncoupled by using the semi-infinite approach. Because the coordinate lines normal to the loaded boundaries are curved in this case, the effects of the boundary forces may be expected to dissipate more rapidly than those for the previously studied case in which in-plane forces $\bar{N}_X(\theta,\bar{X})$ applied at $X = \bar{X}$ can propagate along straight lines. After the integration constants are obtained from a specified combination of static and kinematic edge conditions, the homogeneous open shell solution is completely described. The possible static edge conditions involve

$$
N_\theta(\bar{\theta}), \; \bar{N}_{\theta X}(\bar{\theta}), \; \bar{Q}_\theta(\bar{\theta}), \; \text{and/or} \; \bar{M}_\theta(\bar{\theta})
$$

whereas the kinematic conditions refer to

$$
D_\theta(\bar{\theta}), \; D_X(\bar{\theta}), \; D_n(\bar{\theta}), \; \text{and/or} \; D_{\theta X}(\bar{\theta})
$$

Such solutions are used extensively in the analysis of open cylindrical shell roofs, and extensive tabulations are found in "Design of Cylindrical Concrete Roofs."[14]

We now consider particular solutions for various applied loadings and thermal effects. A general approach is to take

$$\begin{Bmatrix} D_{Xp} \\ \\ D_{\theta p} \\ \\ D_{np} \end{Bmatrix} = \sum_{j=0}^{\infty} \sum_{k=0}^{\infty} \begin{Bmatrix} D_{Xp}^{jk} \cos k\pi \dfrac{X}{L} \cos j\theta \\ \\ D_{\theta p}^{jk} \sin k\pi \dfrac{X}{L} \sin j\theta \\ \\ D_{np}^{jk} \cos k\pi \dfrac{X}{L} \cos j\theta \end{Bmatrix} \qquad (9.80)$$

Such solutions may be combined with the edge load solutions generated from equations (9.76) and (9.77) to solve a wide range of cylindrical shell problems.

Recall from our earlier calculation of the membrane theory displacements for cylindrical shells (section 6.3.4.1) that the solution for the example considered was precisely in the format of equations (9.80), with $j = k = 1$. This strongly suggests that the membrane theory solution will serve as a particular solution in many instances.

Finally, in our general discussion of solution procedures, we outline the technique of Vlasov,[15] which is very attractive for certain problems. Starting with the three coupled equations, (9.68a–c), and using a nondimensional meridional coordinate X/L, he introduces a stress function

$$\Phi = \Phi_h + \Phi_{pX} + \Phi_{p\theta} + \Phi_{pn} \qquad (9.81)$$

from which the three displacements D_X, D_θ, and D_n, as well as the stress resultants and couples can be found by differentiation. The particular solutions correspond to the individual surface loading components p_X, p_θ, and p_n, respectively, while the resulting equation is in the same form as equation (9.75) with Φ replacing D_n. He specializes the equation for a normal loading $p_n(X,\theta)$, and further stipulates that the shell is simply supported on all four sides. Referring to section 6.2.2, this implies that at the boundaries where $X = \bar{X} = $ constant, $N_X = M_X = D_\theta = D_n = 0$; and, at the boundaries where $\theta = \bar{\theta} = $ constant, $N_\theta = M_\theta = D_X = D_n = 0$. Next, he uses the Navier approach, which is widely applied in plate solutions, to write

$$\Phi = \sum_{j=1}^{\infty} \sum_{k=1}^{\infty} \Phi^{jk} \sin j\pi \dfrac{\theta}{\bar{\theta}} \sin k\pi \dfrac{X}{\bar{X}} \qquad (9.82)$$

This stress function automatically satisfies the simply supported boundary conditions. Φ^{jk} is then determined so as to satisfy the governing equation for each harmonic jk.

The solution by Vlasov's technique is fairly simple to implement, although we do not pursue the details here. An interesting and useful result is the solution for a cylindrical shell under a concentrated load which isolates the Green's function for such structures.[16]

9.2.3.2 Column-Supported Cylindrical Shells. We now consider a cylindrical tank resting on a concentric ring of equispaced columns, as shown in figure 9–8. The shell carries an axisymmetric loading $q(X)$, and the transition between the shell and the columns is assumed to be facilitated by a ring beam. To avoid ambiguity in the subsequent development, it is convenient to set the origin for X at the top of the ring beam and to presume that the ring beam depth $B << L$. Thus the tops of the supporting columns are also located at $X = 0$.

This situation is similar to that shown in figure 4–25, except that, of course, here the shell is cylindrical. Whereas we were able to treat the spherical geometry by using the membrane theory, a membrane description is inadequate for the column-supported cylindrical shell. This is illustrated in section 4.3.2 and by figure 4–12(b), which shows that a force $\bar{N}$ applied to one end simply propagates along the straight meridian in the absence of bending.

Now we turn to the problem at hand. The representation of the discrete supports is identical to that shown in figure 4–25, with $\phi_b = \pi/2$ and $R_0(\phi_b) = a$. Here, we have X in place of ϕ for the meridional coordinates. Case [4–25(b)] of the superposition is described by a particular solution, which can be taken as the membrane theory solution, and a homogeneous solution, which satisfies the boundary conditions. The in-plane stress resultants for the particular solution are calculated from equations (4.58) and (4.62a), with the appropriate coordinate modifications:

$$N_{Xbp}(X) = \frac{Q(X)}{2\pi a} \tag{9.83a}$$

$$N_{\theta bp}(X) = q_n(X)a \tag{9.83b}$$

The homogeneous solution for case [4–25(b)] is obtained from the treatment of axisymmetrically loaded semi-infinite cylindrical shells in section 9.2.2.2 as equation (9.21).

If the tank is supported by a circumferentially rigid ring beam, the appropriate boundary conditions would be

$$D_n(0) = 0 \tag{9.84}$$

and either

$$D_{X\theta}(0) = -D_{n,x}(0) = 0 \tag{9.85a}$$

or

$$M_X(0) = M_0 = 0 \tag{9.85b}$$

depending on whether the ring beam is assumed to prevent or to permit

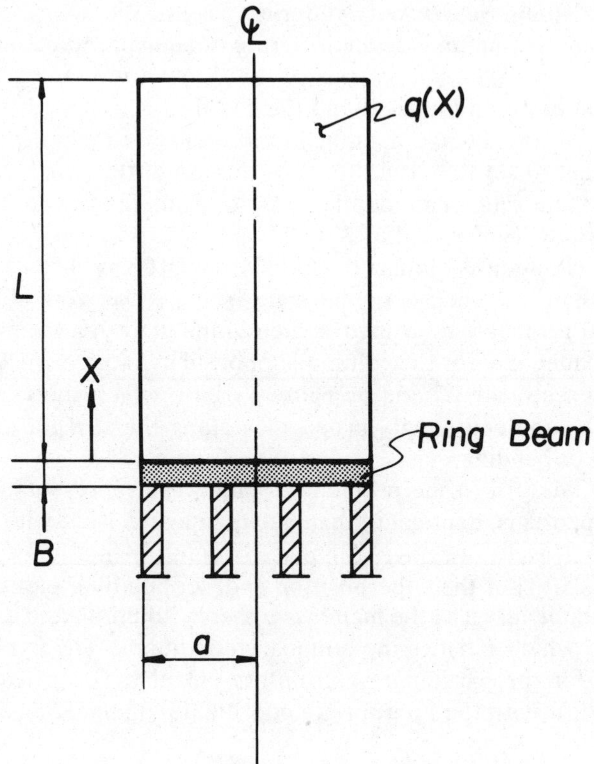

Figure 9–8. Column-Supported Cylindrical Shell

rotations about the circumference. The condition for an actual shell would probably fall somewhere in between the two idealized situations. To find C_1 and C_2, the expressions for D_n and $D_{X\theta}$ evaluated at $X = 0$ are substituted into equations (9.84) and (9.85a), if applicable. In turn, $D_n(0)$ and $D_{X\theta}(0)$ are composed of (a) the particular solution contributions obtained by substituting equations (9.83a) and (9.83b) evaluated at $X = 0$ into equations (6.53) and (6.54); and (b) the homogeneous solution contributions found from equation (9.21). Alternately, if boundary condition (9.85b) is selected, $M_X(0)$ comes from equations (9.22a). Then C_1 and C_2 are evaluated.

To complete case [4–25(b)] of the superposition, $M_{Xb}(X)$, $M_{\theta b}(X)$, and $Q_{Xb}(X)$ are computed from equations (9.25a–c) while $N_{\theta Xb}(X)$ is the sum of equations (9.83b) and (9.25d). $N_{Xb}(X)$ remains as the membrane theory value N_{Xbp}, as given by equation (9.83a).

Case [4–25(c)] of the superposition is a homogenous loading condition that is solved following equation (9.76), with the stress resultants, couples, and displacements having the form of equations (9.77). The boundary conditions at $X = 0$ are given by equations (9.84) and (9.85a) or (9.85b), plus two additional equations. First, the ring beam can be regarded as circumferentially rigid, so that

$$D_\theta(0) = 0 \tag{9.86}$$

The final boundary condition at the base is found from the representation of the meridional stress resultants $N_X(0)$ by (a) the negative of the continuous boundary reaction, and (b) the intensity of the column reactions, as shown on figure 4–25(c). This condition has been expressed earlier by equation (4.129), which is rewritten for the case at hand as

$$N_{Xc}(0,\theta) = [-N_{Xb}(0) + \delta R_{cl}] \tag{9.87}$$

where $\delta = 1$ if $m\left(\dfrac{2\pi}{n_c}\right) - \beta \leq \theta \leq m\left(\dfrac{2\pi}{n_c}\right) + \beta$

$\qquad (m = -n_c/2, \ldots, -1, 0, 1, \ldots, n_c/2$

$\qquad \delta = 0$ for all other θ

$\qquad n_c =$ the total number of columns

$$R_{cl} = \frac{\pi N_{Xb}(0)}{\beta n_c}$$

As before, equation (9.87) is expanded in a Fourier series in θ

$$N_{Xc}(0,\theta) = \sum_{j=1}^{\infty} N_{Xc}^j(0) \cos j\theta \tag{9.88}$$

which gives the Fourier coefficients

$$N_{Xc}^j(0) = \frac{2}{j\beta} N_{Xb}(0) \sin j\beta \quad (j = in_c) \tag{9.89a}$$

and

$$N_{Xc}^j(0) = 0 \quad (j \neq in_c) \tag{9.89b}$$

Thus we have expressed the fourth boundary condition at $X = 0$, [equation (9.87)] in the harmonic form

$$N_{Xc}(0,\theta) = \sum_{j=n_c,2n_c,\ldots}^{\infty} N_{Xc}^j(0) \cos j\theta \tag{9.90}$$

We presume that the semi-infinite assumption is adequate for our purposes for stress analysis, so that we need not be concerned with the boundary conditions at $X = L$. However, from a practical standpoint, note that open cylindrical tanks are generally stiffened at the top to prevent ovaling due to the propagation of the concentrated column reactions $N_{Xc}(0,\theta)$.

For some specific tank structures, the idealized boundary conditions stated in equations (9.84), and (9.85a), and (9.85b) might warrant refinement. One possibility is to incorporate the ring stiffener model introduced in section 9.2.2.7 to generalize equation (9.84). A second possibility concerns the common use of such a tank to contain a liquid or solid. In such vessels, a circular plate or perhaps a shallow cone or other rotational shell bottom would be attached near the ring beam level. Depending on the connection detail, additional radial and, perhaps, rotational restraints would be introduced. The possibilities are many, depending on the actual case in question, and the realistic incorporation of additional constraining elements into the mathematical model may alter the computed stress pattern significantly.[17] The bottom plate or shell, itself, would behave essentially as an axisymmetric member supported by the ring beam. Such structures have been discussed extensively in the preceding chapters.

If the weight of a portion of the contents of the tank is carried directly to the ring beam by a bottom plate or shell instead of being transferred through the shell wall, then the resultant force $\mathbf{Q}(X)$ in equation (9.83a) would *not* include this portion of the loading for the calculation of $N_{Xb}(X)$ for $X > 0$; however, for case [4–25(c)] of the superposition, where $N_{Xb}(0)$ is encountered, the *entire* load from the tank must be included in $Q(0)$ for the calculation of $N_{Xb}(0)$ and the column reaction R_{cl}. For example, if a circular plate bottom were provided at $X = 0$ in the tank shown in figure 9–8, all of the weight of the internal contents would be carried directly to the ring beam if no wall friction was assumed. This would leave only the dead weight of the tank and roof in $\mathbf{Q}(X)$ $X > 0$, but the weight of the contents would be added for $\mathbf{Q}(0)$ in case [4–25(c)].

With the boundary conditions for the idealized problem established as equations (9.84), (9.85a) or (9.85b), (9.86), and (9.90), we set the explicit expressions for $R^j = D_n^j$, $D_{X\theta}^j$ or M_X^j, D_θ^j and N_X^j [which are in the form of equation (9.77)] equal to the boundary values at $X = 0$, which are 0, 0, 0, and $N_{Xc}^j(0)$, respectively. Then the integration constants for harmonic j, C_1^j–C_4^j, are found from the simultaneous solution of the four equations. The stress resultants, couples and displacements for harmonic j are then calculated from the corresponding specific form of equation (9.78).

Convergence is established by the progressive diminution of the results from consecutive participating harmonics ($j = n_c, 2n_c, \ldots$). For the usual cases encountered, four or five harmonics should be adequate. The complete results are obtained by summing the stress resultants and couples and displacements obtained from the specific forms of equation (9.77) over the participating harmonics at various circumferential locations θ, and then adding these values to the case [4–25(b)] results.

Flügge[18] has considered a case with $a/h = 150$ $n_c = 8$, and $\beta = 6°$. This gives an amplification of $360°/(8{\cdot}2\beta) = 360/96 = 3.75$ for N_X at the base; that is, $R_{cl} = 3.75N_{Xb}(0)$. The effect of this amplification and the corresponding change in N_θ dies out at about $X = a$, with the stress resultants above this level practically equal to the membrane theory values.

In Gould[19] the column-supported shell model is generalized to represent a *longitudinally distributed*, as opposed to a *line*, attachment between the column and the shell wall. This study is particularly applicable to elevated storage tanks, such as that shown in figure 2–8(u).

9.2.3.3 Multiple Barrel Shells. A multiple barrel shell composed of adjacent segments of circular cylindrical shells is depicted in figure 9–9(a), with the positive sense of the coordinates taken from figure 4–29. An actual shell roof of this type is shown in figure 2–8(h). The membrane theory analysis of a single shell of this form was conducted in sections 4.4.1 and 6.3.4.1.

First consider the boundary conditions in the X direction. The great majority of shell roofs of this configuration have been designed assuming simply supported boundaries:

$$N_X = M_X = D_\theta = D_n = 0 \text{ at } X = 0 \text{ and } X = L \tag{9.91}$$

If the support is a wall or frame in the Y–Z plane, it must be sufficiently stiff to prevent displacements in the Y or Z directions; yet, it must be sufficiently flexible to allow displacements along the X axis and rotations about the Y axis. Obviously, these requirements are difficult to satisfy exactly in a physical structure, but they are probably the best representation which can be incorporated into a basic analytical solution.

Now consider the boundary conditions in the θ direction. Referring to figure 9–9(a), three situations are identified: exterior, interior-general, and interior-symmetry line. In some cases, the resistance is augmented by stiffening beams spanning the length L; here, we will not consider this situation but only the basic shell. First, the exterior edge may be regarded as free, so that

$$N_\theta = \bar{N}_{\theta X} = M_\theta = \bar{Q}_\theta = 0 \text{ at } \theta = \theta_k \tag{9.92}$$

The interior edges are somewhat more complicated to characterize, since they are continuous with the adjacent shell. Clearly, the displacements D_X, D_θ, and the rotation $D_{\theta X}$ must be compatible between the intersecting shells. Strict enforcement of this condition could result in four simultaneous equations per interior valley line, which greatly complicates the calculations. A simplification, widely used in design, is to treat every interior edge as if it were located on a line of symmetry where the lateral displacement D_Y and the rotation about the longitudinal axis $D_{\theta X}$ vanish. For the coordinate system defined on figure 9–9(a), these conditions become

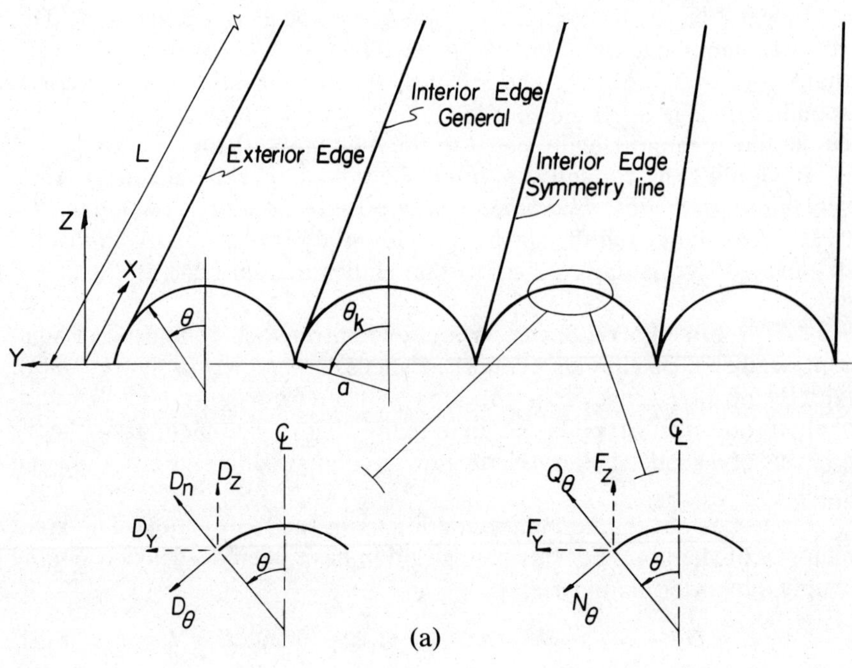

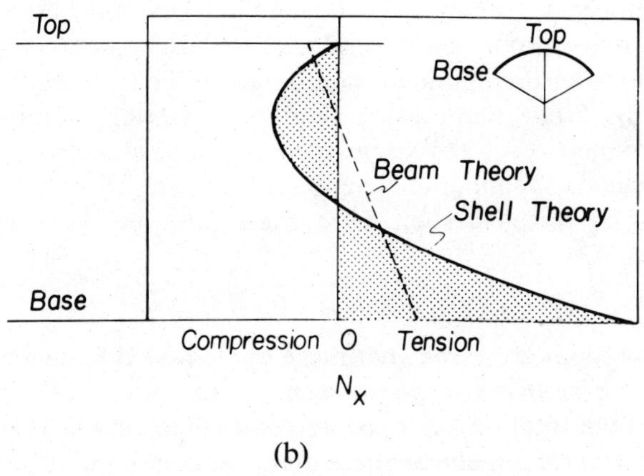

Figure 9–9. Open Cylindrical Shell

$$D_Y(-\theta_k) = 0 ; \quad D_{\theta X}(-\theta_k) = 0 \tag{9.93a}$$

for the interior edge of the outside barrel shell, and

$$D_Y(\pm\theta_k) = 0 ; \quad D_{\theta X}(\pm\theta_k) = 0 \tag{9.93b}$$

for all of the interior barrels, where from figure 9–9(a),

$$D_Y(\theta) = D_\theta \cos\theta + D_n \sin\theta \quad (-\theta_k \le \theta \le \theta_k) \tag{9.93c}$$

If $\gamma_\theta = 0$, the rotation $D_{\theta X}$ is given in terms of D_n and D_θ from equations (9.2h):

$$D_{\theta X} = -\frac{1}{a}(D_{n,\theta} - D_\theta) \tag{9.93d}$$

which is identical to equation (6.55) for the membrane theory rotation.

In addition, we may obtain two static conditions from the symmetry assumption. First, the transverse shearing stress resultants $\bar{N}_{\theta X}$ must be equal and opposite on any two coincident edges. When such edges lie along a symmetry line,

$$\bar{N}_{\theta X}(-\theta_k) = 0 \begin{pmatrix} \text{exterior} \\ \text{barrel} \end{pmatrix} ; \quad \bar{N}_{\theta x}(\pm\theta_k) = 0 \begin{pmatrix} \text{interior} \\ \text{barrels} \end{pmatrix} \tag{9.94a}$$

is the only possibility. Similarly, any resultant force vertical F_Z on one edge must have an equal and opposite counterpart on the other edge. Thus, for edges along symmetry lines,

$$F_Z(-\theta_k) = 0 \begin{pmatrix} \text{exterior} \\ \text{barrel} \end{pmatrix} ; \quad F_Z(\pm\theta_k) = 0 \begin{pmatrix} \text{interior} \\ \text{barrels} \end{pmatrix} \tag{9.94b}$$

where, from figure 9–9(a).

$$F_Z = Q_\theta \cos\theta - N_\theta \sin\theta \quad (-\theta_k \le \theta \le \theta_k) \tag{9.94c}$$

Equations (9.91)–(9.94) prescribe the boundary conditions for a simplified bending theory analysis of cylindrical shells, in which only two cases need be considered—an exterior and a typical interior barrel. Further, the two cases are uncoupled, which expedites the calculations.

We first compare the expressions for the membrane stress resultants (derived in section 4.4.1) and the corresponding displacements (found in section 6.3.4.1) against these boundary conditions. In particular, consider the case of a uniform dead load q_d (force/area) represented by one term of a Fourier series, as given by equation (4.165b):

$$q_d(X) = \frac{4}{\pi}\, q_d \sin \pi \frac{X}{L} \qquad\qquad (9.95)$$

For clarity, all quantities computed from the membrane theory equations are designated by the subscript m. We check the expressions for the stress resultants, equations (4.169a–c), and find that the static boundary conditions in the X direction, $N_X = M_X = 0$ at $X = 0$ and L, are satisfied. An examination of the membrane theory displacements for this case, given by equations (6.56a–c), reveals that the kinematic boundary conditions in the X direction, $D_\theta = D_n = 0$ at $X = 0$ and L, are similarly satisfied. Thus the only possible violation of the membrane theory conditions may occur on the $\theta = \pm\, \theta_k$ boundaries.

Observe from figure 4–31 that the idealized membrane boundary must develop $N_{\theta m}$ and S_m at $\theta = \pm\, \theta_k$. This is in clear conflict with the exterior edge condition, equation (9.92), where N_θ and $\bar{N}_{\theta X}$, which corresponds to S_m, are required to vanish. An interior edge can develop only the Y component of $N_{\theta m}$, $F_{Ym} = N_{\theta m} \cos\theta$, as shown on figure 9–9(a), leaving the Z component, $F_{Zm} = -N_{\theta m} \sin\theta$, unbalanced. With respect to the $\bar{N}_{\theta X}$ or S_m force, however, a careful examination of figure 4–31 reveals that the sense of S_m along two adjacent boundaries as computed from the membrane theory solution would be in the same direction, either in the $(+)$ or $(-)$ X directions. On the other hand, two adjacent shells would be expected to develop *equal* and *opposite* shearing resultants along their common boundary, since no resultant force along the X axis can exist. Moreover, if the interior valley is a symmetry line as assumed in developing equations (9.94a), then $\bar{N}_{\theta X}$ must be equal to 0. It is apparent, then, that the longitudinal boundary conditions implied for the development of the state of stress predicted by the membrane theory are grossly violated.

The basic procedure to rectify these violations consists of starting with the computed membrane stress resultants, which are in equilibrium with the applied loading, and the corresponding displacements. Then, corrective edge loads are applied to satisfy the boundary conditions at $\theta = \pm\, \theta_k$. Mathematically, this is equivalent to adopting the membrane theory solution as the particular solution to the governing equations. This has been shown quantitatively to be a sufficiently close approximation,[20] which is hardly surprising in view of our previous discussions.

With the membrane theory solution taken as the particular solution, we need consider only the homogeneous portion of Donnell's equation for the corrective edge loadings. Specifically, refer to the solutions given by equations (9.78), which were proposed for loading along edges $\theta = $ constant. Notice that these expressions are in the same Fourier series form as our membrane theory solution. This is, of course, no coincidence, but the very reason for taking the membrane theory solutions as Fourier series in the first place. As we have stated earlier, the displacements, stress resultants, and couples corresponding to equations (9.78) take the general form of equation (9.79). Now we further assume that the semi-infinite simplification is applicable, so that the effects at opposite boundaries $\theta = \pm\theta_k$ are *uncoupled*. This leaves only the first half of equation (9.79), with undetermined constants $C_1^k - C_4^k$.

We first consider the exterior edge. The violation of the boundary conditions stated in equation (9.92) are precisely equal to the membrane theory stress resultants evaluated at $\theta = \theta_k$. From equations (4.169a–c),

$$S_m(X, \theta_k) = \frac{8q_d L}{\pi^2} \cos \frac{X}{L} \sin \theta_k \qquad (9.96)$$

and

$$N_{\theta m}(X, \theta_k) = - \frac{4q_d a}{\pi} \sin \pi \frac{X}{L} \cos \theta_k \qquad (9.97)$$

Next, from the semi-infinite form of equations (9.79), with $k = 1$, we select expressions for $\bar{N}_{\theta X}^1$, N_θ^1, M_θ^1, and $\bar{Q}_\theta^1$ in terms of the four integration constants $C_1^1 - C_4^1$. Then we enforce the free edge boundary conditions, equation (9.92), by setting

$$\bar{N}_{\theta X}^1 \left(\frac{X}{L}, \theta_k\right) + S_m\left(\frac{X}{L}, \theta_k\right) = 0 \qquad (9.98a)$$

$$N_\theta^1\left(\frac{X}{L}, \theta_k\right) + N_{\theta m}\left(\frac{X}{L}, \theta_k\right) = 0 \qquad (9.98b)$$

$$M_\theta^1\left(\frac{X}{L}, \theta_k\right) = 0 \qquad (9.98c)$$

$$\bar{Q}_\theta^1\left(\frac{X}{L}, \theta_k\right) = 0 \qquad (9.98d)$$

after which $C_1^1 - C_4^1$ are determined. Note that the functions of (X/L) [$\sin \pi(X/L)$ or $\cos \pi (X/L)$] are the same for both the membrane theory and bending theory solutions, so that once the free edge condition is enforced at *one* point in the interval $0 \le X \le L$, it is satisfied at *every* point. This is another advantage of the Fourier series representation.

Now that the constants of integration have been determined for the exterior edge solution, the stress resultants and couples from the bending theory solution can be evaluated at a sufficient number of coordinates $[(X/L), \theta]$ to obtain the stress distribution. Of course, the membrane theory stress resultants are added to $\bar{N}_{\theta X}^1$, N_θ^1, and N_X^1 to get the complete expression for these functions. Finally, the total displacements may be computed as the sum of the membrane theory values, equations (6.56a–e), and those computed from the bending solutions. This solution is valid in the exterior portion of the shell ($\theta_k \geq \theta \geq 0$).

We have not provided sufficient detail here with respect to the bending solutions to perform calculations on actual shells but detailed procedures and examples are available (see "Design of Cylindrical Concrete Roofs"[21] and Billington[22]). It is often convenient to construct the bending theory corrections to the membrane theory solutions as linear combinations of *unit* load solutions of the type

$$\bar{N}_{\theta X}^1 = 1; \ N_\theta^1 = M_\theta^1 = \bar{Q}_\theta^1 = 0 \text{ at } \left(\frac{X}{L}, \theta_k\right) \tag{9.99a}$$

and

$$N_\theta^1 = 1; \ \bar{N}_{\theta X} = M_\theta^1 = \bar{Q}_\theta^1 = 0 \text{ at } \left(\frac{X}{L}, \theta_k\right) \tag{9.99b}$$

Such solutions are tabulated for a wide range of shell parameters in "Design of Cylindrical Concrete Roofs."[23]

We now take up the idealized interior edge, referring to the boundary conditions given by equations (9.93) and (9.94). We use an analogous procedure, superimposing the membrane theory solution with bending solutions for edge loadings. The constants of integration are evaluated from the conditions

$$D_Y^1\left(\frac{X}{L}, \theta_k\right) + D_{Ym}\left(\frac{X}{L}, \theta_k\right) = 0 \tag{9.100a}$$

$$D_{\theta X}^1\left(\frac{X}{L}, \theta_k\right) + D_{\theta Xm}\left(\frac{X}{L}, \theta_k\right) = 0 \tag{9.100b}$$

$$\bar{N}_{\theta X}^1\left(\frac{X}{L}, \theta_k\right) + S_m\left(\frac{X}{L}, \theta_k\right) = 0 \tag{9.100c}$$

$$F_Z^1\left(\frac{X}{L}, \theta_k\right) + F_{Zm}\left(\frac{X}{L}, \theta_k\right) = 0 \tag{9.100d}$$

where the bending components are found from expressions of the form of equation (9.79), and the membrane parts are given by equations (6.56a–e) and (4.169a–c). Also, D_Y, and $D_{\theta X}$ are expressed in terms of D_θ and D_n by equations (9.93c) and (9.93d), respectively, and F_Z is stated in terms of Q_θ and N_θ by equation (9.94c). Again, these computations may be expedited by the use of the tables in "Design of Cylindrical Concrete Roofs."[24]

Finally, with respect to simply supported circular cylindrical shells without intermediate supports, we should assess the relevance of the rather complex analysis we have outlined. We made a start in this direction at the conclusion of section 4.4.1. Now we can amplify some of these remarks, since we have, in principle, satisfied the boundary conditions on all edges.

Basically, at every cross section, the shell must resist the statical bending moment due to the applied loading, regardless of the theory used. For a uniform load w (force/length) the statical moment at the center line $X = L/2$ is the well-known

$$\bar{M}_X = w\,\frac{L^2}{8} \tag{9.101a}$$

or, for the first harmonic Fourier series approximation

$$\bar{M}_X = w\,\frac{L^2}{\Pi^2} \tag{9.101b}$$

It is of interest to compare the magnitude and distribution of the stress resultant N_X over the cross section calculated by using the bending theory of shells, to the like quantity found from elementary beam theory

$$N_X = \frac{\bar{M}_X Z}{I} \tag{9.102}$$

where $I =$ the moment of inertia of the cross section with unit thickness about the centroidal axis. This may be regarded as a gross measure of the relevance of the shell theory, as opposed to beam theory, calculations. In figure 9–9(b), we show the distribution of N_X over the cross section for a single barrel shell analyzed in "Design of Cylindrical Concrete Roofs,"[25] along with the straight line distribution given by equation (9.102). Note that both the maximum tensile and compressive stress resultants are more than double the values computed from the linear strain, beam theory formula. It is readily verified that both distributions of N_X yield practically the same statical moment $\bar{M}_X$ about the centroid of the cross section.[26]

9.3 Shells of Revolution

9.3.1 General

The bending of shells of revolution is a realtively complex subject from a mathematical standpoint. Also, the availability of very efficient and powerful numerically based computer programs has made many of the classical means of solving such problems somewhat archaic. Moreover, the classical solutions are developed in considerable detail in a number of popular monographs (Novozhilov,[27] Flügge,[28] and Kraus[29]), and little would be accomplished here by restating these lengthy derivations. Instead, we explore only some of the important classical findings, and then present an energy formulation suitable for a finite element solution.

9.3.2 Axisymmetrical Loading—Analytical Solutions

9.3.2.1. Governing Equations. An early classical formulation of the shell bending problem was presented by H. Reissner (the father of the often cited E. Reissner), who studied the spherical shell.[30] This formulation, generalized by E. Meissner,[31] represents the fourth order system by a coupled set of two second order equations in terms of the meridional rotation $D_{\phi\theta}$ and the transverse shearing resultant Q_ϕ; here, the derivation is generalized to include specifically orthotropic materials.

We begin by specializing the equilibrium equations, equations (3.24a, c, e), for the case of axisymmetric bending:

$$(R_0 N_\phi)_{,\phi} - R_\phi \cos \phi\, N_\theta + R_0 Q_\phi + q_\phi R_\phi R_0 = 0 \tag{9.103a}$$

$$(R_0 Q_\phi)_{,\phi} - R_0 N_\phi - R_\phi \sin \phi\, N_\theta + q_n R_\phi R_0 = 0 \tag{9.103b}$$

$$(R_0 M_\phi)_{,\phi} - R_\phi \cos \phi\, M_\theta - R_\phi R_0 Q_\phi = 0 \tag{9.103c}$$

We next combine the strain-displacement relationship for axisymmetric shells, equations (5.48a, c, d, g), and the constitutive law, equation (6.12), into a set of stress resultant and stress couple-displacement equations:

$$N_\phi = D_{or11}\left[\frac{1}{R_\phi}(D_{\phi,\phi} + D_n)\right]$$

$$+ D_{or12}\left[\frac{1}{R}(\cos \phi\, D_\phi + \sin \phi\, D_n)\right] - N_{\phi Tor} \tag{9.104a}$$

$$N_\theta = D_{or12}\left[\frac{1}{R_\phi}(D_{\phi,\phi} + D_n)\right]$$

$$+ D_{or22}\left[\frac{1}{R_0}(\cos \phi\, D_\phi + \sin \phi\, D_n)\right] - N_{\theta Tor} \tag{9.104b}$$

$$M_\phi = D_{or44}\left[\frac{1}{R_\phi}D_{\phi\theta,\phi}\right] + D_{or45}\left[\frac{\cos \phi}{R_0}D_{\phi\theta}\right] - M_{\phi Tor} \tag{9.104c}$$

$$M_\theta = D_{or45}\left[\frac{1}{R_\phi}D_{\phi\theta,\phi}\right] + D_{or55}\left[\frac{\cos \phi}{R_0}D_{\phi\theta}\right] - M_{\theta Tor} \tag{9.104d}$$

Symmetry of the constitutive matrix is presumed, so that $D_{or12} = D_{or21}$ and $D_{or45} = D_{or54}$. Also, as before, the thermal terms are presumed to be known at the outset.

The boundary conditions corresponding to the axisymmetric formulation are specialized from figure 6–3 as $N_\phi = \bar{N}_\phi$ or $D_\phi = \bar{D}_\phi$; $Q_\phi = \bar{Q}_\phi$ or $D_n = \bar{D}_n$; and $M_\phi = \bar{M}_\phi$ or $D_{\phi\theta} = \bar{D}_{\phi\theta}$.

We now substitute equations (9.104c) and (9.104d) into equation (9.103c). To simplify the derivation here somewhat, the elements of the constitutive matrix $[\mathbf{D}_{or}]$ are taken as constant, but they can be included as variables for shells with continuously varying material properties and/or thickness.

Continuing, we have

$$D_{\phi\theta,\phi\phi} + \frac{R_\phi}{R_0}\left(\frac{R_0}{R_\phi}\right)_{,\phi} D_{\phi\theta,\phi}$$

$$-\frac{1}{D_{or44}}\left(\frac{R_\phi}{R_0}\right)\left[D_{or45}\sin\phi + D_{or55}\frac{R_\phi\cos^2\phi}{R_0}\right]D_{\phi\theta} - \frac{R_\phi^2}{D_{or44}}Q_\phi$$

$$= \frac{1}{D_{or44}}\left(\frac{R_\phi}{R_0}\right)[(R_0 M_{\phi Tor}),_\phi - R_\phi\cos\phi\, M_{\theta Tor}] \qquad (9.105)$$

which is the first of two governing equations in $D_{\phi\theta}$ and Q_ϕ.

To obtain the second equation, we first consider equations (9.104a) and (9.104b). We form the combinations

$$\frac{R_\phi}{D_{or11}}\left[(9.104a) - \frac{D_{or12}}{D_{or22}}(9.104b)\right]$$

$$\text{and}\quad \frac{R_\theta}{D_{or22}}\left[(9.104b) - \frac{D_{or12}}{D_{or11}}(9.104a)\right]$$

and rewrite the expression $(1/R_0) \cdot (\cos\phi\, D_\phi + \sin\phi\, D_n)$ as $(1/R_\theta) \cdot (\cot\phi\, D_\phi + D_n)$ to get

$$D_{\phi,\phi} + D_n = \frac{R_\phi}{D_{or11}(1 - \mu_1\mu_2)}[(N_\phi + N_{\phi Tor}) - \mu_2(N_\theta + N_{\theta Tor})]$$

$$(9.106a)$$

and

$$\cot\phi\, D_\phi + D_n = \frac{R_\theta}{D_{or22}(1 - \mu_1\mu_2)}[-\mu_1(N_\phi + N_{\phi Tor}) + (N_\theta + N_{\phi Tor})]$$

$$(9.106b)$$

where

$$\mu_1 = \frac{D_{or12}}{D_{or11}} \quad \text{and} \quad \mu_2 = \frac{D_{or12}}{D_{or22}} \tag{9.106c}$$

Next, we eliminate D_n between equations (9.106a) and (9.106b) to find

$$D_{\phi,\phi} - \cot \phi \, D_\phi = \frac{1}{(1 + \mu_1\mu_2)} \left[\left(\frac{R_\phi}{D_{or11}} + \frac{\mu_1 R_\theta}{D_{or22}} \right) (N_\phi + N_{\phi Tor}) \right.$$

$$\left. - \left(\frac{R_\theta}{D_{or22}} + \frac{\mu_2 R_\phi}{D_{or11}} \right) (N_\theta + N_{\theta Tor}) \right] \tag{9.107a}$$

and differentiate equation (9.106b) with respect to ϕ, which gives

$$\cot \phi \, D_{\phi,\phi} - \csc^2 \phi D_\phi + D_{n,\phi} = \frac{1}{D_{or22}(1 - \mu_1\mu_2)}$$

$$\times \{R_\theta[-\mu_1(N_\phi + N_{\phi Tor}) + (N_\theta + N_{\theta Tor})]\}_{,\phi} \tag{9.107b}$$

We then eliminate $D_{\phi,\phi}$ from equations (9.107a) and (9.107b) by taking $[(9.107b) - \cot \phi \, (9.197a)]$. The l.h.s. becomes

$$(\cot^2 \phi - \csc^2 \phi)D_\phi + D_{n,\phi} = D_{n,\phi} - D_\phi \tag{9.108}$$

It is now convenient to suppress the transverse shearing strains. From equation (5.48g), note that for $\gamma_\phi = 0$,

$$D_{\phi\theta} = -\frac{1}{R_\phi} (D_{n,\phi} - D_\phi) \tag{9.109}$$

so that $-(1/R_\phi) \cdot (\text{l.h.s.}) \, (9.108) = D_{\phi\theta}$, which is one of our dependent variables.

To complete the derivation, the stress resultants N_ϕ and N_θ on the r.h.s. of equation (9.107b) must be expressed in terms of the other variable Q_ϕ. This is most easily accomplished by using the overall equilibrium concept introduced in section 4.3.3.1. With Q_ϕ now included, equation (4.58) generalizes to

$$\mathbf{Q}(\phi) = [N_\phi(\phi) \sin \phi - Q_\phi(\phi) \cos \phi]2\pi R_0(\phi) \tag{9.110}$$

where $\mathbf{Q}(\phi)$ is the resultant vertical load in the *negative Z* direction, expressed as a function of the surface loading by equation (4.59). Therefore,

$$N_\phi = \frac{\mathbf{Q}(\phi)}{2\pi R_0 \sin \phi} + Q_\phi \cot \phi \tag{9.111a}$$

Now, from equation (9.103b),

$$N_\theta = \frac{1}{R_\phi \sin \phi} \left[(R_0 Q_\phi)_{,\phi} - R_0 N_\phi + q_n R_\phi R_0 \right]$$

$$= \frac{1}{R_\phi \sin \phi} \left[(R_0 Q_\phi)_{,\phi} - R_0 \cot \phi \, Q_\phi - \frac{Q(\phi)}{2\Pi \sin \phi} + q_n R_\phi R_0 \right] \quad (9.111b)$$

Equations (9.111a) and (9.111b) may be used to remove N_ϕ and N_θ from the combined equations (9.107a) and (9.107b), i.e. [(9.107b) $-$ cot ϕ(9.107a)], which then becomes a second equation in $D_{\phi\theta}$ and Q_ϕ. This is tedious to carry through in the general form; consequently, at this point, the specialization for a commonly encountered case seems prudent.

9.3.2.2 Homogeneous Solution for Spherical Shells. Consider an isotropic shell and no surface loading or thermal effects. Then, $q_n = Q(\phi) = 0$; $\mu_1 = \mu_2 = \mu$; $D_{or11} = D_{or22} = Eh/(1 - \mu^2)$; $D_{or44} = D_{or55} = D$; and $D_{or45} = \mu D$, where D has been previously defined as $Eh^3/[12(1 - \mu^2)]$. Also, for the spherical geometry, $R_\phi = R_\theta = a$ and $R_0 = a \sin \phi$. We may anticipate that the forthcoming solution will represent the influence of edge forces and moments which, in combination with a membrane theory solution for the surface loading, constitutes a complete solution to the spherical shell problem.

Now we proceed with the aforementioned [(9.107b) $-$ cot ϕ (9.107a)] combination of equations (9.107a) and and (9.107b), which, in view of equation (9.109), gives

$$D_{\phi\theta} = \frac{1}{Eh} \left[(1 + \mu) \cot \phi \, (N_\phi - N_\theta) - N_{\theta,\phi} + \mu N_{\phi,\phi} \right] \quad (9.112)$$

Equations (9.111a) and (9.111b) simplify to

$$N_\phi = Q_\phi \cot \phi \qquad\qquad (9.113a)$$

$$N_\theta = Q_{\phi,\phi} \qquad\qquad (9.113b)$$

where we have employed the Gauss–Codazzi relationship $R_{0,\phi} = R_\phi \cos \phi$ via equation (2.47).

Substituting equations (9.113a) and (9.113b) into (9.112), we have

$$D_{\phi\theta} = -\frac{1}{Eh} \left[Q_{\phi,\phi\phi} + \cot \phi Q_{\phi,\phi} - (\cot^2 \phi - \mu) Q_\phi \right] \quad (9.114a)$$

The specialized form of the first of our governing equations, equations (9.105), is

$$\frac{a^2}{D} Q_\phi = D_{\phi\theta,\phi\phi} + \cot \phi D_{\phi\theta,\phi} - (\cot^2 \phi + \mu) D_{\phi\theta} \quad (9.114b)$$

Noting the similarity between the proceding two equations, we may define the operator

$$\mathbf{L}(\ ) = (\ )_{,\phi\phi} + \cot\phi\ (\ )_{,\phi} - \cot^2\phi\ (\ ) \tag{9.115}$$

(See Flügge[32]), whereupon the equations become

$$\mathbf{L}(Q_\phi) + \mu Q_\phi = -EhD_{\phi\theta} \tag{9.116a}$$

$$\mathbf{L}(D_{\phi\theta}) - \mu D_{\phi\theta} = \frac{a^2}{D} Q_\phi \tag{9.116b}$$

We may substitute equation (9.116b) into (9.116a), or (9.116a) into (9.116b), to get, respectively,

$$\mathbf{LL}(D_{\phi\theta}) - \mu^2 D_{\phi\theta} = -\frac{a^2 Eh}{D} D_{\phi\theta} \tag{9.117a}$$

$$\mathbf{LL}(Q_\phi) - \mu^2 Q_\phi = -\frac{a^2 Eh}{D} Q_\phi \tag{9.117b}$$

so that the equations are uncoupled. If one variable is determined, the other is immediately known from equations (9.116a) and (9.116b), and all other terms follow from the previous formulae.

It is of interest to rewrite equation (9.117b) in the form

$$\mathbf{LL}(Q_\phi) + 4k^4 Q_\phi = 0 \tag{9.118a}$$

where

$$k^4 = \frac{1}{4}\left(\frac{a^2 Eh}{D} - \mu^2\right)$$

$$= 3(1 - \mu^2)\frac{a^2}{h^2} - \frac{\mu^2}{4} \tag{9.118b}$$

The resulting equation (9.118a) is quite similar to equation (9.16), which describes cylindrical shells. Here, of course, the operator $\mathbf{L}(\ )$ causes more complication in the eventual solution.

It is also convenient to express equation (9.118a) in complex form as

$$\mathbf{L}[\mathbf{L}(Q_\phi) + 2ik^2 Q_\phi] - 2ik^2[\mathbf{L}(Q_\phi) + 2ik^2 Q_\phi] = 0 \tag{9.119a}$$

$$\mathbf{L}[\mathbf{L}(Q_\phi) - 2ik^2 Q_\phi] + 2ik^2[\mathbf{L}(Q_\phi) - 2ik^2 Q_\phi] = 0 \tag{9.119b}$$

(See Flügge[33]), which can be easily verified to be equivalent to the original equation.

Thus the solution of the two second order equations:

$$\mathbf{L}(Q_\phi) + 2ik^2 Q_\phi = 0 \qquad\qquad (9.120a)$$

$$\mathbf{L}(Q_\phi) - 2ik^2 Q_\phi = 0 \qquad\qquad (9.120b)$$

will constitute the solution to the fourth order equation (9.118a).

The formulation of the governing equations for other geometrics follow similar but sometimes more complicated steps. The interested reader is referred to Novozhilov[34] Flügge,[35] and Kraus.[36]

An exact solution to the governing equations for the spherical geometry exists. If one of the equations, e.g., equation (9.120a), is fully expanded, we have

$$Q_{\phi,\phi\phi} + \cot\phi \, Q_{\phi,\phi} - \cot^2\phi \; Q_\phi + 2ik^2 Q_\phi = 0 \qquad (9.121)$$

which is a second order linear differential equation with variable coefficients.

The further transformations

$$\cos^2\phi = \psi \qquad\qquad (9.122a)$$

and

$$Q_\phi = \tilde{Q}_\phi \sin\phi \qquad\qquad (9.122b)$$

put the equation into the form

$$\tilde{Q}_{\phi,\psi\psi} + \frac{1 - 5\psi}{2\psi(1 - \psi)} \, \tilde{Q}_{\phi,\psi} - \frac{1 - 2i\psi^2}{4\psi(1 - \psi)} \, \tilde{Q}_\phi = 0 \qquad (9.123)$$

which is known as the *hypergeometric equation*. Such equations may be solved with some rather slowly converging power series expressions.[37]

9.3.2.3 Asymptotic Solutions for Spherical Shells.

One of the most powerful methods for solving the complicated differential equations that arise in the bending theory of shells is that of *asymptotic integration*. This method is developed in Kraus[38] and Novozhilov[39] and applied to both symmetrically and asymmetrically loaded shells of revolution in the latter reference, but a detailed development is beyond our objectives in this book. It is sufficient to mention here that the asymptotic integration technique, as applied to second order equations such as equation (9.121), transforms the equation into another form in which the first derivative of the dependent variable is eliminated.

The attractiveness of representing the solution of the bending theory equations for rotational shells in a relatively simple form has given rise to some useful approximate methods, in which it is postulated that the solutions to the homogeneous equations have the form

$$\left. \begin{array}{l} e^{\pm k\phi} \sin k\phi \\[12pt] e^{\pm k\phi} \cos k\phi \end{array} \right\} \qquad (9.124)$$

Although expressions of this type may be rigorously obtained using asymptotic integration, such solutions can be directly derived for shells in which the geometric parameters remain essentially constant over a significant portion of the shell by simply neglecting certain terms in the original differential equation. For example, note in equation (9.121) that the second and third terms are multiplied by $\cot \phi$ and $\cot^2 \phi$, which are relatively small when $\phi \geq \pi/4$, whereas the fourth term has a coefficient of k^2, which, from equation (9.118b), is proportional to a/h. Thus it is reasonable to use the simplified equation

$$Q_{\phi,\phi\phi} \pm 2ik^2 Q_\phi = 0 \qquad (9.125)$$

to find approximate solutions for the bending of spherical shells under edge loads, provided that ϕ is not too small. If we are interested in solutions in the vicinity of the lower boundary of spherical shells due to edge loading, as shown in figure 9–10, it is only necessary that ϕ_b not be too small. This reasoning is also applicable for other shells with smoothly varying geometry, such as conical shells, and is sometimes referred to as the Geckeler approximation.[40] Despite the lack of mathematical rigor, such "quick and dirty" solutions are often the only relatively simple analytical means available to check highly complex, computer-based analyses.

We may write the solution of equation (9.125) from the auxiliary equations

$$m^2 \pm 2ik^2 = 0 \qquad (9.126)$$

which give

$$m = \pm(1 \pm i)k \qquad (9.127)$$

so that the complete solution is

$$Q_\phi = B_1 e^{(1+i)k\phi} + B_2 e^{(1-i)k\phi} + B_3 e^{-(1+i)k\phi} + B_4 e^{-(1-i)k\phi} \qquad (9.128)$$

where B_1–B_4 are integration constants.

We now wish to redefine the integration constants so that Q_ϕ is expressed in terms of real functions. First, we use the identities

$$\left. \begin{array}{l} e^{ik\phi} = \cos k\phi + i\sin k\phi \\[12pt] e^{-ik\phi} = \cos k\phi - i\sin k\phi \end{array} \right\} \qquad (9.129)$$

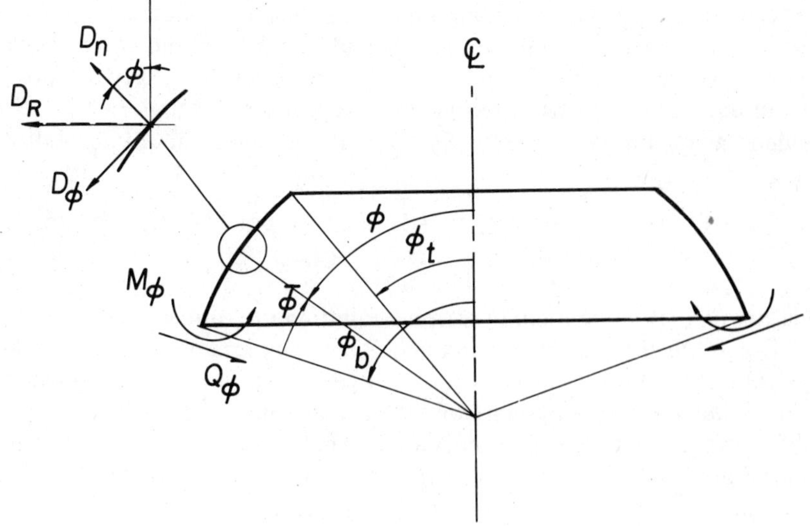

Figure 9-10. Edge-Loaded Spherical Shell

to rewrite equation (9.128) as

$$Q_\phi = (B_1 + B_2)e^{k\phi}\cos k\phi + i(B_1 - B_2)e^{k\phi}\sin k\phi$$

$$+ (B_3 + B_4)e^{-k\phi}\cos k\phi - i(B_3 - B_4)e^{-k\phi}\sin k\phi \tag{9.130a}$$

It is convenient to express this solution as a function of the auxiliary coordinate

$$\bar{\phi} = \phi_b - \phi \tag{9.130b}$$

which we introduce into the first term of equation (9.130a):

$$(B_1 + B_2)e^{k\phi}\cos k\phi = (B_1 + B_2)e^{k(\phi_b - \bar{\phi})}\cos k(\phi_b - \bar{\phi})$$

$$= (B_1 + B_2)e^{k\phi_b}e^{-k\bar{\phi}}(\cos k\phi_b \cos k\bar{\phi} + \sin k\phi_b \sin k\bar{\phi})$$

$$= [(B_1 + B_2)e^{k\phi_b}\cos k\phi_b]e^{-k\bar{\phi}}\cos k\bar{\phi}$$

$$+ [(B_1 + B_2)e^{k\phi_b}\sin k\phi_b]e^{-k\bar{\phi}}\sin k\bar{\phi} \tag{9.130c}$$

Similarly, the second term becomes

$$i(B_1 - B_2)e^{k\phi}\sin k\phi = [i(B_1 - B_2)e^{k\phi_b}\sin k\phi_b]e^{-k\bar{\phi}}\cos k\phi_b$$

$$- [i(B_1 - B_2)e^{k\phi_b}\cos k\phi_b]e^{-k\bar{\phi}}\sin k\bar{\phi} \tag{9.130d}$$

The terms in [ ] in the preceding equations may be collected as coefficients of the variable terms $e^{-k\bar{\phi}}\cos k\bar{\phi}$ and $e^{-k\bar{\phi}}\sin k\bar{\phi}$ and can be redefined as new arbitrary constants C_1 and C_2. We retain the last two expressions in equation (9.130a) in terms of ϕ, with $(B_3 + B_4)$ and $-i(B_3 - B_4)$ redefined as C_3 and C_4, respectively, so that the new form of the solution is

$$Q_\phi = e^{-k\bar{\phi}}(C_1 \cos k\bar{\phi} + C_2 \sin k\bar{\phi}) + e^{-k\phi}(C_3 \cos k\phi + C_4 \sin k\phi)$$

$$(9.131)$$

Equation (9.131) is identical in form to equation (9.20) if $\bar{\phi}$ is replaced by $\phi_b - \phi$ and the constants are redefined. This comparison gives rise to the physical interpretation of the Geckeler approximation as the replacement of the actual shell by a cylindrical shell of suitable dimensions.

The foregoing transformation is an alternate means of introducing the semi-infinite concept that we utilized extensively for cylindrical shells. The solution is uncoupled into parts that decay from the boundaries $\phi = \phi_b$ and $\phi = \phi_t$, respectively. If the influence of the loading at one boundary does not penetrate into the shell affected by the loading at the other boundary, we have a semi-infinite shell. Considering edge loading originating at $\phi = \phi_b$, we retain

$$Q_\phi(\bar{\phi}) = e^{-k\bar{\phi}}(C_1 \cos k\bar{\phi} + C_2 \sin k\bar{\phi}) \qquad (9.132)$$

We now evaluate the corresponding stress resultants, couples, and displacements. In performing the required differentiations, note that

$$(\ ),_\phi = -(\ ),_{\bar{\phi}} \qquad (9.133)$$

From equation (9.111a), with $\mathbf{Q}(\phi) = 0$,

$$N_\phi(\bar{\phi}) = Q_\phi \cot \phi \qquad (9.134a)$$

$$= e^{-k\bar{\phi}} \cot \phi (C_1 \cos k\bar{\phi} + C_2 \sin k\bar{\phi}) \qquad (9.134b)$$

$$= e^{-k\bar{\phi}} \cot (\phi_b - \bar{\phi})(C_1 \cos k\bar{\phi} + C_2 \sin k\bar{\phi}) \qquad (9.134c)$$

Expanding equation (9.103b) with $q_n = 0$, we have

$$a(\sin \phi Q_\phi),_\phi - a \sin \phi\, N_\phi - a \sin \phi\, N_\theta = 0 \qquad (9.135a)$$

Solving for N_θ in view of equation (9.134a), we find

$$N_\theta(\bar{\phi}) = Q_{\phi,\phi}$$

$$= -Q_{\phi,\bar{\phi}}$$

$$= (C_1 - C_2)ke^{-k\bar{\phi}} \cos k\bar{\phi} + (C_1 + C_2)ke^{-k\bar{\phi}} \sin k\bar{\phi} \qquad (9.135b)$$

Next, we turn to the rotation $D_{\phi\theta}$, which is expressed by equation (9.116a). It is consistent with our previous approximations to take $\mathbf{L}(\ ) \simeq (\ )_{,\phi\phi}$ and also to neglect the μQ_ϕ term, whereby

$$D_{\phi\theta}(\bar\phi) = -\frac{1}{Eh}Q_{\phi,\phi\phi}$$

$$= -\frac{1}{Eh}Q_{\phi,\bar\phi\bar\phi}$$

$$= -\frac{2}{Eh}k^2 e^{-k\bar\phi}(C_1 \sin k\bar\phi - C_2 \cos k\bar\phi) \qquad (9.136)$$

The actual displacements are sometimes of interest. D_ϕ and D_n may be individually found from the simultaneous solution of equations (6.34a) and (6.34b). Of more use for our purposes is the radial displacement D_R, which is defined in the inset of figure 9–10 as

$$D_R = D_\phi \cos \phi + D_n \sin \phi \qquad (9.137a)$$

Referring to equation (6.34b),

$$D_R(\bar\phi) = a\epsilon_\theta = \frac{a}{Eh}(N_\theta - \mu N_\phi) \qquad (9.137b)$$

which is easily evaluated from equations (9.134) and (9.135).

We now turn to the stress couples which can be obtained from equations (9.104c) and (9.104d). Again, it is consistent with our previous arguments to neglect the second terms in these equations in comparison to the first, whereby

$$M_\phi = \frac{D}{a}D_{\phi\theta,\phi} \qquad (9.138a)$$

which is also conveniently written, using equations (9.136) and (9.118b), as

$$M_\phi = -\frac{a}{4k^4}Q_{\phi,\phi\phi\phi} = \frac{a}{4k^4}Q_{\phi,\bar\phi\bar\phi\bar\phi} \qquad (9.138b)$$

Evaluating this expression,

$$M_\phi(\bar\phi) = \frac{a}{2k}e^{-k\bar\phi}[(C_1 + C_2)\cos k\bar\phi - (C_1 - C_2)\sin k\bar\phi] \qquad (9.138c)$$

Also, from equation (9.104d), with $D_{or\,45} = \mu$,

$$M_\theta(\bar\phi) = \mu M_\phi(\bar\phi) \qquad (9.139)$$

We have thus derived approximate but explicit expressions for the forces, moments, and displacements in an edge-loaded spherical shell. From these equations, it is obvious that the assumption leading to equation (9.125)—the neglect of Q_ϕ and $Q_{\phi,\phi}$ as compared to $Q_{\phi,\phi\phi}$—is reasonable since each subsequent differentiation gives a factor k that is of $O(a/h)$. Moreover, the presence of the cot ϕ and $\cot^2\phi$ coefficients of $Q_{\phi,\phi}$ and Q_ϕ, respectively, in equation (9.121) further reduce the influence of the latter terms when $\phi > \Pi/4$. The limits of application of the semi-infinite solution can be established using figure 9–2, with kX replaced by $k\bar{\phi}$ where k is given by equation (9.118b).

For practical application, is is helpful to represent the previous solution in the same format as we used for the edge-loaded cylindrical shell. Consider the loading shown in figure 9–10, which is analogous to figure 9–1 for the cylindrical shell. We take

$$M_0 = M_\phi(\phi_b) \tag{9.140a}$$

$$Q_0 = Q_\phi(\phi_b) \tag{9.140b}$$

whereupon, from equations (9.138c) and (9.132) with $\bar{\phi} = 0$,

$$Q_0 = C_1 \tag{9.141a}$$

$$M_0 = \frac{a}{2k} (C_1 + C_2) \tag{9.141b}$$

from which

$$C_1 = Q_0 \tag{9.142a}$$

$$C_2 = \frac{2k}{a} M_0 - Q_0 \tag{9.142b}$$

A matrix of derivatives of Q_ϕ, similar to that written for D_n in matrix 9–1, is useful for further computations.

9.3.2.4 Compound Shells. We are now able to resume our consideration of the cylindrical shell with a spherical head, as shown in figure 4–14 and discussed in section 4.3.4. Previously, we found that the membrane theory by itself was inadequate, since (a) radial deformations between the two shells were incompatible, and (b) except for the complete hemisphere, the radial component of the meridional force in the spherical shell was unbalanced. We may now fully analyze the shell by the superposition method described in section 6.3.1, using the notation introduced in equation (6.30).

On figure 9–11(a), we show the pressurized shell from figure 4–14, with the axial coordinate taken as X, and the origin placed at the end of

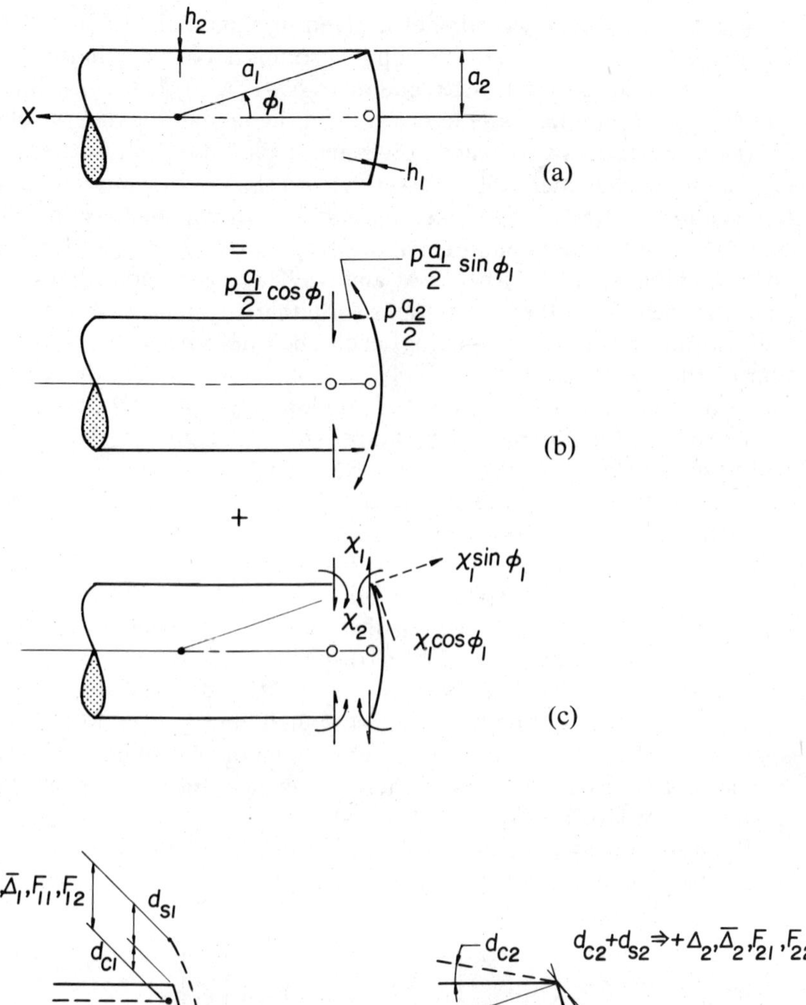

Figure 9–11. Compound Shell

the cylinder. The respective thicknesses of the spherical and cylindrical segments are defined as h_1 and h_2, and the Young's moduli and Poisson's ratios as E_1 and E_2 and μ_1 and μ_2.

Loading case [9–11(b)] requires the membrane theory solutions for the cylindrical and the spherical shells, given in equations (4.77)–(4.80) and equations (6.38a–e), as well as a bending solution for the cylindrical shell, given by equations (9.22)–(9.29) and matrix 9–1, with $Q_0 = p\,(a_1/2) \cos \phi_1$ and $M_0 = 0$. As we have noted previously, the membrane theory solution for the internal pressure loading gives incompatibilities for both the *relative* radial displacement [figure 9–11(d)] and the *relative* meridional rotation [figure 9–11(e)] of the spherical and cylindrical shells at their junction. We denote these quantities in the original shell, case [9–11(a)], as Δ_1 and Δ_2, and in case [9–11(b)] as $\bar{\Delta}_1$ and $\bar{\Delta}_2$. There is no contribution to $\bar{\Delta}_2$ from the membrane theory solutions for either segment, since, under the uniform internal pressure loading, each shell deforms without any rotation of the meridian. This is easily visualized. However, there still remains a contribution of $\bar{\Delta}_2$ from the line load $Q_0 = p(a_1/2) \cos \phi_1$.

To restore the compatibility, we provide loading case [9–11(c)], consisting of the *static correspondents* of Δ_1 and Δ_2—namely, a radial transverse shearing force χ_1 and a meridional moment χ_2, equal and opposite on each segment. Solutions for these loading systems are given for the cylindrical shell by equations (9.22)–(9.29) and matrix 9–1, with $Q_0 = \chi_1$ and $M_0 = \chi_2$, and for the spherical shell by equations (9.133)–(9.141), with $Q_0 = -\chi_1 \sin \phi$, and $M_0 = \chi_2$. For the flexibility formulation that we are using, it is expedient to solve the following loading cases: (i) $Q_0 = 1$, $M_0 = 0$ on the cylindrical shell, and $Q_0 = -\sin \phi_1$, $M_0 = 0$ on the spherical shell; and (ii) $Q_0 = 0$, $M_0 = 1$ on the cylindrical shell, and $Q_0 = 0$, $M_0 = 1$ on the spherical shell. For case (i), the relative radial displacement and rotation are designated as F_{11} and F_{21}, whereas for case (ii), the corresponding quantities are labeled F_{21} and F_{22}.

We now consider the details of the evaluation of the individual terms in equation (6.30) for this problem:

$$\left\{ \begin{array}{c} \Delta_1 \\ \Delta_2 \end{array} \right\} = \left\{ \begin{array}{c} \bar{\Delta}_1 \\ \bar{\Delta}_2 \end{array} \right\} + \left[\begin{array}{cc} F_{11} & F_{12} \\ F_{21} & F_{22} \end{array} \right] \left\{ \begin{array}{c} \chi_1 \\ \chi_2 \end{array} \right\} \tag{9.143}$$

First, note that we are dealing with *relative* displacements and rotations, which are computed from the absolute displacements of the two segments of the shell. Therefore, we must carefully define the sign convention for these quantities. Corresponding to the senses assumed for the redundants χ_1 and χ_2 in figure 9–11(c), it is consistent to take the positive sense of the *relative* radial displacement as inward on the cylinder and outward on the sphere. Similarly, a positive *relative* meridional rotation will open the internal angle between the cylinder and the sphere, originally $\Pi/2 + \phi_1$. These are illustrated on figure 9–11(d) and (e). For example, if we compute the radial displacement in the cylinder D_n due to one of the loading cases, and it is *positive* (outward) by the cylindrical shell convention, it takes a *negative* sign in the relative displacement expression, since it is opposite to in sense to d_{c1}. On the other hand, a *positive* computed value for the radial displacement of the spherical shell D_R is consistent with d_{s1}, so that it takes a *positive* sign in the relative displacement expression. Similarly, a *positive* meridional rotation $D_{X\theta}$ in the cylindrical shell may be seen from figure 5–2(a) (with $\alpha = X$ and $\beta = \theta$) to be opposite in sense to d_{c2}, so that it takes a *negative* sign. The rotation of the spherical shell must be examined carefully. From figure 5–2(a) (with $\alpha = \phi$ and $\beta = \theta$), it appears as if a *positive* meridional rotation $D_{\phi\theta}(\phi)$, corresponds to a *positive* d_{s2}. However, referring to equation (9.136), $D_{\phi\theta}$ is expressed as a function of the variable $\bar{\phi} = \phi_b - \phi = \phi_1 - \phi$. Consequently, figure 5–2(a) should be superimposed on the spherical shell in figure 9–11(e), with $\mathbf{t}_\alpha$ in the direction of increasing $\bar{\phi}$. This means that a *positive* $D_{\phi\theta}(\bar{\phi})$ corresponds to a *negative* d_{s2}.

Next, note that the overall compatibility condition requires that the relative radial displacement and relative meridional rotation between the two segments vanishes, or

$$\{\Delta\} = \{\Delta_1\,\Delta_2\} = \{0\} \tag{9.144}$$

For the evaluation of the elements of $\{\bar{\Delta}\}$ and $[\mathbf{F}]$, we refer to the chart of basic solutions given in Table 9–1. Note that the cylindrical and spherical segments may each have different constant thicknesses and material properties. In the last column, the most probable algebraic sign of the computed displacement or rotation on the shell segment is assumed, and the corresponding algebraic sign of the contribution of this quantity to the relative displacement is given. To illustrate how we perform some of the computations, we investigate the symmetry of the flexibility matrix by evaluating F_{12} and F_{21}.

For the cylindrical shell, we compute $D_{X\theta}$ for $Q_0 = 1$ and $M_0 = 0$. From equation (9.6d), with $\gamma_x = 0$, $D_{X\theta} = -D_{n,X}$. Also from matrix 9–1(b)

$$D_{n,X}(0) = \frac{1}{2k_2^2 D_2}$$

so that

$$D_{X\theta} = -\frac{1}{2k_2^2 D_2}$$

and

$$d_{c2} = +\frac{1}{2k_2^2 D_2} \tag{9.145a}$$

Table 9–1
Basic Solutions for a Compound Shell

Segment	Loading	Compute	Equations	Contributes to	Sign Convention
	p	D_n $(D_{X\theta} = 0)$	(9.42a) with $N_x = p(a_1/2)\sin\phi_1$ and $N_\theta = pa_2$	$\bar{\Delta}_1$	$+D_n \rightarrow -d_{c1}$
	$Q_0 = p\,(a_1/2)\cos\phi_1$	D_n	matrix 9–1, row (a)	$\bar{\Delta}_1$	$-D_n \rightarrow +d_{c1}$
Cylindrical	$M_0 = 0$	$D_{X\theta}$	(9.6d) with $\gamma_X = 0$; then matrix 9–1, row (b)	$\bar{\Delta}_2$	$-D_{X\theta} = +d_{c2}$
$a = a_2$		D_n	matrix 9–1, row (a)	F_{11}	$-D_n \rightarrow +d_{c1}$
$h = h_2,\ \mu = \mu_2$ $E = E_2,\ D = D_2$ $X = 0$	$Q_0 = 1$ $M_0 = 0$	$D_{X\theta}$	(9.6d) with $\gamma_X = 0$; then matrix 9–1, row (b)	F_{21}	$-D_{X\theta} \rightarrow +d_{c2}$
$k_2^4 = \dfrac{3(1 - \mu_2^2)}{(a_2 h_2)^2}$		D_n	matrix 9–1, row (a)	F_{12}	$-D_n \rightarrow +d_{c1}$
	$Q_0 = 0$ $M_0 = 1$	$D_{X\theta}$	(9.6d) with $\gamma_X = 0$; then matrix 9–1, row (b)	F_{22}	$-D_{X\theta} \rightarrow +d_{c2}$

	p				
		D_R $(D_{\phi\theta}=0)$	(9.137b) with $N_\phi = N_\theta = p(a_1/2)$	$\bar{\Delta}_1$	$+D_R \to +d_{s1}$
Spherical	$Q_0 = -\sin\phi_1$	D_R	(9.137b) with N_ϕ and N_θ from (9.134c) and (9.135b)	F_{11}	$+D_R \to +d_{s1}$
$R_\phi = R_\theta = a$	$M_0 = 0$				
$a = a_1$	or				
$h = h_1,\ \mu = \mu_1$	$C_1 = -\sin\phi_1$	$D_{\phi\theta}$	(9.136)	F_{21}	$+D_{\phi\theta} \to -d_{s2}$
$E = E_1,\ D = D_1$	$C_2 = \sin\phi_1$				
$\phi = \phi_b = \phi_1$			(9.137b) with		
$\bar{\phi} = 0$	$Q_0 = 0$	D_R	N_ϕ and N_θ from (9.134c) and (9.135b)	F_{12}	$-D_R \to -d_{s1}$
	$M_0 = 1$				
$k_1^4 = 3(1 - \mu_1^2)\left(\dfrac{a_1}{h_1}\right)^2$	or				
	$C_1 = 0$	$D_{\phi\theta}$	(9.136)	F_{22}	$+D_{\phi\theta} \to -d_{s2}$
	$C_2 = 2k_1/a_1$				

is the contribution of the cylindrical shell to F_{21}. Next, we evaluate D_n for $Q_0 = 0$ and $M_0 = 1$. From matrix 9–1(a),

$$D_n(0) = -\frac{1}{2k_2^2 D_2}$$

giving

$$d_{c1} = +\frac{1}{2k_2^2 D_2} \qquad (9.145\text{b})$$

as the contribution of the cylindrical shell to F_{12}. The symmetry is thus verified for the contributions of the cylindrical shell.

For the spherical shell, we calculate $D_{\phi\theta}$ for $Q_0 = -\sin\phi_1$ and $M_0 = 0$. This corresponds to $C_1 = -\sin\phi_1$ and $C_2 = \sin\phi_1$ in equation (9.136), so that

$$D_{\phi\theta}(0) = \frac{2}{E_1 h_1} k_1^2 \sin\phi_1$$

giving

$$d_{s2} = \frac{-2}{E_1 h_1} k_1^2 \sin\phi_1 \qquad (9.146)$$

as the contribution of the spherical shell to F_{21}. Finally, we find D_R for $Q_0 = 0$ and $M_0 = 1$. Taking $C_1 = 0$ and $C_2 = 2(k_1/a_1)$, we have from equations (9.134c) and (9.135b) evaluated at $\bar{\phi} = 0$

$$\left.\begin{aligned} N_\phi(0) &= 0 \\[2mm] N_\theta(0) &= -2\frac{k_1^2}{a_1} \end{aligned}\right\} \qquad (9.147\text{a})$$

and then from equation (9.137b),

$$D_R(0) = -2\frac{k_1^2}{E_1 h_1} \qquad (9.147\text{b})$$

which gives

$$d_{s1} = -2\frac{k_1^2}{E_1 h_1} \qquad (9.147\text{c})$$

as the contribution of the spherical shell to F_{12}.

Now F_{12} and F_{21} are found from equations (9.145) and (9.146) as

$$F_{12} = d_{c1} + d_{s1} = \frac{1}{2k_2^2 D_2} - \frac{2k_1^2}{E_1 h_1} \tag{9.148a}$$

$$F_{21} = d_{c2} + d_{s2} = \frac{1}{2k_2^2 D_2} - \frac{2k_1^2}{E_1 h_1} \sin \phi_1 \tag{9.148b}$$

Thus we see that the contribution of the spherical shell solution causes some unsymmetry in [$\mathbf{F}$]. This may be somewhat disconcerting at first, but when we reflect on the nature of the approximate solution, the discrepancy may be explained.

Recalling from section 9.3.2.3 that the spherical segment has been represented, in effect, as an equivalent cylinder, we write

$$d_{s1} = -\frac{2k_1^2}{E_1 h_1} = -\frac{1}{2\tilde{k}_1^2 D_1} \tag{9.149a}$$

and

$$d_{s2} = -\frac{1}{2\tilde{k}_1^2 D_1} \sin \phi_1 \tag{9.149b}$$

where $\tilde{k}_1$ is defined in the same way as k for the cylindrical shell:

$$\tilde{k}_1^4 = \frac{3(1 - \mu_1^2)}{(a_1 h_1)^2} \tag{9.149c}$$

We recognize that the solutions for d_{s1} and d_{s2} as given in the above form are indeed solutions for equivalent cylindrical shells. Therefore, the discrepancy, $\sin \phi_1$, would not be reflected, since $\phi_1 = (\pi/2)$ for a cylindrical shell. Obviously, as $\phi_1 \to (\pi/2)$, the discrepancy diminishes. Moreover, the approximate solution was based on $\csc \phi \leq 1$, so that this trend is consistent. For the case of a hemispherical head with both shells having the same material properties and thicknesses, $a_1 = a_2$ and $F_{12} = F_{21} = 0$.

Once all of the terms in equation (9.143) are evaluated,

$$\begin{Bmatrix} \chi_1 \\ \chi_2 \end{Bmatrix} = - \begin{bmatrix} \mathbf{F}^{-1} \end{bmatrix} \begin{Bmatrix} \bar{\Delta}_1 \\ \bar{\Delta}_2 \end{Bmatrix} \tag{9.150}$$

in view of equation (9.145). After determining χ_1 and χ_2, the stress resultants, couples, and displacements in the cylindrical and spherical segments are found as linear combinations of the solutions for the loading cases shown in figure 9–11(b) and (c).

We shall not pursue further details of the calculations here, but refer to a numerical study in Flügge.[41] There, two cases are considered, a hemispherical head and a shallower cap. For both cases, $h_1 = h_2 = h$, $h/a =$

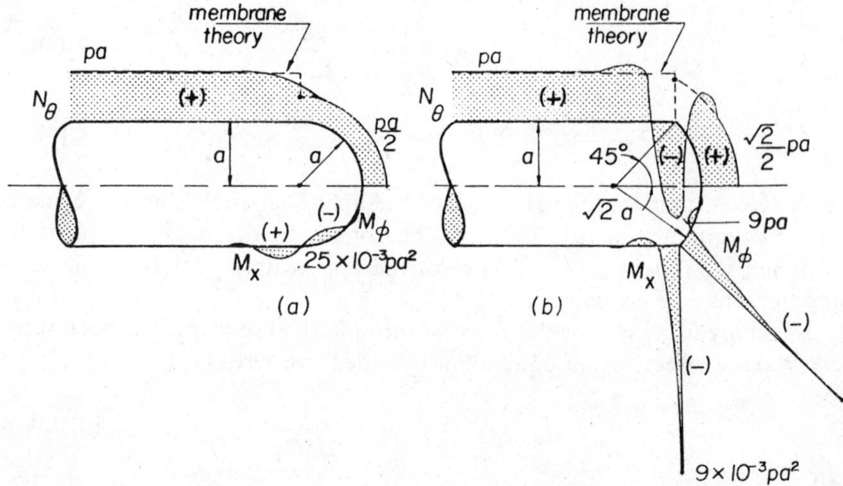

Figure 9–12. Stress Resultants and Couples for a Cylindrical Shell with a Spherical Head

0.01, and $\mu = 0.3$. In Flügge the shallow cap is chosen with $\phi_1 = 45°$, whereas we have studied $\phi_1 = 60°$ and $75°$ as well. On figure 9–12, comparative values of the meridional stress couple M_X or M_ϕ, and the circumferential or hoop stress resultant N_θ are shown for two cases.

The meridional moments increase greatly as ϕ_1 decreases. Similarly, for the circumferential stress resultant, the hemispherical head provides a smooth transition for the hoop stresses in the two segments, while the shallow caps result in increasing magnitudes of N_θ as well as a change in sense to compression, which gives rise to the possibility of circumferential buckling or wrinkling. Note that moderate discontinuities appear in the plotted values for the shallow cap cases at the intersection. These are attributed to the approximations contained in the spherical shell solution.

While a hemispherical cap might seem more attractive on the basis of this comparison, the fabrication advantages of shallow heads have been pointed out previously, and the practical solution is often to include a circumferential ring stiffener, as we see in figure 2–8(r). Details regarding the design of such stiffeners and other practical considerations are found in Pirok and Wozniak.[42] When an elastic stiffener is used, it is possible to generalize the model to include the stiffener, following the same reasoning we employed in section 9.2.2.7. A comprehensive study of ring stiffened cylindrical-conical shells under hydrostatic loading was performed by the author and his co-workers, and the results are reported in Gould et al.[43] and Wang and Gould.[44] In general, an adequate circumferential stiffener greatly moderates the extreme amplifications caused by the geometric discontinuity as shown in figure 9–12.[45]

Another possibility for reducing the stress amplification due to geometric discontinuities is the use of a transition segment such as a torospherical head, briefly mentioned in section 4.3.4 and illustrated in figure 7–16(b). Ranjan and Steele[46] have examined the range of applicability of approximate analytical solutions derived from asymptotic expansions of the dependent variables to the solution of torospherical shells. From their study, one can draw some conclusions as to the optimum radius of the toroidal knuckle as a function of the cylindrical and spherical radii and the shell thickness. Also, their study suggests a lower bound approximation for the critical internal pressure corresponding to circumferential wrinkling. Although they obtain fairly accurate solutions with their analytical formulae, based on comparisons with numerical and experimental results, it would appear that the best present technology for an actual design case would include a finite element analysis using doubly curved rotational shell elements. This technique is introduced in section 9.3.4 and is widely documented in the technical literature.

9.3.3. Asymmetrical Loading—Analytical Solutions

When the loading on a shell of revolution is not axisymmetric, the Fourier series technique that we employed extensively in chapter 4 is again useful to separate the independent variables. The problem then reduces to the solution of a set of ordinary differential equations for each harmonic j.

General solutions for spherical and conical shells are given by Flügge.[47] These solutions are based on the classical displacement method, and are exact in the sense that only quantities of $O(h^2/R^2):1$ are neglected. These solutions are of value in studying such cases as static wind loading and discrete column supports, which we discussed in sections 4.3.6.2 and 4.3.7.3., respectively.

For geometries other than spherical and conical and, of course, cylindrical (which we have treated separately), exact solutions are scarce.

As a general approach, the formulation of Novozhilov[48]; is very attractive. He follows the classical *force* method rather than the *displacement* procedure that we have employed almost exclusively in our study of plate bending in chapter 8 and of cylindrical shells in this chapter. We did use a force formulation for the membrane theory in chapter 4; but this involved only the equilibrium equations, since that problem is statically determinate. In the bending theory, however, the situation is more complicated, and it is necessary to express the strain-displacement or compatibility equations, (5.52) and (5.53), in terms of the stress resultants and couples. Once this is accomplished via the constitutive law, the resulting expressions are combined with the equilibrium relations to provide a consistent set of equations in which the stress resultants and couples are the unknowns. In this context, it is of interest to note that the H. Reissner–Meissner approach presented in section 9.3.2.1 is a combination of a *force* and a *displacement* formulation, since the dependent variables are Q_ϕ and $D_{\phi\theta}$. This is often termed a *mixed* formulation. We should also point out an interesting feature of the Novozhilov formulation of the shell bending problem, whereby the dependent variables are grouped into complex variables. This results in a halving of the order of the ultimate set of equations to be solved, and thereby simplifies the algebra. The solution of the ensuing equations is accomplished by the asymptotic integration technique, to which we have alluded previously. For shells of revolution, it generally is necessary to neglect terms of $O[j^2(h/R)]$, so that, for higher harmonics, the accuracy diminishes. A large number of practical problems can be described rather closely in terms of the lower harmonics—such as the wind loading shown on figure 4–28—so that the Novozhilov method has wide applicability. The author has extended this approach to include cases where terms of $O[j^2(h/R)]$ may be significant, such as in the column-supported boundary problem discussed in section 4.3.7.3. (See two articles by Gould and Lee.[49])

It is felt, however, that the most feasible approach to the general solution of shells of revolution under asymmetrical as well as symmetrical loading today is to use an energy-based numerical solution. We develop the required energy expressions for shells of revolution in the next section.

9.3.4 Energy Formulation

9.3.4.1 General Considerations. The total potential energy functional U_τ defined in section 7.4.2 is specialized here for a shell of revolution. It is convenient to consider an arbitrary segment of a shell bound by two parallel circles defined by the meridional angles ϕ_i and ϕ_{i+1}, as shown in figure 9–13. Here, the Z axis is oriented so that $Z = 0$ corresponds to the upper boundary of the shell. This segment may be a portion of a shell or even an entire shell. Also, ϕ_i may be 0°, which indicates a closed shell or dome. It is evident that the total potential energy for a shell composed of the assemblage of a number of such segments is the sum of the energies of all the segments. Therefore, the development, for the most part, may focus on a single general segment, such as that shown in figure 9–13. Of course, the segments adjoining system boundaries require some specialization to

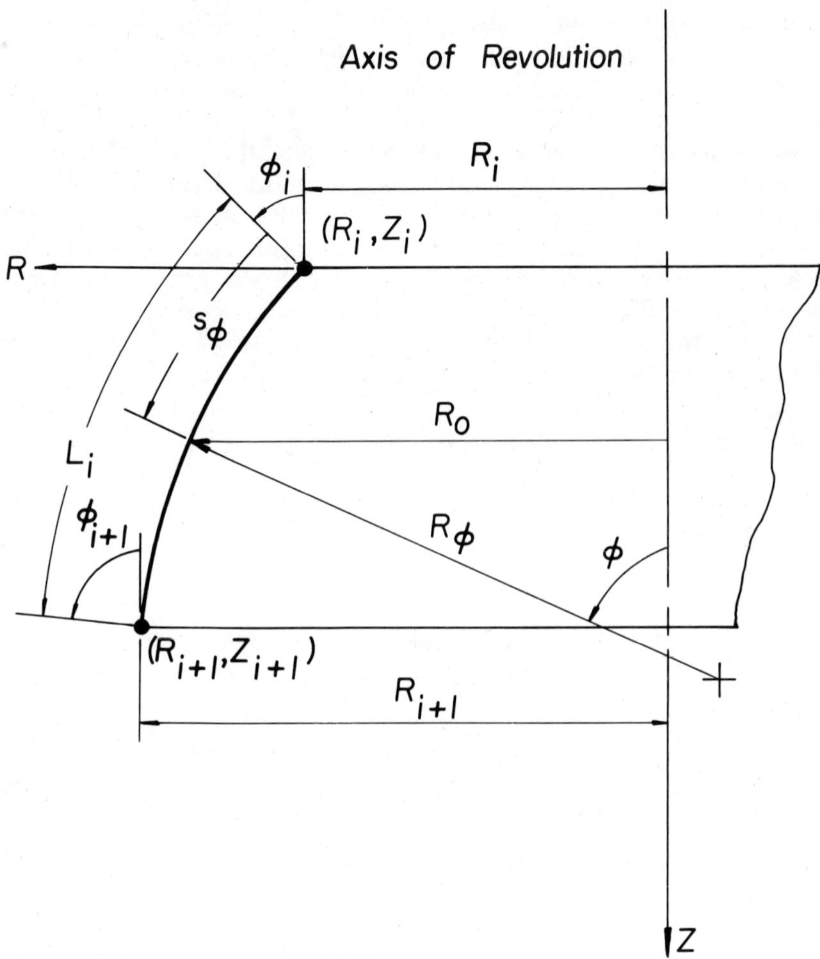

Source: L.J. Brombolich and P.L. Gould, "A High-Precision Curved Shell Finite Element," *AIAA Journal*, 10, no. 6 (June 1972): 727.

Figure 9–13. Shell of Revolution Geometry (Reprinted with permission of American Institute of Aeronautics and Astronautics).

accommodate prescribed boundary conditions. We also should remember that the stationary condition, stated previously as equation (7.14),

$$\delta U_\tau = 0 \qquad (9.151)$$

implies a global extremum and should be applied to the entire assembled shell. Nevertheless, the numerical techniques generally applied to this problem permit almost all of the calculations to be performed at the segment or local, as opposed to the global level.

9.3.4.2 Geometry. Recall from section 2.8.2 that three possibilities were presented for the meridional coordinate: the meridional angle ϕ, the axial coordinate Z, and the meridional arc length s_ϕ. In fact, the various computations that are required in an actual problem sometimes involve functions of two, or even all three of these possibilities. For this topic, it is convenient to use the $R-Z$ Cartesian coordinates to locate stations on the shell, but to take the arc length as the primary dependent variable. It is convenient to define for *each* segment a nondimensional arc length variable

$$s = \frac{s_\phi}{L_i} \qquad (0 \le s \le 1) \tag{9.152a}$$

where s_ϕ is measured from (R_i, Z_i), and where, from equation (2.46),

$$L_i = \int_{Z_i}^{Z_{i+1}} [1 + (R_{0,z})^2]^{1/2}\, dZ \tag{9.152b}$$

We also find it convenient to express the following geometric relations that are required in the formulation as functions of Z:[50]

$$R_0(Z) = R \tag{9.153a}$$

$$R_\phi(Z) = \frac{[1 + (R_{0,z})^2]^{3/2}}{R_{0,zz}} \tag{9.153b}$$

$$\sin \phi\,(Z) = [1 + (R_{0,z})^2]^{-1/2} \tag{9.153c}$$

$$\cos \phi\,(Z) = R_{0,z}[1 + (R_{0,z})^2]^{-1/2} \tag{9.153d}$$

$$R_\theta(Z) = \frac{R_0}{\sin \phi} \tag{9.153e}$$

9.3.4.3 Fourier Series Representation. As is customary for shells of revolution, all external loading, stress resultants and couples, strains and changes in curvature, and displacements and rotations are represented by Fourier series in the circumferential variable θ. These expressions are given as equations (4.84a) and (4.84b) for the loading and in-plane stress resultants, and by equations (5.50) and (5.51) for the strains and changes in curvature, and for the displacements and rotations, respectively. To make our discussion complete, we write the corresponding expressions for the bending stress resultants:

$$\begin{Bmatrix} Q_\phi \\ Q_\theta \\ M_\phi \\ M_\theta \\ M_{\phi\theta} \end{Bmatrix} = \sum_{j=0}^{\infty} \begin{Bmatrix} Q_\phi^j \cos j\theta \\ Q_\theta^j \sin j\theta \\ M_\phi^j \cos j\theta \\ M_\theta^j \cos j\theta \\ M_{\phi\theta}^j \sin j\theta \end{Bmatrix} \qquad (9.154)$$

We may then further restrict our development to typical harmonic j of segment i, for which the stress resultants and couples and the strains and changes in curvature are represented by vectors $\{N^j\}$ and $\{\epsilon^j\}$, respectively, as defined in matrix 6–2 and equation (6.10). Typical elements of these vectors are $N_\phi^j \cos j\theta$, $\omega^j \sin j\theta$, etc. Also, we have the displacement vector $\{\Delta_i^j\}$, defined in accordance with equation (5.51), as

$$\{\Delta_i^j\} = \{D_\phi^j \cos j\theta \; D_\theta^j \sin j\theta \; D_n^j \cos j\theta \; D_{\phi\theta}^j \cos j\theta \; D_{\theta\phi}^j \sin j\theta\} \qquad (9.155)$$

9.3.4.4 Strain-Displacement Relationships. If we anticipate a displacement formulation, we will have need for the strains and changes in curvature expressed as functions of the displacements and rotations. We have derived such compatibility relationships for harmonic j as equations (5.52) and (5.53). We may transform these equations to the nondimensional arc length variable s by noting, from figure 9–13, that

$$R_\phi \, d\phi = ds_\phi = L_i \, ds \qquad (9.156a)$$

from which we may express

$$( )_{,\phi} = ( )_{,s} \frac{ds}{d\phi} \qquad (9.156b)$$

as

$$( )_{,\phi} = ( )_{,s} \frac{R_\phi}{L_i} \qquad (9.157)$$

The resulting relations are conveniently written in the matrix form

$$\{\epsilon^j\} = [B_i^j]\{\Delta_i^j\} \qquad (9.158)$$

where $[B_i^j]$ is given by matrix 9–3.

Note that $[B_i^j]$ is an operator matrix and must always premultiply $\{\Delta_i^j\}$. Modified expressions for κ_ϕ^j, κ_θ^j, and τ^j, with transverse shearing strains suppressed, are easily written, but are not particularly advantageous in the formulation.

Matrix 9–3

$$[\mathbf{B}^j{}_i] = \begin{bmatrix}
\dfrac{1}{L_i}(\ )_{,s} & 0 & \dfrac{1}{R_\phi} & 0 & 0 \\[2ex]
\dfrac{\cos\phi}{R_0} & \dfrac{j}{R_0} & \dfrac{\sin\phi}{R_0} & 0 & 0 \\[2ex]
\dfrac{-j}{R_0} & \dfrac{1}{L_i}(\ )_{,s} - \dfrac{\cos\phi}{R_0} & 0 & 0 & 0 \\[2ex]
0 & 0 & 0 & \dfrac{1}{L_i}(\ )_{,s} & 0 \\[2ex]
0 & 0 & 0 & \dfrac{\cos\phi}{R_0} & \dfrac{j}{R_0} \\[2ex]
0 & 0 & 0 & \dfrac{-j}{2R_0} & \dfrac{1}{2L_i}(\ )_{,s} - \dfrac{\cos\phi}{2R_0} \\[2ex]
\dfrac{1}{R_\phi} & 0 & -\dfrac{1}{L_i}(\ )_{,s} & -1 & 0 \\[2ex]
0 & \dfrac{\sin\phi}{R_0} & \dfrac{j}{R_0} & 0 & -1
\end{bmatrix}$$

We also require a modification of the strain-displacement relationships if a closed shell is encountered. We have shown two cases in figure 4–1 and have recorded the pole condition for both possibilities in table 6–1. Here, the origin for the coordinate system is generally selected, so that $s = 0$ when $R_0 = 0$.

To establish the pole conditions and to suitably modify the strain-displacement relationships, we simply postulate that all strains and changes in curvature remain *finite* as $s \to 0$. Consider, for example, the expression for ϵ_θ^j, which we write from equation (9.158) and matrix 9–3 as

$$\epsilon_\theta^j = \frac{\cos \phi}{R_0} D_\phi^j + \frac{j}{R_0} D_\theta^j + \frac{\sin \phi}{R_0} D_n^j \tag{9.159}$$

For both shells shown in figure 4–1, $R_0(0) = 0$, so that for the first term to remain finite as $s \to 0$, $(\cos \phi \, D_\phi^j)_{s=0} = 0$, which creates an indeterminate form. The second term will remain finite only if $(jD_\theta^j)_{s=0} = 0$, and the third term requires that $(\sin \phi \, D_n^j)_{s=0} = 0$. Hence, we have the general pole condition

$$(\cos \phi \, D_\phi^j + jD_\theta^j + \sin \phi \, D_n^j)_{s=0} = 0 \tag{9.160}$$

for which we may investigate the individual pole conditions. We have six separate possibilities to study, consisting of two types of closed shells, $(\phi)_{s=0} = 0$ and $(\phi)_{s=0} = \phi_t$, for $j = 0$, $j = 1$, and $j > 1$. It is easiest to treat the harmonics individually.

$j = 0$. For this case, equation (9.160) reduces to

$$(\cos \phi \, D_\phi^j + \sin \phi \, D_n^j)_{s=0} = 0 \tag{9.161a}$$

If $(\phi)_{s=0} = 0$, the second term drops out, so that no restriction is needed on D_n^j; whereas if $(\phi)_{s=0} = \phi_t$,

$$D_\phi^j(0) = 0 \tag{9.161b}$$

as recorded in the second line, second column of table 6–1. Also, for the latter case, equation (9.161a) gives

$$D_\phi^j(0) = -\tan \phi_t \, D_n^j(0) \tag{9.161c}$$

as listed in the second line, third column of table 6–1.

$j = 1$. Here, equation (9.160) becomes

$$(\cos \phi \, D_\phi^j + D_\theta^j + \sin \phi \, D_n^j)_{s=0} = 0 \tag{9.162a}$$

If $(\phi)_{s=0} = 0$,

$$D_\phi^j(0) + D_\theta^j(0) = 0 \tag{9.162b}$$

as recorded in the third line, second column of table 6–1. Note that $D_n^j(0) = 0$ is also listed in the table. This arises out of consideration of γ_θ and is not required here—but neither is it contradicted. Now, if $(\phi)_{s=0} = \phi_t$, we have

$$(\cos \phi_t\, D_\phi^j + D_\theta^j + \sin \phi_t\, D_n^j)_{s=0} = 0 \qquad (9.163a)$$

Examining the corresponding entry in row 3, column 3 of table 6–1, we find

$$D_\phi(0) = -\cos \phi_t\, D_\theta(0)\,; \quad D_n(0) = -\sin \phi_t\, D_\theta(0) \qquad (9.163b)$$

which obviously cannot be completely obtained from equation (9.163a). However, equation (9.163b) does satisfy equation (9.163a), so that no contradiction is encountered.

$j > 1$. In this case,

$$D_\phi^j(0) = D_\theta^j(0) = D_n^j(0) = 0 \qquad (9.164)$$

as listed in line 4 of table 6–1 for both types of closed shells, obviously satisfies equation (9.160), but, as before, consideration of some of the other individual relationships are required to certify the correctness of these conditions.

Proceeding with similar arguments for the other strains that have singularities at $Z = 0$—i.e., ω^j, κ_θ^j, $\tau_{\theta\phi}^j$, and γ_θ^i—the remaining entries in table 6–1 may be verified.

Having isolated the necessary constraints on the displacements to avoid singularities, we now turn to the actual modified strain-displacement relationships. We again use ϵ_θ^j as given by equation (9.159) to illustrate the approach. In view of equation (9.160), equation (9.159) evaluated at $s = 0$ becomes

$$(\epsilon_\theta^j)_{s=0} = \left[\frac{\cos \phi\, D_\phi^j + jD_\theta^j + \sin \phi\, D_n^j}{R_0}\right]_{s=0} = \frac{0}{0} \qquad (9.165)$$

which is an indeterminate form. Using L'Hospital's rule, we may investigate the $\lim_{s \to 0} \epsilon_\theta^j$. It is convenient to differentiate with respect to ϕ, and then to convert to s using equation (9.157). Proceeding, we have

$$\lim_{s \to 0} \epsilon_\theta^j =$$

$$\left[\frac{-\sin \phi\, D_\phi^j(L_i/R_\phi) + \cos \phi\, D_{\phi,s}^j + jD_{\phi,s}^j}{R_\phi \cos \phi(L_i/R_\phi)}\right]_{s \to 0}$$

$$+ \left[\frac{\cos \phi\, D_n^j(L_i/R_\phi) + \sin \phi\, D_{n,s}^j}{R_\phi \cos \phi(L_i/R_\phi)}\right]_{s \to 0} \qquad (9.166)$$

$$= \left[\frac{1}{L_i}\, D_{\phi,s}^j + \frac{j}{L_i \cos \phi}\, D_{\theta,s}^j + \frac{1}{R_\phi}\, D_n^j - \frac{\tan \phi}{R_\phi}\, D_\phi^j + \frac{\tan \phi}{L_i}\, D_{n,s}^j\right]_{s \to 0}$$

For the more common closed shell, the dome, $\phi \to 0$ as $s \to 0$ and the last two terms drop out, giving

$$\epsilon_\theta^j(0) = \frac{1}{L_i} D_{\phi,s}^j(0) + \frac{j}{L_i} D_{\theta,s}^j(0) + \frac{1}{R_\phi(0)} D_n^j(0) \qquad (9.167)$$

In practice, when a dome is modeled by multiple segments, equation (9.167) is used for $\epsilon_\theta^j(s)$ in the uppermost segment where $i = 0$. To make this discussion complete, we present the entire set of modified strain-displacement equations for a *dome*:[51]

$$\{\epsilon_j^0\} = [\mathbf{B}_0^j] \{\mathbf{\Delta}_0^j\} \qquad (9.168)$$

where rows 2, 3, 5, 6, and 8 (corresponding to ϵ_θ, ω, κ_ϕ, τ, and γ_θ) of $[\mathbf{B}_i^j]$ are replaced by the modified equations, such as equation (9.167), as depicted in matrix 9–4.

Additionally, there are circumstances in which the pole cannot conveniently be located at $s = 0$, but coincides with $s = L_i$. For example, a shell that is closed at both ends will present such a situation at one end unless complete loading and geometric symmetry is present, whereupon only half the shell must be considered. Similar modified strain-displacement equations may be developed by enforcing the finite strain conditions at $s = L_i$ instead of $s = 0$.

An analogous set of equations can be developed for the case shown in figure 4–1(b) where $\phi = \phi_t$.

Thus we have the strain-displacement relationships given as a function of the nondimensional arc length variable s by matrix 9–3 and equation (9.158), augmented by special equations for pole segments.

9.3.4.5 Total Potential Energy Functional. U_T, as defined by equation (7.14), is the sum of U_ϵ, the strain energy, and U_q, the potential energy of the applied loads.

First, considering the strain energy U_ϵ, as given by equation (7.4), we may integrate through the thickness to obtain the strain energy for segment i in harmonic j:

$$U_{\epsilon i}^j = \frac{1}{2} \int_0^1 \int_{-\pi}^{\pi} [\mathbf{N}^j] \{\epsilon^j\} \, dA_i \qquad (9.169a)$$

where the differential surface area

$$dA_i = RL_i \, d\theta \, ds \qquad (9.169b)$$

We proceed with the displacement formulation, generalizing for specifically orthotropic shells, by replacing $\{\mathbf{N}^j\}$ with $\{\epsilon^j\}$ using equation (6.12):

Matrix 9–4

$$[\mathbf{B}_0^j] = \begin{bmatrix} \dfrac{1}{L_0}(\ )_{,s} & 0 & \dfrac{1}{R_\phi} & 0 & 0 \\[2ex] \dfrac{1}{L_0}(\ )_{,s} & \dfrac{j}{L_0}(\ )_{,s} & \dfrac{1}{R_\phi} & 0 & 0 \\[2ex] \dfrac{-j}{L_0}(\ )_{,s} & 0 & 0 & 0 & 0 \\[2ex] 0 & 0 & 0 & \dfrac{1}{L_0}(\ )_{,s} & 0 \\[2ex] 0 & 0 & 0 & \dfrac{1}{L_0}(\ )_{,s} & \dfrac{j}{L_0}(\ )_{,s} \\[2ex] 0 & 0 & 0 & \dfrac{-j}{2L_0}(\ )_{,s} & 0 \\[2ex] \dfrac{1}{R_\phi} & 0 & \dfrac{-1}{L_0}(\ )_{,s} & -1 & 0 \\[2ex] 0 & \dfrac{1}{R_\phi} & \dfrac{j}{L_0}(\ )_{,s} & 0 & -1 \end{bmatrix}$$

$$U^j_{\epsilon i} = \frac{1}{2} \int_0^1 \int_{-\pi}^{\pi} \left[[\mathbf{D}_{or}] \{\boldsymbol{\epsilon}^{\,j}\} - \{\mathbf{N}_{Tor}\} \right]^T \{\boldsymbol{\epsilon}^{\,j}\}\, dA_i \qquad (9.170a)$$

$$= \frac{1}{2} \int_0^1 \int_{-\pi}^{\pi} \left([\boldsymbol{\epsilon}^{\,j}][\mathbf{D}_{or}] \{\boldsymbol{\epsilon}^{\,j}\} - [\mathbf{N}_{Tor}]\{\boldsymbol{\epsilon}^{\,j}\} \right) dA_i \qquad (9.170b)$$

noting that $[\mathbf{D}_{or}]$ is symmetric. Finally, we replace the strains by the displacements using matrix 9–3 and equating (9.158) and get

$$U^j_{\epsilon i} = \frac{1}{2} \int_0^1 \int_{-\pi}^{\pi} \left(\left[[\mathbf{B}^{\,j}_i] \{\boldsymbol{\Delta}^{\,j}_i\} \right]^T [\mathbf{D}_{or}] \left[[\mathbf{B}^{\,j}_i] \{\boldsymbol{\Delta}^{\,j}_i\} \right] \right.$$
$$\left. - \lfloor \mathbf{N}_{Tor} \rfloor \left[[\mathbf{B}^{\,j}_i] \{\boldsymbol{\Delta}^{\,j}_i\} \right] \right) dA_i \qquad (9.171)$$

Now we turn to the potential energy of the applied loads U_q, as defined in equation (7.10). Of the three terms contained in the general expression equation (7.10a) that accommodate distributed, line, and concentrated loadings, only the first term and the second integral of the second term are pertinent here. Using Fourier series, line loads along the s_α or s_ϕ coordinate line, as represented by the first integral in the second term, may be included in the first term, and any concentrated loads, as represented by the third term, may be included in the second integral of the second term. Therefore, in terms of the present notation,

$$U^j_{qi} = - \int_0^1 \int_{-\pi}^{\pi} \lfloor \boldsymbol{\Delta}^{\,j}_i \rfloor \{\mathbf{q}^{\,j}_i\}\, dA_i$$
$$- \int_{-\pi}^{\pi} \lfloor \boldsymbol{\Delta}^{\,j}_i(Z_i) \rfloor \{\hat{\mathbf{q}}^{\,j}_i(Z_i)\}\, R(Z_i)\, d\theta$$
$$- \int_{-\pi}^{\pi} \lfloor \boldsymbol{\Delta}^{\,j}_i(Z_{i+1}) \rfloor \{\hat{\mathbf{q}}^{\,j}_i(Z_{i+1})\}\, R(Z_{i+1})\, d\theta \qquad (9.172)$$

where $\{\mathbf{q}^{\,j}_i\}$ $\{\boldsymbol{\Delta}^{\,j}(Z_i)\}$, and $\{\hat{\mathbf{q}}^{\,j}_i(Z_i)\}$ are defined corresponding to equation (7.10a). Note that in a numerical calculation, the second and the third terms of equation (9.172) would not ordinarily both appear. The second term might be included with segment $i - 1$, or the third with segment $i + 1$, depending on the computational format.

For the entire shell,

$$U^j_T = \sum_{i=1}^{p} (U^j_{\epsilon i} + U^j_{qi}) \qquad (9.173)$$

where the shell is assumed to be comprised of p segments. The Principle of Minimum Total Potential Energy [Equation (7.14)] gives the equilibrium condition

$$\delta U^{j}_{\tau} = 0 \tag{9.174}$$

For asymmetric loading, the complete solution is composed of the superposition of the solutions to equation (9.174) for the relevant participating harmonics.

9.3.4.6 Rayleigh–Ritz Solution.

The Rayleigh–Ritz method was presented in section 7.4.3 as a promising numerical approach to the solution of equations such as equation (9.174). In the context of the present development for a single general harmonic j, the vector $\{\Delta\}$, as intrcduced in equation (7.15), is the assemblage of the vectors $\{\Delta^{j}_{i}\}$ for all ot the segments comprising the shell. Recall that $\{\Delta^{j}_{i}\}$ was defined in equation (9.155) as the vector of displacements for segment i. The index m introduced in equation (7.15) is equal to 5 in this case, since there are five displacement components in $\{\Delta^{j}_{i}\}$ except, of course, for the $j = 0$ case, when D_{θ} and $D_{\theta\phi} = 0$ and m is reduced to 3. But, for the general case, there will be five so-called *comparison or trial* functions of the form of equation (7.16) for each of the p segments of the shell. Polynomials of the form

$$\Delta^{j}_{ik} = c^{j}_{iko} + c^{j}_{ik1}s + \ldots c^{j}_{ikl}s^{l} \ldots + c^{j}_{ikn}s^{n} \quad \begin{pmatrix} k = 1, 5 \\ i = 1, p \end{pmatrix} \tag{9.175}$$

where $\Delta^{j}_{i1} = D^{j}_{\phi}$, $\Delta^{j}_{i2} = D^{j}_{\theta}$, $\Delta^{j}_{i3} = D^{j}_{n}$, $\Delta^{j}_{i4} = D^{j}_{\phi\theta}$, and $\Delta^{j}_{i5} = D^{j}_{\theta\phi}$ are quite popular for this application. It is important to note that n, the order of the approximation for each Δ^{j}_{ik}, may vary for each displacement k, segment i, and harmonic j.

Upon substitution of the set of $5p$ comparison functions of the form of equation (9.175) for $\{\Delta^{j}_{i}\}$ into equations (9.171) and (9.172), and the operation by $[\mathbf{B}^{j}_{i}]$ on the polynomials, the extremum problem represented by equation (9.174), is transformed into the maximum–minimum problem of the calculus

$$U^{j}_{\tau,c^{j}_{ikl}} = 0 \quad \begin{pmatrix} l = 0,n \\ k = 1,5 \\ i = 1,p \end{pmatrix} \tag{9.176}$$

The maximum–minimum conditions, equation (9.176), are necessary, but not sufficient, to determine the n coefficients (c^{j}_{kl}) that are present for each of the five comparison functions Δ^{j}_{ik} in each segment $i = 1, p$. In addition to equation (9.176), there are *intersegment continuity conditions*, which recognize that the displacements are continuous along the common boundary of adjacent segments. Also, *kinematic boundary conditions* may be prescribed on the external boundaries that fix the values of the corresponding displacements at those points. Thus, in summary, the maximum–minimum conditions, the intersegment continuity conditions, and the kinematic boundary conditions comprise the necessary and sufficient conditions to approximately satisfy the Principle of Minimum Total Potential Energy. The further development of this approach is beyond our objectives here and is left to the province of the finite element method.[52]

One further point should be noted, however. Static boundary conditions are also encountered. These conditions enter as the *natural boundary conditions* of the variational problem; this means, essentially, that if the kinematic correspondent of a certain force or moment is not prescribed on the boundary, that force or moment will take the value of the corresponding external loading component along the boundary. As an example, consider a typical stress resultant N_ϕ on a given external boundary $\phi = \phi'$. If the displacement $D_\phi(\phi')$ is not prescribed *a priori*, along that boundary, then $N_\phi(\phi')$ is equal to the corresponding element of the external loading vector $\hat{q}_\phi(\phi')$. If $\hat{q}_\phi(\phi') = 0$, then $N_\phi(\phi') = 0$.

We have outlined a quite general, energy-based approach that may be developed and extended to solve many complex problems in the field of shells of revolution on a fairly routine, automated basis.

9.4 Shells of Translation

9.4.1 General

The general theory of shells of translation, apart from cylindrical shells, is lightly treated in contemporary English language textbooks. Although there are many papers in the literature that describe analysis methods for such shells, most are based on idealized boundary conditions that do not necessarily match the physical situation very closely. For example, the inclusion of elastic edge members, which are often present in reality, greatly complicates the analysis. From a practical standpoint, however, numerous translational shells having spectacularly large spans have been constructed. As an example, we may take the shell shown in figure 2–8(a). Such major structures certainly warrant a complete investigation, beyond the scope of the membrane theory. The means for carrying out such investigations are often computer-based numerical solutions, such as finite element or finite difference techniques, or carefully executed experimental studies using physical models. The current availability and capabilities of numerical analysis procedures for translational shells is described in Schnobrich.[53] Here, we introduce two theories upon which many of the currently popular numerical algorithms are founded.

9.4.2 Mushtari–Donnell–Vlasov (MDV) Theory

This theory is a simplification of the general theory of shells in orthogonal curvilinear coordinates and is based on the assumptions that (a) the effect of the transverse shearing forces Q_α and Q_β in the in-plane equilibrium equations is negligible; and (b) the influence of the normal displacement D_n will predominate over the influences of the in-plane displacements D_α and D_β in the bending response of the shell. We have discussed the first assumption at length in connection with other derivations, and it should be acceptable provided the membrane theory boundary conditions are not grossly violated. The second assumption is most feasible as the shell becomes relatively flatter, approaching a plate. Such a shell is said to be *shallow*. Shallow shells were introduced in section 2.8.1, and we build on this geometric concept in the following paragraphs. Some quantitative criteria for a shell to be classified as shallow are also stated later.

Proceeding, we consider first the equilibrium equations in general form, equations (3.17). We omit the transverse shearing stress terms as stated in assumption (a) and take $N_{\alpha\beta} = N_{\beta\alpha} = S$:

$$(BN_\alpha)_{,\alpha} + (AS)_{,\beta} + A_{,\beta}S - B_{,\alpha}N_\beta + q_\alpha AB = 0 \qquad (9.177a)$$

$$(BS)_{,\alpha} + (AN_\beta)_{,\beta} + B_{,\alpha}S - A_{,\beta}N_\alpha + q_\beta AB = 0 \qquad (9.177b)$$

$$(BQ_\alpha)_{,\alpha} + (AQ_\beta)_{,\beta} - N_\alpha\frac{AB}{R_\alpha} - N_\beta\frac{AB}{R_\beta} + q_n AB = 0 \qquad (9.177c)$$

We first wish to express the transverse shearing forces in terms of the normal displacement D_n. With transverse shearing strains suppressed, Q_α and Q_β are given in terms of the bending stress resultants through equations (3.22a) and (3.22b):

$$Q_\beta = \frac{1}{AB}\left[(BM_{\alpha\beta})_{,\alpha} + (AM_\beta)_{,\beta} + B_{,\alpha}M_{\beta\alpha} - A_{,\beta}M_\alpha\right] \qquad (9.178a)$$

$$Q_\alpha = \frac{1}{AB}\left[(BM_\alpha)_{,\alpha} + (AM_{\beta\alpha})_{,\beta} + A_{,\beta}M_{\alpha\beta} - B_{,\alpha}M_\beta\right] \qquad (9.178b)$$

Next, we apply our constitutive law matrix 6–2, omitting the thermal terms, and take $M_{\alpha\beta} = M_{\beta\alpha}$ to write the stress couples as functions of κ_α, κ_β, and τ:

$$M_\alpha = D(\kappa_\alpha + \mu\kappa_\beta) \tag{9.179a}$$

$$M_\beta = D(\mu\kappa_\alpha + \kappa_\beta) \tag{9.179b}$$

$$M_{\alpha\beta} = D(1 - \mu)\tau \tag{9.179c}$$

Before substituting equations (9.179a–c) into (9.178a) and (9.178b), we express the changes in curvature as functions of the displacements. These relationships are given, in general, by equations (5.47a–c), but we invoke assumption (b) and eliminate the contributions of D_α and D_β. The simplified compatibility equations are

$$\kappa_\alpha = -\frac{1}{A}\left(\frac{1}{A}\,D_{n,\alpha}\right)_{,\alpha} - \frac{1}{AB^2}\,A_{,\beta}\,D_{n,\beta} \tag{9.180a}$$

$$\kappa_\beta = -\frac{1}{B}\left(\frac{1}{B}\,D_{n,\beta}\right)_{,\beta} - \frac{1}{A^2B}\,B_{,\alpha}\,D_{n,\alpha} \tag{9.180b}$$

$$\tau = -\frac{1}{AB}\left(D_{n,\alpha\beta} - \frac{1}{A}\,A_{,\beta}D_{n,\alpha} - \frac{1}{B}\,B_{,\alpha}\,D_{n,\beta}\right) \tag{9.180c}$$

The next step is to substitute equations (9.180a–c) into equations (9.179a–c) and then into equations (9.178a) and (9.178b). The algebra is considerable, and we defer to Novozhilov[54] to write

$$Q_\alpha = -D\frac{1}{A}(\nabla^2 D_n)_{,\alpha} \tag{9.181a}$$

$$Q_\beta = -D\frac{1}{B}(\nabla^2 D_n)_{,\beta} \tag{9.181b}$$

where the Laplacian, in general terms, is given by

$$\nabla^2(\ ) = \frac{1}{AB}\left\{\left[\frac{B}{A}(\ )_{,\alpha}\right]_{,\alpha} + \left[\frac{A}{B}(\ )_{,\beta}\right]_{,\beta}\right\} \tag{9.181c}$$

Finally, substituting equations (9.181a) and (9.181b) into equation (9.177c) and dividing through by AB, we find

$$\frac{-D}{AB}\left\{\left[\frac{B}{A}(\nabla^2 D_n)_{,\alpha}\right]_{,\alpha} + \left[\frac{A}{B}(\nabla^2 D_n)_{,\beta}\right]_{,\beta}\right\} - \frac{N_\alpha}{R_\alpha} - \frac{N_\beta}{R_\beta} + q_n = 0 \tag{9.182}$$

or, in view of equation (9.181c),

$$D\nabla^4 D_n + \frac{N_\alpha}{R_\alpha} + \frac{N_\beta}{R_\beta} - q_n = 0 \tag{9.183}$$

which we recognize as being very similar to the equilibrium equation for plates. Thus we have equations (9.177a) and (9.177b) together with equations (9.183) and the requisite boundary conditions constituting the shell theory which was apparently derived independently by Mushtari in the Soviet Union and Donnell in the United States.[55]

The governing equations at this stage of the development are somewhat similar to those encountered in section 8.3.3 following the derivation of equation (8.52), which, together with equation (8.35) formed a coupled set of equations for the finite displacement theory of plates. The dependent variables in those equations were the *normal displacement* and the *in-plane stress resultants*, which is essentially the same situation we now have with equations (9.177a) and (9.177b) and equation (9.183). Therefore, the introduction of a stress function F, defined such that

$$N_\alpha = \frac{1}{B}\left(\frac{1}{B}F_{,\beta}\right)_{,\beta} + \frac{1}{AB}B_{,\alpha}\frac{1}{A}F_{,\alpha} \tag{9.184a}$$

$$N_\beta = \frac{1}{A}\left(\frac{1}{A}F_{,\alpha}\right)_{,\alpha} + \frac{1}{AB}A_{,\beta}\frac{1}{B}F_{,\beta} \tag{9.184b}$$

$$S = -\frac{1}{AB}\left(F_{,\alpha\beta} - \frac{1}{A}A_{,\beta}F_{,\alpha} - \frac{1}{B}B_{,\alpha}F_{,\beta}\right) \tag{9.184c}$$

is quite logical.[56]

The details of the back substitution are tedious and basically are similar to our previous exercises in the development of the von Kármán equations in section 8.3.3. We omit the details and follow the procedure suggested by Vlasov.[57] Also, we consider the special case where the in-plane components of the surface loading q_α and q_β are equal to 0. Thus, we have

$$\frac{1}{A^2B}\left[\left(\frac{1}{A}B_{,\alpha}\right)_{,\alpha} + \left(\frac{1}{B}A_{,\beta}\right)_{,\beta}\right]F_{,\alpha} = -\frac{1}{R_\alpha R_\beta}\frac{1}{A}F_{,\alpha} = 0 \tag{9.185a}$$

$$\frac{1}{AB^2}\left[\left(\frac{1}{A}B_{,\alpha}\right)_{,\alpha} + \left(\frac{1}{B}A_{,\beta}\right)_{,\beta}\right]F_{,\beta} = -\frac{1}{R_\alpha R_\beta}\frac{1}{B}F_{,\beta} = 0 \tag{9.185b}$$

and

$$D\nabla^4 D_n + \bar{\nabla}^2 F - q_n = 0 \tag{9.186a}$$

where

$$\bar{\nabla}^2(\ ) = \frac{1}{AB}\left\{\left[\frac{1}{R_\beta}\frac{B}{A}(\ )_{,\alpha}\right]_{,\alpha} + \left[\frac{1}{R_\alpha}\frac{A}{B}(\ )_{,\beta}\right]_{,\beta}\right\} \tag{9.186b}$$

The equalities in equations (9.185a) and (9.185b) have been obtained by invoking equation (2.37), the third Gauss–Codazzi relationship, while the second term in equation (9.186a) is expressed in terms of the operator $\bar{\nabla}^2(\ )$ by using the first two Gauss–Codazzi relations, equations (2.35) and (2.36).

It may now be argued that for certain classes of problems, equations (9.185a) and (9.185b) may be considered to be satisfied. First, for shells of zero Gaussian curvature, such as cylindrical or conical shells, equations (9.185a) and (9.185b) are *identically* satisfied. Also, if we think of shells that are relatively flat or shallow, the Gaussian curvature will be large and the coefficients of $F,_\alpha$ and $F,_\beta$ in equations (9.185a) and (9.185b) will be an order of magnitude smaller than the coefficients of the same terms in equations (9.186a) and (9.186b), so the former equations can be regarded to be *approximately* satisfied. A third class of problems for which the first two equations may be ignored are those cases in which the stress resultants are rapidly varying functions of α and β. Then, the higher derivatives of F predominate, and the contribution of the terms associated with $F,_\alpha$ and $F,_\beta$ will be relatively small as compared to those multiplying $F,_{\alpha\alpha}$ and $F,_{\beta\beta}$. Similar reasoning was used to motivate the Geckeler approximation in section 9.3.2.3.

We now seek a second relationship between D_n and F. This accomplishment is attributed to Vlasov by Novozhilov[58] and requires two general steps.

The first operation is the derivation of the compatibility equations for the deformation of the middle surface in terms of strains and curvatures only. These equations are analogous to the St. Venant equations of the Theory of Elasticity, and may be established following the same steps employed in the derivation of the Gauss–Codazzi relations in section 2.6 if the tangent vectors to the *deformed* middle surface are used in place of the tangent vectors to the *undeformed* middle surface. Specifically, $\mathbf{t}_{n,\alpha}$ and $\mathbf{t}_{n,\beta}$ in equation (2.33) are replaced by $\mathbf{r}',_\alpha$ and $\mathbf{r}',_\beta$, as given by equations (5.9) and (5.11), respectively. Then the identical steps are carried out on these tangent vectors to derive the compatibility equations, which may be viewed as Gauss–Codazzi relations for the *deformed* middle surface.

The second operation required is the substitution of equations (9.180a–c) for the changes in curvature, the replacement of the in-plane strains by the corresponding stress resultants using matrix 6–2, and the subsequent introduction of the stress function F via equations (9.184a–c).

The final result, after eliminating terms similar to those previously dropped in equations (9.185a) and (9.185b), is

$$\frac{1}{Eh}\,\nabla^4 F - \bar{\nabla}^2 D_n = 0 \qquad (9.187)$$

Thus, the solution of the coupled equations (9.186a) and (9.187) is sufficient to describe the complete pattern of stress and deformation within the limitations of the approximations introduced in the course of the derivation. Novozhilov observed that these equations form the basis for the investigation of many practical problems.[59] Also, explicit solutions are provided by Vlasov[60] based on some further approximations: (a) the shell is described by Cartesian coordinates, with

$$A = B \cong 1 \tag{9.188}$$

and (b) R_α and R_β are taken as constant (average values) and are given by

$$R_\alpha(X,Y) = \bar{R}_X \tag{9.189}$$

$$R_\beta(X, Y) = \bar{R}_Y \tag{9.190}$$

These assumptions are specifically tied to the shallow shell notion and, according to Vlasov, can be utilized for shells that cover rectangular plan areas if the rise is no more than 1/5 of the smaller side of the rectangle.

The introduction of the simplifications given in equations (9.188)–(9.190) into the previous equations reduces the operators $\nabla^2(\ )$ and $\bar{\nabla}^2(\ )$ to

$$\nabla^2(\ ) = (\ )_{,XX} + (\ )_{,YY} \tag{9.191}$$

and

$$\bar{\nabla}^2(\ ) = -\frac{1}{\bar{R}_Y}(\ )_{,XX} - \frac{1}{\bar{R}_X}(\ )_{,YY} \tag{9.192}$$

where $\bar{R}_Y$ and $\bar{R}_X$ are average values of the radii of curvature along the respective coordinate lines. The solution then proceeds with the introduction of a further potential function G, such that

$$D_n = \nabla^4 G \tag{9.193a}$$

and

$$F = Eh\bar{\nabla}^2 G \tag{9.193b}$$

When introduced into equation (9.186a), equations (9.193) give[61]

$$\nabla^8 G + \frac{12(1 - \mu^2)}{h^2} \bar{\nabla}^4 G - \frac{q_n}{D} = 0 \tag{9.194}$$

There is a marked similarity between equation (9.194) and equation (9.75), which governs the general bending of cylindrical shells, and the solution strategies are similar. Once G is found, F is determined from equation (9.193b). The interested reader is referred to Vlasov[62] for the complete development of this solution. As an example, the bending of an elliptic paraboloid shell of a similar configuration to that shown on figure 4–41 can be investigated using equation (9.194). Also, these equations can be readily generalized to the study of the instability of certain shells, as we see later.

9.4.3 Theory of Shallow Shells

Another theory, roughly equivalent to the simplification of the MDV theory facilitated by equations (9.189) and (9.190), is widely used in shell analysis. The approach is to employ some simplifying assumptions at the outset to derive the governing equations in a form which is particularly suited for shell roofs covering rectangular plan areas. Although the theory is commonly termed the Theory of Shallow Shells, it has been used to analyze shells that are only locally shallow when the original shell is divided into finite segments or elements—a procedure to which we have often alluded in our previous discussions.

Refer to the basic geometric relationships introduced in section 2.8.1. Recalling the assumption that $(Z_{,X})^2$ and $(Z_{,Y})^2$ may be neglected in comparison to unity, we may evaluate the principal radii of curvature from equations (2.29a) and (2.29b) as

$$R_Y = -\frac{1}{Z_{,YY}} \tag{9.195a}$$

and

$$R_X = -\frac{1}{Z_{,XX}} \tag{9.195b}$$

The expressions for the in-plane strains ϵ_X and ϵ_Y are obtained directly from equations (5.46a) and (5.46b) as

$$\epsilon_X = D_{X,X} - \frac{1}{Z_{,XX}} D_n \tag{9.196a}$$

and

$$\epsilon_Y = D_{Y,Y} - \frac{1}{Z_{,YY}} D_n \tag{9.196c}$$

The equation for the in-plane shearing strain could similarly be written from equation (5.46c). However, it is more correct to use a somewhat different form that arises because the Cartesian coordinates are not orthogonal when projected on the surface. The expanded form of ω is[63]

$$\omega = D_{Y,X} + D_{X,Y} - 2Z_{,XY}D_n \tag{9.197}$$

The term $Z_{,XY} = -(1/R_{XY})$ may be called the twisting curvature, which we have encountered previously in section 4.4.2 when planar X–Y coordinates were also used. Additionally, we have the expressions for κ_α, κ_β, and τ, which, with $A = B = 1$, reduce to

$$\kappa_X = -D_{n,XX} \tag{9.198a}$$

$$\kappa_X = -D_{n,YY} \tag{9.198b}$$

$$\tau = -D_{n,XY} \tag{9.198c}$$

Equations (9.196)–(9.198) are the compatibility equations for this theory.

We now proceed as in the previous section. The analogue to the normal equilibrium equation, (9.186a) and (9.186b), is[64]

$$D\nabla^4 D_n - (Z_{,YY}F_{,XX} - 2Z_{,XY}F_{,XY} + Z_{,XX}F_{,YY}) - q_n = 0 \tag{9.199}$$

Where the stress function is related to the stress resultants by the simplified form of equations (9.194a–c),

$$N_X = F_{,YY} \tag{9.200a}$$

$$N_Y = F_{,XX} \tag{9.200b}$$

$$S = -F_{,XY} \tag{9.200c}$$

Similarly, an analogous equation to equation (9.187) is developed. This equation is written as[65]

$$\frac{1}{Eh}\nabla^4 F + (Z_{,YY}D_{n,XX} - 2Z_{,XY}D_{n,XY} + Z_{,XX}D_{n,YY}) = EhP \tag{9.201a}$$

where P is derived from the applied surface loading as

$$P = (1 + \mu)(P_{X,YY} + P_{Y,XX}) - \mu\nabla^2(P_X + P_Y) \tag{9.201b}$$

$$P_X(X, Y) = \int q_X \, dX \tag{9.201c}$$

$$P_Y(X, Y) = \int q_Y \, dY \tag{9.201d}$$

and

$$\nabla^2(\ ) = (\ )_{,XX} + (\ )_{,YY} \tag{9.201e}$$

P_X and P_Y in equations (9.201c) and (9.201d) are cumulative in-plane forces and are set to 0 on boundaries that are unrestrained in the X and Y directions, respectively. Note that this theory is slightly complicated by starting with nonorthogonal coordinates, but retains the generality of including the in-plane components of the surface loading q_X and q_Y. For most shells encountered in practice, the MDV theory and the Theory of Shallow Shells are fairly equivalent. From an applications standpoint, the most significant difference is probably in the nature of the external boundaries. If the boundaries are coincident with the coordinate lines of the middle surface, the general form of the Mushtari–Donnell–Vlasov theory is appropriate; if the boundaries fall along the Cartesian coordinate lines, the Theory of Shallow Shells or the even simpler specialization of the MDV theory seems to be more attractive.

9.4.4 Strain Energy for Shallow Shells

The energy expressions and principles derived in chapter 7 are directly adaptable to shells of translation, but we have not carried out the details of the specializations based on the theories presented in the last two sections. However, it is of general interest to examine the expression for strain energy, equation (7.6), using the simplified strain-displacement relationship, equations (9.196)–(9.198). With the transverse shearing strains neglected, term [I] of equation (7.7) becomes

$$[\mathrm{I}] = \epsilon_X^2 + \epsilon_Y^2 + 2\mu\epsilon_X\epsilon_Y + \frac{1-\mu}{2}\omega^2$$

$$= \left(D_{X,X} - \frac{1}{Z_{,XX}}\,D_n\right)^2 + \left(D_{Y,Y} - \frac{1}{Z_{,YY}}\,D_n\right)^2$$

$$+ 2\mu\left(D_{X,X} - \frac{1}{Z_{,XX}}\,D_n\right)\left(D_{Y,Y} - \frac{1}{Z_{,YY}}\,D_n\right)$$

$$+ \frac{1-\mu}{2}\,(D_{Y,X} + D_{X,Y} - 2Z_{,XY}D_n)^2 \tag{9.202}$$

Term [II] is dropped in the linear theory, while

$$[\mathrm{III}] = (\kappa_X + \kappa_Y)^2 - 2(1-\mu)(\kappa_X\kappa_Y - \tau^2)$$

$$= (D_{n,XX} + D_{n,YY})^2 - 2(1-\mu)(D_{n,XX}D_{n,YY} - D_{n,XY}^2) \tag{9.203}$$

The resulting expression,

$$U_\epsilon = \frac{E}{2(1-\mu^2)} \int_S \left\{ h[\mathrm{I}] + \frac{h^3}{12}\,[\mathrm{III}] \right\} dX\,dY \tag{9.204}$$

is frequently encountered in the literature. The thermal terms, as defined in equation (7.6), can easily be added if required.

9.5 Instability and Finite Deformations

9.5.1 General

Local instabilities characterized by waves of comparatively small wave length, similar to those occurring in plate structures, are of interest in many thin-walled vessels where compressive stresses are present. This type of instability is delineated from general instability, which could correspond to the buckling of an entire stack or tower as a beam-column. Since the amplitudes encountered in local instabilities are comparatively small, this phenomenon can be analyzed in many cases by using the assumption of the shallow shell theories presented in the previous sections.

There are several possible approaches to the *elastic* instability analysis of shells. These are illustrated schematically on figure 9–14, where we have plotted the normal displacement D_n against a loading parameter $\bar{N}$, which represents some combination of the in-plane loading.

First, if we can arrive at a general solution for the normal displacement in the presence of transverse and in-plane loading, and then observe that certain discrete values of $\bar{N}$ cause the displacements to become excessively large, we term these values of $\bar{N}$ the critical loads $\bar{N}_{cr}$. This is illustrated by the curves labeled *1* in figure 9–14, with the different branches indicating different values of the known transverse loading. This approach is widely used in the study of plate instability. Note that the same $\bar{N}_{cr}$ is obtained regardless of the transverse loading, which is q_n for shells.

The second approach follows from the realization that $\bar{N}_{cr}$ is independent of q_n. This suggests that the transverse loading can be ignored in the stability analysis, which means that for $\bar{N} < \bar{N}_{cr}$, $D_n = 0$ as indicated by curve 2, which is the special case of *1* for $q_n = 0$. Then at $N = \bar{N}_{cr}$, the displacements become unbounded as before. This approach is attractive because the complete solution for D_n as a function of q_n and $\bar{N}$ is *not* required. Rather, the states of equilibrium just prior to and just after instability can be examined without concern for the remainder of the load deflection curve. At the transition point $(\bar{N}_{cr}, 0)$, a *bifurcation* of equilibrium is said to occur, since there are at least two possible states of equilibrium, $D_n = 0$ and $D_n \to \infty$. This approach is of course familiar from the classical solution of the Euler column buckling problem, and we employ it here as well. Although it is implied by curve 2 that the equilibrium path prior to buckling is linear, bifurcation buckling of shells may also follow a nonlinear prebuckling path.[66]

A third approach which should be mentioned is that of using *geometrically nonlinear* strain-displacement relationships to determine the complete load-deflection curve—unloading as well as loading. As illustrated by curve *3*, a critical load $\bar{N}_{cr(nl)}$, which occurs at a limit point, may be found somewhat below $\bar{N}_{cr}$. A classical illustration is an axially com-

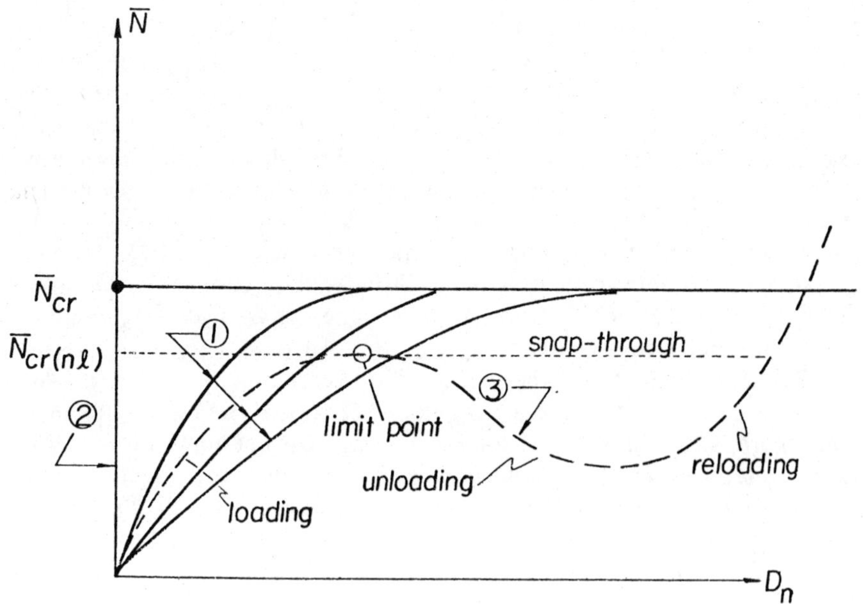

Figure 9–14. Types of Shell Instability

pressed conical shell, such as that shown in figure 4–13, where the axial shortening and accompanying rotational deformation would cause an immediate degradation of the stiffness.[67] Then the curve would show unloading and perhaps reloading as other elements of resistance are mobilized by the finite deformations. If this latter resistance is adequate, the load may eventually surpass $\bar{N}_{cr(nl)}$ and even $\bar{N}_{cr}$, provided the material capacity is not exceeded. An actual shell cannot unload in most cases, so there may be a sudden transition from the *loading* to the *reloading* branches of the curve at $\bar{N} = \bar{N}_{cr(nl)}$, with a corresponding jump in D_n. This is termed *snap-through* buckling, and is characteristic of initially curved structures but beyond our scope here.

Note that all three of the general approaches discussed have been predicated on elastic buckling. If material nonlinearities are present for $\bar{N} \leq \bar{N}_{cr}$, then additional complications arise. Sometimes these can be handled conveniently by the tangent modulus theory.

9.5.2 Equilibrium Method

As we mentioned in our discussion of the bifurcation of equilibrium approach to elastic instability in the previous section, we need only consider the immediate neighborhood of the transition point $(\bar{N}_{cr}, 0)$. Up to this point, the shell will be regarded as being subject to a membrane state of stress N_{X0}, N_{Y0}, and S_0.[68] At the transition point, the greatly increased normal displacements result in the development of *normal* components of these in-plane stress resultants. These normal components may be treated as surface loading in the t_n direction, since the starting point for the analysis is the state of stress N_{X0}, N_{Y0}, and S_0. With this in mind, the normal components of the membrane stress resultants are used for q_n in equations (9.183), (9.186a) and (9.186b), if the MDV theory is used; in equations (9.193) and (9.194) if the simplified version of the MDV theory is used; and in equations (9.199) and (9.201a–e) if we choose the Theory of Shallow Shells. It now remains to determine the normal components of the membrane stress resultants.

Refer to figure 8–4, where the normal components of the in-plane stress resultants for a plate are apparent. (The analysis corresponding to this figure is contained in section 8.3.1.) Since the net normal force due to any single stress resultant is the *difference* of the normal projections at the two ends of the differential segment, this analysis is valid for shells as well. Following equation (8.33), with the transverse shearing forces and in-plane loads omitted, we have for Cartesian coordinates

$$q_{n1} = -N_{X0}D_{XY,X} - N_{Y0}D_{YX,Y} - N_{XY0}(D_{YX,X} + D_{XY,Y}) \qquad (9.205)$$

With transverse shearing strains suppressed and $N_{XY0} = S_0$, and with D_X and D_Y neglected in comparison with D_n in the Cartesian coordinate version of equations (5.46g) and (5.46h) in the spirit of the shallow shell assumptions, we have

$$\left. \begin{array}{c} D_{XY} = -D_{n,X} \\[1em] D_{YX} = -D_{n,Y} \end{array} \right\} \qquad (9.206)$$

so that

$$q_{n1} = N_{X0}D_{n,XX} + N_{Y0}D_{n,YY} + 2S_0D_{n,XY} \qquad (9.207a)$$

which is sometimes written as

$$q_{n1} = -[N_{X0}\kappa_X + N_{Y0}\kappa_Y + 2S_0\tau] \qquad (9.207b)$$

in view of equation (9.198). For orthogonal curvilinear coordinates, the same analysis is valid, and

$$q_{n1} = -[N_{\alpha0}\kappa_\alpha + N_{\beta0}\kappa_\beta + 2S_0\tau] \qquad (9.207c)$$

We proceed with the simplest of the three theories, the MDV theory in Cartesian coordinates. The governing equation is written in terms of the potential function G from equation (9.194) and q_n is taken as q_{n1}, as given by equation (9.207a):[69]

$$\nabla^8 G + \frac{12(1-\mu^2)}{h^2} \, \bar{\nabla}^4 G =$$

$$\frac{1}{D}(N_{X0}D_{n,XX} + N_{Y0}D_{n,YY} + 2S_0 D_{n,XY}) \qquad (9.208a)$$

where again

$$\nabla^2(\ ) = (\ )_{,XX} + (\ )_{,YY} \qquad (9.208b)$$

$$\bar{\nabla}^2(\ ) = \frac{1}{R_Y}(\ )_{,XX} + \frac{1}{R_X}(\ )_{,YY} \qquad (9.208c)$$

and G is related to D_n and F by

$$D_n = \nabla^4 G \qquad (9.208d)$$

and

$$F = Eh\bar{\nabla}^2 G \qquad (9.208e)$$

If we write the r.h.s. of equation (9.208a) in terms of G, we have

$$\nabla^8 G + \frac{12(1-\mu^2)}{h^2} \, \bar{\nabla}^4 G =$$

$$\frac{1}{D}[N_{X0}(\nabla^4 G)_{,XX} + N_{Y0}(\nabla^4 G)_{,YY} + 2S_0(\nabla^4 G)_{,XY}] \qquad (9.209)$$

We apply this equation to the stability analysis of spherical shells in a later section.

9.5.3 Energy Method for Cylindrical Shells

From an energy standpoint, the transition between the prebuckled and postbuckled states may be represented by the following stability criteria:[70] (a) there is no bending prior to the onset of buckling, so that the total strain energy is due to the in-plane stress resultants; (b) at the onset of instability, there are additional contributions to the strain energy due to middle surface straining and bending; and (c) the increase in strain energy as buckling occurs must be equal to the work done by the external loading and by the components of the in-plane forces that act through *normal* displacements. The latter source of "external" work is analogous to the load component q_{n1} introduced in the previous section, and is not present during infinitesimal deformations. We illustrate this for a cylindrical shell under uniform axial loading, as shown in figure 9–15.

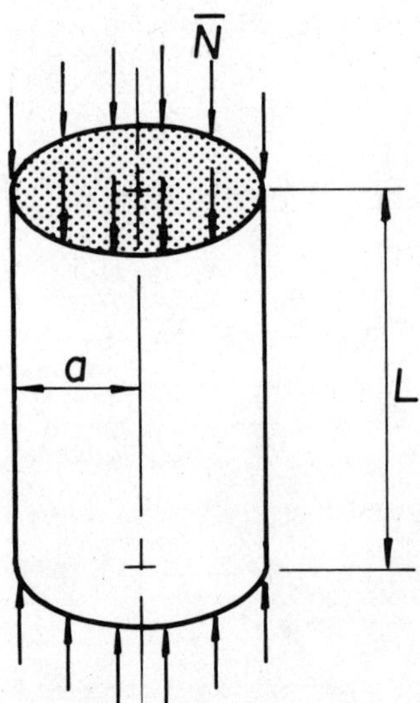

Figure 9–15. Axially Loaded Cylindrical Shell

We first consider the strains and charges in curvature prior to and following buckling.

In the prebuckled state, which we designated by the subscript 0, we assume that there is no bending. There is no normal loading acting, so that the third equilibrium equation, (9.5b), with $Q_x = q_{n0} = 0$, gives $N_{\theta 0} = 0$. We now refer to the constitutive law, matrix 6–2 with $\alpha = X$, $\beta = \theta$, and no thermal terms. The axial strain just before buckling is found from the first equation in matrix 6–2 as

$$\epsilon_{X0} = -\frac{\bar{N}_{cr}}{Eh} \tag{9.210a}$$

The corresponding circumferential strain follows from the second equation in matrix 6–2 as

$$\epsilon_{\theta 0} = -\mu \epsilon_{X0} \tag{9.210b}$$

Due to buckling, the normal displacement becomes D_n, producing a circumferential strain $\epsilon_{\theta 1}$. This strain is proportional to D_n through equation (9.6b), so that

$$\epsilon_{\theta 1} = \frac{D_n}{a} \tag{9.211}$$

Therefore, the total circumferential strain after the shell has buckled is

$$\epsilon_\theta = \epsilon_{\theta 0} + \epsilon_{\theta 1}$$

$$= \frac{D_n}{a} - \mu \epsilon_{X0} \tag{9.212}$$

To find the corresponding meridional strain, we again refer to the first equation of matrix 6–2 with $N_X = -\bar{N}_{cr}$:

$$-\bar{N}_{cr} = \frac{Eh}{1 - \mu^2} (\epsilon_X + \mu \epsilon_\theta) \tag{9.213a}$$

from which

$$\epsilon_X = -\frac{(1 - \mu^2)}{Eh} \bar{N}_{cr} - \mu \epsilon_\theta \tag{9.213b}$$

Substituting for $\bar{N}_{cr}/Eh$ from equation (9.210a), and for ϵ_θ from equation (9.212), we find

$$\epsilon_X = (1 - \mu^2)\epsilon_{X0} - \mu \left(\frac{D_n}{a} - \mu \epsilon_{X0} \right)$$

$$= \epsilon_{X0} - \mu \frac{D_n}{a} \tag{9.214}$$

Also, we have the change in curvature after buckling

$$\kappa_X = -D_{n,XX} \tag{9.215}$$

Of course, $\kappa_X = 0$ prior to buckling.

Now we may evaluate the change in strain energy during buckling. We may specialize equation (7.6) for the case of cylindrical shells with axisymmetric loading to

$$U_\epsilon = \frac{Ea\pi}{1 - \mu^2} \int_0^L \left\{ h(\epsilon_X^2 + \epsilon_\theta^2 + 2\mu \epsilon_X \epsilon_\theta) + \frac{h^3}{12}\kappa_X^2 \right\} dX \tag{9.216}$$

The change in strain energy is given by

$$\Delta U_\epsilon = U_\epsilon(\epsilon_X, \epsilon_\theta, \kappa_X) - U_{\epsilon 0}(\epsilon_{X0}, \epsilon_{\theta 0}, 0) \tag{9.217}$$

We substitute equations (9.214), (9.212), and (9.215) for ϵ_X, ϵ_θ, and κ_X into equation (9.216) to compute U_ϵ; also, using equations (9.210a) and (9.210b), we take ϵ_{X0} and $-\mu \epsilon_{X0}$ for ϵ_X and ϵ_θ to evaluate $U_{\epsilon 0}$. Equation (9.217) then becomes

$$\Delta U_\epsilon = a\pi \int_0^L \left[Eh\left(\frac{D_n}{a}\right)^2 - 2\mu Eh\epsilon_{X0}\left(\frac{D_n}{a}\right) + D \cdot D_{n,XX}^2 \right] dX \tag{9.218}$$

Next, we consider the work done by the external loading and the nor-

376

mal components of the in-plane forces during buckling. For the external loading, $\bar{N} = -\bar{N}_{cr}$

$$\Delta U_{q\bar{N}} = 2\pi a \int_0^L -\bar{N}_{cr}(\epsilon_X - \epsilon_{X0})\, dX$$

$$= 2\mu\pi\bar{N}_{cr} \int_0^L D_n\, dX \qquad (9.219)$$

in view of equation (9.214). Note that there is no 1/2 coefficient in this expression, since the full $\bar{N}_{cr}$ is already present prior to the onset of buckling. For the normal components of the in-plane forces, refer to the loading q_{n1} given by equation (9.207a). The work done over the circumference by q_{n1} acting through D_n is

$$\Delta U_{qn} = \tfrac{1}{2} \int_0^L q_{n1}D_n \cdot 2\pi a\, dX \qquad (9.220a)$$

In equation (9.207a) for q_{n1}, we take $N_{X0} = -\bar{N}_{cr}$ and $N_{Y0} = S_0 = 0$, whereupon equation (9.220a) becomes

$$\Delta U_{qn} = \int_0^L \pi a \bar{N}_{cr} D_{n,XX} D_n\, dX \qquad (9.220b)$$

Adding equations (9.219) and (9.220b), we have

$$\Delta U_q = \Delta U_{q\bar{N}} + \Delta U_{qn}$$

$$= 2\pi\bar{N}_{cr}\left\{\mu \int_0^L \left[D_n - \frac{a}{2} D_{n,XX}D_n\right] dX\right\} \qquad (9.221)$$

Finally, equating equations (9.218) and (9.221) completes the application of the stability criteria:

$$\Delta U_\epsilon = \Delta U_q \qquad (9.222)$$

It is apparent that these equations can be treated most easily with a Rayleigh–Ritz-type approximation for D_n. We provide an example in a later section.

9.5.4 Buckling of Pressurized Spherical Shells

Vlasov has investigated the stability of pressurized spherical shells by using equations (9.208a–e) and (9.209).[71] For a sphere of radius a, equation (9.208c) becomes

$$\bar{\nabla}^2(\ ) = \frac{1}{a}\nabla^2(\ ) \qquad (9.223)$$

The membrane theory of axisymmetrically loaded spherical shells has been presented in section 4.3.2.2. For a uniform internal suction or external pressure p, $q_n = -p$, and equations (4.27) and (4.28) give

$$N_\phi = N_\theta = -\frac{pa}{2} \tag{9.224}$$

from which the prebuckling stresses

$$N_{X0} = N_{Y0} = -\frac{pa}{2} \tag{9.225a}$$

and

$$S_0 = 0 \tag{9.225b}$$

Then the r.h.s. of equation (9.209) becomes

$$\frac{pa}{2D}\left[(\nabla^4 G)_{,XX} + (\nabla^4 G)_{,YY}\right] = \frac{-pa}{2D}\nabla^6 G \tag{9.226}$$

Now, in view of equations (9.224)–(9.226), equation (9.209) reduces to

$$\nabla^8 G + \frac{12(1-\mu^2)}{h^2 a^2}\nabla^4 G + \frac{pa}{2D}\nabla^6 G = 0 \tag{9.227}$$

We seek solutions of the form

$$\nabla^2 G = \lambda G \tag{9.228}$$

where λ is a characteristic or eigenvalue related to $p = p_{cr}$, and obtain

$$\lambda^2\left[\lambda^2 + \frac{p_{cr}a}{2D}\lambda + \frac{12(1-\mu^2)}{h^2 a^2}\right] = 0 \tag{9.229}$$

The trivial solutions $\lambda = 0$ are discarded, and we have remaining the quadratic equation

$$\lambda^2 + \frac{p_{cr}a}{2D}\lambda + \frac{12(1-\mu^2)}{h^2 a^2} = 0 \tag{9.230}$$

which we write in terms of p_{cr} as

$$p_{cr} = -\frac{2D}{a}\left[\lambda + \frac{12(1-\mu^2)}{h^2 a^2 \lambda}\right] \tag{9.231}$$

It is unclear as to which value of λ corresponds to the minimum buckling load.

We may proceed by invoking the minimization condition to find the lowest value of p_{cr}, p_{cr1}. This condition is

$$p_{cr,\lambda} = 0 \tag{9.232a}$$

from which

$$1 - \frac{12(1 - \mu^2)}{h^2 a^2 \lambda^2} = 0 \tag{9.232b}$$

or

$$\lambda = \pm \frac{2\sqrt{[3(1 - \mu^2)]}}{ha} \tag{9.232c}$$

The negative root of λ is meaningful, since it results in a positive p_{cr}, which has already been defined as an internal suction or external pressure.

With

$$\lambda = \frac{-2\sqrt{[3(1 - \mu^2)]}}{ha} \tag{9.233}$$

$$p_{cr1} = \frac{2E}{\sqrt{[3(1 - \mu^2)]}} \left(\frac{h}{a}\right)^2 \tag{9.234}$$

is the lowest buckling pressure for a spherical shell. This equation is applicable for complete spheres as well as spherical shells that have close to ideal membrane boundary conditions. A direct treatment of spherical shell buckling in polar coordinates is developed in Brush and Almroth.[72]

We have arrived at the lowest critical load without first finding a general solution for the normal displacement D_n and may subsequently investigate the normal displacement D_n, which is the *mode shape* of the buckle. If we return to equation (9.208d),

$$D_n = \nabla^4 G$$

$$= \nabla^2(\lambda G)$$

$$= \lambda \nabla^2 G \tag{9.235}$$

The potential function G, in turn, is found from equation (9.228)

$$\nabla^2 G - \lambda G = 0 \tag{9.236}$$

where λ is given by equation (9.233). The solutions of equation (9.236) are the eigenfunctions of the problem. Recall, however, that we have used the simplest form of the MDV theory based on Cartesian coordinates. Consequently, solutions of equations (9.235) and (9.236) that satisfy the kinematic boundary conditions are only feasible if the shell covers a rectangular plan area. For boundaries coincident with the coordinate lines, the original form of the theory, equations (9.186a), (9.186b), and (9.187), should be used in place of the subsequently simplified equation (9.194).

This is carried out in some detail in Vlasov,[73] where it is shown that the lowest critical pressure only differs from p_{cr1}, as computed from equation (9.234), by a term of $O[\mu(h/a)]:1$. Nevertheless, the more rigorous solution provides a basis for determining the eigenfunctions and subsequent displacement or mode shapes. In many applications, only p_{cr1} is of interest, and the solution obtained from equation (9.234) is sufficient to estimate the buckling pressure. Note that, in this regard, quite large factors of safety against *elastic* buckling—perhaps three to five or even more—are commonly specified for the design of thin shells. Since the bifurcation analysis does not account for initial imperfections or for nonlinear behavior, the critical pressure obtained theoretically is very difficult to approach in reality, even in carefully controlled experiments,[74] and a very ample factor of safety is prudent unless a more sophisticated analysis can be made. A means for introducing additional refinements into the analysis is described in section 9.5.6.

9.5.5 Buckling of Axially Loaded Cylindrical Shells

We now investigate the buckling of the shell shown in figure 9–15. We use the energy criterion, equation (9.222), and assume a one-term Rayleigh–Ritz-type solution for D_n

$$D_n = C \sin m\pi \frac{X}{L} \tag{9.237}$$

Substituting equation (9.237) into ΔU_ϵ and ΔU_q as given by equations (9.218) and (9.221), respectively, gives

$$\Delta U_\epsilon = a\pi \left\{ \frac{EhC^2L}{2a^2} - \frac{2\mu Eh\epsilon_{X0}C}{a} \int_0^L \sin m\pi \frac{X}{L} dX + D \frac{m^4\pi^4C^2}{2L^3} \right\} \tag{9.238}$$

and

$$\Delta U_q = 2\pi \bar{N}_{cr} \left\{ \mu C \int_0^L \sin m\pi \frac{X}{L} dX - \frac{a}{4} \cdot \frac{m^2\pi^2}{L} C^2 \right\} \tag{9.239}$$

Then, after substituting equations (9.238) and (9.239) into equation (9.222), and noting from equation (9.210a) that $\epsilon_{X0} = (-N_{cr}/Eh)$, we find all C and C^2 terms cancelling and

$$\bar{N}_{cr} = D \left[\frac{m^2\pi^2}{L^2} + \frac{Eh}{a^2D} \cdot \frac{L^2}{m^2\pi^2} \right] \tag{9.240}$$

The minimum $\bar{N}_{cr}$ is found from

$$\bar{N}_{cr,m} = 0 \tag{9.241a}$$

giving

$$\frac{m\pi}{L} = \sqrt[4]{\left(\frac{Eh}{a^2 D}\right)} \tag{9.241b}$$

and

$$\bar{N}_{cr1} = 2D \sqrt{\left(\frac{Eh}{a^2 D}\right)}$$

$$= \frac{Eh^2}{a\sqrt{[3(1 - \mu^2)]}} \tag{9.242}$$

The length of the buckled waves is

$$\frac{L}{m} = \pi \sqrt[4]{\left(\frac{a^2 D}{Eh}\right)} \approx 1.72 \sqrt{(ah)} \tag{9.243}$$

The identical result may be obtained by the equilibrium equation approach, equation (9.209).

Various other specific solutions based on the two approaches developed in this chapter are available in the literature for spherical and cylindrical shells. Using either approach, the solutions are quite simple and may be used for other geometries with some judicious geometric approximations (see Vlasov[75] and Flügge[76]). Also, an extensive array of more complex situations, including body forces, fluid pressure, torsion, orthotropic shells, and initial imperfections, is treated in Brush and Almroth.[77]

9.5.6 Finite Displacement of Shallow Shells

Recall from section 8.3.3 that the introduction of the geometrically nonlinear strain-displacement equations, (8.47a–c) lead to the von Kármán equations describing the finite displacements for plates. Vlasov has derived a similar set of equations for shallow shells.[78]

In this case, the nonlinear strain-displacement relations are the same as for plates, except that $\bar{\epsilon}_X$ and $\bar{\epsilon}_Y$ each have an additional term proportional to the radii of curvature, as we may verify from equations (5.46a) and (5.46b). Therefore, the appropriate equations in Cartesian coordinates are

$$\bar{\epsilon}_X = D_{X,X} + \frac{D_n}{R_X} + \frac{1}{2} D_{n,X}^2 \tag{9.244a}$$

$$\bar{\epsilon}_Y = D_{Y,Y} + \frac{D_n}{R_Y} + \frac{1}{2} D_{n,Y}^2 \tag{9.244b}$$

$$\bar{\omega} = D_{Y,X} + D_{X,Y} + D_{n,X}D_{n,Y} \tag{9.244c}$$

If we follow the identical steps leading to equation (8.54), we get[79]

$$\nabla^4 F = Eh\left[(D_{n,XY})^2 - D_{n,XX}D_{n,YY} - \frac{1}{R_X}D_{n,YY} - \frac{1}{R_Y} D_{n,XX}\right] \tag{9.245}$$

as the generalized form of that compatibility equation. As before, F is the stress function defined by equation (8.53).

Now, considering the second equation which was derived for plates as equation (8.55), note that this equation followed from equation (8.35), which is the finite displacement version of the basic biharmonic equation for plate bending, equation (8.12). Following the same reasoning, the generalization of equation (8.12) to shallow shell theory is equation (9.183), where the term

$$\frac{N_\alpha}{R_\alpha} + \frac{N_\beta}{R_\beta}$$

is added to the l.h.s. This term enters into the shallow shell equivalent of equation (8.35) as

$$\frac{N_X}{R_X} + \frac{N_Y}{R_Y}$$

and subsequently into the shallow shell form of equation (8.55) as

$$- \left(\frac{1}{R_X} F_{,YY} + \frac{1}{R_Y}F_{,XX}\right)$$

on the r.h.s. Finally, we have the generalization of equilibrium equation (8.55):

$$\nabla^4 D_n =$$

$$\frac{1}{D}\left[q_n + F_{,YY}D_{n,XX} + F_{,XX}D_{n,YY} - 2F_{,XY}D_{n,XY} - \frac{1}{R_X} F_{,YY} - \frac{1}{R_Y} F_{,XX}\right]$$

$$\tag{9.246}$$

Equations (9.245) and (9.246) constitute the generalized von Kármán equations, which form a finite displacement theory for shallow shells. In these equations, the in-plane stress resultants N_Y and N_Y are assumed as positive. The corresponding plate equations (8.54) and (8.55), of course, follow for $R_X = R_Y = \infty$; also, for a cylindrical shell of radius a, $R_X = \infty$

382

and $R_Y = a$. Similarly, for a spherical shell with radius a, $R_X = R_Y = a$.

We have noted in the previous section that the elastic buckling load based on infinitesimal deformation theory as presented in sections 9.1.1 through 9.1.5, may only be a rough indication of the load at which an actual shell may buckle. We have also suggested that a more elaborate analysis may provide theoretical results which are in reasonable agreement with physical experiments. Such an analysis has been provided by two of the most distinguished mechanicians and engineers of the twentieth century, T. von Kármán and H. S. Tsein, in their study of the postbuckling behavior of cylindrical shells based on the finite deformation equations presented in this section.[80] They used a Ritz approach and assumed an approximate expression for D_n, selecting the unspecified parameters based on minimization of the potential energy functional and on some test results. This analysis provides a buckling equation similar to equation (9.238) modified by a factor proportional to the buckled normal deflection to thickness ratio D_n/h. For $D_n/h = 0$, the critical load is given by equation (9.238), while as D_n/h increases, the critical load decreases rapidly to only about 1/3 of this value.[81]

A frequently used equation for the buckling of axially loaded cylindrical shells based on this analysis is

$$\bar{N}_{cr1} = \frac{Eh^2}{3a\sqrt{[3(1 - \mu^2)]}}$$

$$\simeq 0.2\frac{Eh^2}{a} \tag{9.247}$$

By way of confirmation of this theory, an experimental study of open aluminum cylindrical shells with simply supported, uniformly compressed ends and free longitudinal edges gave measured buckling loads that correlated quite well with numerical predictions based on equations (9.241a), (9.241b), and (9.242).[82] It was observed that the correlation between analytically predicted and experimentally measured buckling loads was considerably better than similar comparisons for closed cylindrical shells. It was also noted that open shells have a theoretical critical stress of only 10 to 15 percent of the corresponding values for comparable closed shells in the range of parameters investigated. A relatively complete compilation of the rather extensive literature in this area is contained in the study.

Notes

1. M. Henényi, *Beams on Elastic Foundations* (Ann Arbor: University of Michigan Press, 1946).

2. K. C. Nielsen, *Differential Equations* (New York: Barnes & Noble, 1962), p. 95.

3. S. Timoshenko and S. Woinowski-Krieger, *Theory of Plates and Shells* (New York: McGraw-Hill, 1959), pp. 472–473; E. Y. W. Tsui, *Stresses in Shells of Revolution* (Menlo Park, Calif.: Pacific Coast Publishers, 1968), p. 63.

4. H. Kraus, *Thin Elastic Shells* (New York: Wiley, 1967), pp. 136–139.

5. Timoshenko and Woinowsky-Krieger, *Theory of Plates and Shells*, p. 481.

6. W. Flügge, *Stresses in Shells*, 2nd ed. (Berlin: Springer-Verlag, 1973), chap. 5 and pp. 513–515.

7. V. V. Novozhilov, *Thin Shell Theory* (translated from 2nd Russian ed. by P. G. Lowe (Groningen, The Netherlands: Noordhoff, 1964), pp. 215–218).

8. V. Z. Vlasov, *General Theory of Shells and Its Application in Engineering*, NASA TTF–99 (Washington, D.C.: National Aeronautics and Space Administration, April 1964).

9. Kraus, *Thin Elastic Shells*, pp. 221–229.

10. Ibid., pp. 200–204.

11. Ibid.

12. Ibid., pp. 297–314.

13. Flügge, *Stresses in Shells*, pp. 217–250.

14. "Design of Cylindrical Concrete Roofs," *ASCE Manual of Engineering Practice* no. 31 (New York: American Society of Civil Engineers, 1952).

15. Vlasov, *General Theory of Shells*, pp. 359–394.

16. Ibid., pp. 376–394.

17. P. L. Gould et al., "Column Supported Cylindrical-Conical Tanks," *Journal of the Structural Division, ASCE* 102. no. ST2 (February 1976):429–447.

18. Flügge, *Stresses in Shells*, pp. 231–235.

19. Gould et al., "Column Supported Cylindrical-Conical Tanks," pp. 429–447.

20. D. P. Billington, *Thin Shell Concrete Structures* (New York: McGraw-Hill, 1965), pp. 171–172.

21. "Design of Cylindrical Concrete Roofs."

22. Billington, *Thin Shell Concrete Design*, chap. 6.

23. "Design of Cylindrical Concrete Roofs."

24. Ibid.

384

25. Ibid., pp. 48–57.

26. Ibid., pp. 48–57.

27. Novozhilov, *Thin Shell Theory*, chap. IV.

28. Flügge, *Stresses in Shells*, chap. 6.

29. Kraus, *Thin Elastic Shells*, chaps. 5, 6, and 7.

30. H. Reissner, *Spannungen in Kugelschalen* (Leipzig: Muller-Breslau-Fetschrift, 1912), pp. 181–193.

31. E. Meissner, "Das Elastizitatsproblem für dünne Schalen von Ringflächen-, Kugel-, and Kegelform," *Physik Zeit.* 14(1913):343–349.

32. Flügge, *Stresses in Shells*, pp. 326–344.

33. Ibid.

34. Novozhilov, *Thin Shell Theory*, chaps. III–IV.

35. Flügge, *Stresses in Shells*, chap. 6.

36. Kraus, *Thin Elastic Shells*, chaps. 5, 6 and 7.

37. Flügge, *Stresses in Shells*, pp. 326–334.

38. Kraus, *Thin Elastic Shells*, pp. 271–283.

39. Novozhilov, *Thin Shell Theory*, pp. 315–319.

40. Kraus, *Thin Elastic Shells*, pp. 261–271.

41. Flügge, *Stresses in Shells*, pp. 341–351.

42. J. N. Pirok and R. S. Wozniak, "Steel Tanks," in *Structural Engineering Handbook*, E. H. Gaylord, Jr., and C. N. Gaylord, eds. (New York: McGraw-Hill, 1968), chap. 23.

43. Gould et al., "Column Supported Cylindrical-Conical Tanks."

44. R. S. C. Wang and P. L. Gould, "Continuously Supported Cylindrical-Conical Tanks," *Journal of the Structural Division*, ASCE 100, no. ST10 (October 1974): 2037–2052.

45. Gould et al., "Column Supported Cylindrical-Conical Tanks."

46. G. V. Ranjan and C. R. Steele, "Analysis of Torospherical Pressure Vessels," *Journal of the Engineering Mechanics Division, ASCE* 102, no. EM4 (August 1976): 643–657.

47. Flügge, *Stresses in Shells*, pp. 386–413.

48. Novozhilov, *Thin Shell Theory*, chap. 4.

49. P. L. Gould and S. L. Lee, "Hyperboloids of Revolution Supported on Columns," *Journal of the Engineering Mechanics Division, ASCE* 95, no. EM5 (October 1969): 1083–1100; P. L. Gould and S. L. Lee, "Column-Supported Hyperboloids under Wind Load" (Zurich, Switzerland: Publications, International Association for Bridge and Structural Engineering, 1971), vol. 31–11, pp. 47–64.

50. L. J. Brombolich and P. L. Gould, "Finite Element Analysis of Shells of Revolution by Minimization of the Potential Energy Functional," Proceedings of the Symposium on Applications of Finite Element Methods in Civil Engineering (Nashville, Tenn.: Vanderbilt University, 1969, pp. 279–307.

51. Ibid.

52. Ibid. and L. J. Brombolich and P. L. Gould, "A High-Precision Curved Shell Finite Element," *Synoptic, AIAA Journal* 10, no. 6 (June 1972): 727–728.

53. W. C. Schnobrich, "Analysis of Hyperbolic Paraboloid Shells," *Concrete Thin Shells*, publication SP–28 (Detroit: American Concrete Institute, 1971), pp. 275–311.

54. Novozhilov, *Thin Shell Theory*, pp. 88–94.

55. Ibid.

56. Ibid.

57. Vlasov, *General Theory of Shells*, p. 345.

58. Novozhilov, *Thin Shell Theory*, p. 92.

59. Ibid.

60. Vlasov, *General Theory of Shells*, pp. 495–514.

61. Ibid.

62. Ibid.

63. Novozhilov, *Thin Shell Theory*, pp. 94–99.

64. Ibid.

65. Ibid.

66. B. O. Almroth and J. H. Starnes, "The Computer in Shell Stability Analysis," *Journal of the Engineering Mechanics Division, ASCE* 101, no. EM6 (December 1975): 873–888.

67. Ibid.

68. Vlasov, *General Theory of Shells*, pp. 521–525.

69. Ibid.

70. S. Timoshenko and J. Gere, *Theory of Elastic Stability* (New York: McGraw-Hill, 1961), pp. 457–519.

71. Vlasov, *General Theory of Shells*, pp. 525–529.

72. D. O. Brush and B. O. Almroth, *Buckling of Bars, Plates, and Shells* (New York: McGraw-Hill, 1975), pp. 142–258.

73. Vlasov, *General Theory of Shells*, pp. 525–529.

74. Timoshenko and Gere, *Theory of Elastic Stability*, pp. 468–473.

75. Vlasov, *General Theory of Shells*, pp. 529–538.

76. Flügge, *Stresses in Shells*, chap. 7.

77. Brush and Almroth, *Buckling of Bars, Plates, and Shells*.

78. Vlasov, *General Theory of Shells*, pp. 538–543.

79. Ibid.

80. T. von Kármán and H. S. Tsien, "The Buckling of Thin Cylindrical Shells under Elastic Compression," *Journal of Aeronautical Sciences* 8 (1941): 303.

81. Timoshenko and Gere, *Theory of Elastic Stability*, pp. 47–173.

82. T. H. Yang and S. A. Guralnick, "Buckling of Axially Loaded Open Shells," *Journal of the Engineering Mechanics Division, ASCE* 102, no. EM2 (April 1976): 199–211.

10 Conclusion

10.1 General

The preceding chapters have treated the analysis of thin elastic plates and shells from a unified viewpoint insofar as possible. Specializations to specific forms—e.g., membrane shells, flat plates—are introduced in a logical sequence after the general aspects of the theory are set forth. The emphasis has been on explaining the mechanics of surface structures, with ample examples to illustrate the most important aspects. In general, relatively simple and idealized cases that could readily be solved analytically were used to illustrate the salient points. Additionally, fairly complete compilations of known analytical solutions for both plates[1] and thin shells[2] are available to supplement the examples given in the text. The solution of more complex structures is left to the domain of numerical analysis, and several citations to the literature in this area were provided. Although the calculations for those situations may be more involved, it is felt that such structures should reflect the same basic behavioral characteristics as the relatively simple illustrations contained in this book.

10.2 Proportioning

One topic that has not been extensively addressed in this text is the *proportioning* component of the shell design process. Although the emphasis here has been on the *analysis* component, one must have something to analyze, and it is our hope that some interest in this question has been aroused. It is clear that the specific form of the surface structure is often determined by the anticipated function: flat plates for carrying pedestrian and vehicular traffic, shells of revolution for containment, and shallow shells for long, clear span roofs. Also, the construction material—metal and concrete—plays a very important role in the proportioning process.

Although the proportioning of surface structures is obviously very much related to the function and the construction material, it is possible to offer a few general comments on this topic. We consider separately (a) regions of the structure where the stresses are primarily in-plane or membrane; and (b) regions where there is significant bending action.

In the first case, tensile stresses should be resisted entirely by reinforcing steel in concrete shells, or by ample wall thicknesses for metal shells. Regions with compressive direct stresses are generally controlled by stability requirements, particularly in metal plate structures.

In the second case, the bending moments or stress couples may be resisted by considering a concrete section to act as a wide flexural member. However, the relatively small effective depth available may complicate the provision of ample reinforcing steel. For thin metal structures, it is often impractical to resist high local bending stresses with the basic shell thickness, and stiffening members are often attached.

Beyond these brief generalizations, the proportioning of thin surface structures is a somewhat subjective and always challenging branch of structural design that has become somewhat specialized for various structural forms and materials. For thin concrete shells, D. Billington has considered the proportioning problem in a somewhat comprehensive manner,[3] and for steel tanks, J. Pirok and R. Wozniak have illustrated several practical techniques of design.[4] Beyond these rather general articles, there are a number of authoritative works for concrete shells,[5] for steel tanks[6] and pressure vessels,[7] and for cooling towers.[8]

Although many of the procedures for proportioning surface structures may appear to be involved, the basic rules of structural design apply, augmented by special requirements due to the relative thinness of the system. This is vividly illustrated in figure 10–1, where the various types of reinforcement required in an umbrella shell are shown. The orthogonal grid for the shell proper, the additional shell steel in the vicinity of the columns, and the edge beam reinforcement are difficult to accommodate in a relatively thin section, and careful planning and supervision are necessary to ensure a sound structure.

Notes

1. R. Szilard, *Theory and Analysis of Plates* (Englewood Cliffs, N.J.: Prentice-Hall, 1974).

2. E. H. Baker, L. Kovalevsky, and F. L. Rish, *Structural Analysis of Shells* (New York: McGraw-Hill, 1972).

3. D. B. Billington, "Thin-Shell Concrete Structures," in *Structural Engineering Handbook*, E. H. Gaylord, Jr., and C. N. Gaylord, eds. (New York: McGraw-Hill, 1968), sec. 20.

4. J. N. Pirok and R. Wozniak, "Steel Tanks," in *Structural Engineering Handbook*, E. H. Gaylord, Jr., and C. N. Gaylord, eds. (New York: McGraw-Hill, 1968), sec. 23.

Figure 10–1. Reinforcement for an Umbrella Shell (Courtesy Mr. A. Monsey)

5. "Concrete Shell Structures: Practice and Commentary," *Journal of the ACI* 61, no. 9 (September 1964): 1091–1108.

6. "Recommended Rules for Design and Construction of Large, Welded, Low Pressure Storage Tanks," API Standard 620 American Petroleum Institute, 1966, "Welded Steel Tanks for Oil Storage," API Standard 650, American Petroleum Institute, 1973; "AWWA Standard for Steel Tanks—Standpipes, Reserving and Elevated Tanks—for Water Storage," AWWA D100–67, American Water Works Association, or AWS D5.2–67, American Welding Society.

7. "Rules for Construction of Pressure Vessels," div. 1, 1971 ed., ASME Boiler and Pressure Vessel Code, sec. VIII, American Society of Mechanical Engineers, 1971; "Alternative Rules for Pressure Vessels," div. 2, 1971 ed., ASME Boiler and Pressure Vessel Code, sec. VIII, American Society of Mechanical Engineers, 1971.

8. "Reinforced Concrete Cooling Tower Shells: Practice and Commentary," *Journal of the ACI* 74, no. 1 (January 1977): 22–31.

Index

Index

About the Author

Phillip L. Gould is a professor of civil engineering at Washington University, a registered structural and professional engineer, and a member of several professional societies. He received the B.S. and M.S. degrees at the University of Illinois at Urbana and worked as a structural engineer, engaging in the design of institutional and multistory buildings and highway bridges. He received the Ph.D. from Northwestern University.

Dr. Gould's research activities have centered on thin shell structures with applications to finite element analysis, biomedical engineering, and the design of hyperbolic cooling towers. Numerous technical papers have accrued from these studies. He has also served as a consultant to various industries and governmental agencies. A Senior US Scientist Award enabled Dr. Gould to spend a year as a guest professor in the German Federal Republic and he is engaged in various cooperative research projects in Europe and Asia.